AMERICAN GENESIS

THOMAS P. HUGHES

AMERICAN GENESIS

A CENTURY OF INVENTION AND TECHNOLOGICAL ENTHUSIASM 1870–1970

The University of Chicago Press
Chicago and London

Thomas Hughes is the Mellon Professor Emeritus in the Department of the History and Sociology of Science at the University of Pennsylvania and Distinguished Visiting Professor at the Massachusetts Institute of Technology.

The University of Chicago Press, Chicago 60637
 Published 2004
Originally published by Viking Penguin, a division of Penguin Books USA Inc.
Printed in the United States of America

13 12 11 10 09 08 07 06 05 04 1 2 3 4 5
ISBN: 0-226-35927-1 (paper)

Library of Congress Cataloging-in-Publication Data

Hughes, Thomas Parke.
American genesis : a century of invention and technological enthusiasm, 1870–1970 / Thomas P. Hughes.
p. cm.
Reprint. Originally published: New York : Viking, 1989.
Includes bibliographical references and index.
ISBN 0-226-35927-1 (pbk. : alk. paper)
1. Technology—United States—History. I. Title.

T21 .H82 2004
609.73–dc22
2003067221

∞ The paper used in this publication meets the minimum requirements of the American National Standard for Information Sciences—Permanence of Paper for Printed Library Materials, ANSI Z39.48-1992.

For my father, Hunter Russell Hughes,
who believed in the America
of Edison and Ford

CONTENTS

CONTENTS

FOREWORD

The last events mentioned in the earlier edition of *American Genesis* were the nuclear reactor catastrophe at Three-Mile Island in 1979, the space shuttle *Challenger* tragedy of 1986, and the Chernobyl nuclear plant disaster in the Soviet Union in the same year. I characterized all three as large technological-system accidents. These events were on my mind as well as counterculture ideas when I suggested in my concluding remarks that there might be a reaction in the United States against the further spread of large technological systems.

Counterculture values of the 1960s especially impressed college students and a college-educated generation under thirty years of age. The popularity of *The Whole Earth Catalog* (1968), with its call for the use of simple tools and a return to a near-primitive, environmentally friendly style of life demonstrated an enthusiasm for small-scale, appropriate technology and a hostility towards the large manufacturing and marketing systems of industrial capitalism and the weapons systems deployed in Vietnam.

Influenced by my awareness of system accidents and counterculture values, I suggested in my concluding remarks that the momentum or inertia of large systems might be countered in the future by a new set of values. I went so far as to suggest that the next large technological development or revolution might take place outside of the United States because it was saddled by a commitment to large systems. Because the Vietnam

War stimulated anti-war sentiments, I doubted that the United States would continue to develop increasing numbers of large weapons systems.

As I look back upon my flawed anticipations, I recall my criticism in *American Genesis* of public intellectuals, such as Lewis Mumford, who assumed that the second industrial revolution would bring a new technologically shaped society that would reflect their desire for green, or environmentally friendly, regionalism dependent upon electric power and the automobile. I argued that they allowed their values to shape their predictions and that they ignored the conservative power of vested interests. I allowed my values, attuned as they were to the counterculture, to shape my anticipations as well.

I no longer believe that counterculture values will diminish the influence of large technological systems, and I no longer expect major technological change to occur outside the United States. Instead of a demise of large technological systems and a loss of U.S. technological preeminence since the 1980s, we have witnessed what appears to be another technological revolution, the so-called information revolution. It has taken place largely in the United States. Instead of a retreat from weapons systems, we have recently seen a sizeable increase in the Pentagon's budget to the level where U.S. military expenditures exceed those of Russia, Britain, France, and Germany combined.

I now consider the information revolution to be the major technological and societal development of the past two decades, and I envisage it shaping technical, political, economic, social, and cultural developments for the foreseeable future, the collapse of the dotcom bubble notwithstanding. To bring *American Genesis* up-to-date, therefore, I shall describe the origins and development of the information revolution, fully aware that a historian lacks perspective and will err in dealing with the recent past. I hope, nevertheless, that my survey of the information revolution will have the depth and scope of the survey of the second industrial revolution that I provided in the earlier edition of *American Genesis*. At this time there are few comparable accounts of the information revolution.

My approach will be to assume that the information revolution has developed, and will develop, in a way comparable in scope and character to earlier industrial revolutions. I shall summarize the characteristics of the British industrial revolution of the late eighteenth and early nineteenth centuries and the second industrial revolution of the late nineteenth and early twentieth centuries that occurred in the United States and Germany. Having identified the major characteristics of the earlier revolutions, I shall describe the information revolution seeking similar characteristics in this new preface.

In general, today's accounts of the information revolution focus upon artifacts, such as computers and the Internet. This approach is myopic. We should broaden our concept of the information revolution. The other industrial revolutions involved far more than hardware. Beside technical artifacts and systems, these earlier revolutions involved political, economic, social, organizational, and cultural changes that greatly influenced people's lives, as will the information revolution.

THE BRITISH INDUSTRIAL REVOLUTION

Take, for example, the broad scope of the British industrial revolution, to which the information revolution is often compared. With ease and familiarity, historians tell of inventors of humble background and limited means introducing a wave of inventions that interacted systematically. Newly invented steam engines drove recently mechanized textile factories; furnaces produced iron using abundant coke instead of scarce charcoal; iron machine tools turned out steam engine components; and canals made possible the flow of coal, iron, cotton, and other commodities throughout the evolving system. A rapidly growing chemical industry supplied bleaches for the textile mills. When deciding whether there is an information revolution comparable to the British industrial revolution, we should look for such systematic technical interactions.

Thoughtful accounts of the British industrial revolution do not ignore demographic, social, and political changes accompanying technical ones. Before the revolution, commerce and industry concentrated along navigable rivers and coastal ports. In the eighteenth century, however, engineers laid out canals that facilitated the relocation of industry and population to English Midland cities situated along these canals. The demographic shift changed the face of England. Are there comparable demographic shifts associated with the information revolution?

On the political level, the revolution stimulated the rise of an urban middle class whose wealth came from industry rather than from land. Leading industrialists and their banking and commercial associates cultivated laissez-faire liberalism, advocated free trade, opposed the corn laws, or tariffs, and demanded suffrage be extended to male, middle class voters in the industrial cities. The industrial revolution lastingly changed British politics. Has the information revolution brought similar changes?

In the realm of culture, steam driven presses and typesetting machines increased the circulation of newspapers and books. The availability of new or cheaper materials and the spread of railways brought cathedral-like urban train stations whose designers innovated with glass and iron. The increased production of consumer goods stimulated the middle class to laden their bodies with textiles and their homes with gewgaws. A Victorian cultural style emerged. Can we speak of an information revolution culture and style?

THE SECOND INDUSTRIAL REVOLUTION

The second industrial revolution also involved a wave of interconnected inventions, as well as demographic, political, and cultural changes. Innovative modes of management and new organizational forms also appeared. Professional inventor-entrepreneurs launched major energy, communication, and production systems that increasingly structured the industrialized world. Composed of both technical and organizational components, these systems spreading across the landscape became a second, human-built creation where people lived and worked.

Independent inventor-entrepreneurs presided over the invention, development, and innovation of large technological systems. Outstanding among the Americans were Thomas Edison, Alexander Graham Bell, Nikola Tesla, who introduced an alternating current electric power system, Edwin Armstrong, inventor of FM radio, and Elmer Sperry, inventor of automatic airplane and ship pilots. They made alliances with capitalists to form companies based on their patents. The technological systems introduced by inventor-entrepreneurs evolved according to a pattern. Shortly we shall observe that the internet evolved in a similar way.

Managerial style changed during the second industrial revolution. The availability of electric motors allowed a flexible layout of factories. Henry Ford and his engineers organized material and energy flows so that raw materials entered his automobile factories in an unending stream and Model T automobiles flowed from moving assembly lines. Frederick W. Taylor, father of scientific management, rationalized plant layout and human labor, which he treated as a machine component. On the organizational level, giant manufacturing firms, like General Motors, organized hierarchical, multi-layered, managerial structures.

The second industrial revolution also brought momentous changes in warfare. The internal combustion engine made possible the tank and the airplane. Electric motors moved stealthy submarines. Steam turbines powered all big gun battleships. Advances in industrial chemistry contributed substantially to the production of explosives and poison gases.

The second industrial revolution also brought dramatic cultural changes. Futurist and constructivist art mirrored technology's system, order, and control values. Machine art found a prominent place in museums and art galleries. Designers and enthusiastic celebrants of the International Style of architecture believed that this modern, technology-based style was comparable in scope and influence to the medieval, renaissance and baroque styles.

Demographic shifts and technical innovations resulted in new industrial cities. Berlin, with its world-renowned electrical industry, became known as the city of light. Essen, Germany, signified the awesome industrial power of Germany's Ruhr region. Manhattan became a center of commerce and innovation attracting inventors like a magnet. Philadelphia nurtured light and heavy manufacturers and called itself a workshop of the world; coal, coke, and iron poured forth from smoky Pittsburgh; and Chicago showed the prodigious power of an industrial city by reversing the flow of the Chicago River to carry off the city's wastes.

Skyscrapers symbolized the constructive power of industrial cities. In

Minneapolis, Minnesota, and Buffalo, New York, the stark forms of the great grain silos represented the might of mechanized agriculture. At roaring Niagara Falls, electrical engineers and chemists built an integrated hydroelectric and industrial complex that became a model for technological development throughout the world. On the outskirts of Detroit rose the Ford River Rouge Plant, history's largest production machine. Will Silicon Valley find a place in history comparable to the one established by the industrial cities of the second industrial revolution?

THE INFORMATION REVOLUTION[1]

Assuming that the information revolution will resemble the earlier revolutions, we should seek interacting artifacts and systems at its technical core. The role of inventor-entrepreneurs and the evolutionary pattern of large technical systems, especially the internet, should be considered. We should look for political and social changes, including demographic ones. Of particular importance are the cultural changes stimulated by the information revolution, including those in architecture, art, and science.

A TECHNICAL CORE

Like the British industrial revolution and the second industrial revolution, the information revolution has interacting artifacts and systems at its technical core. The interactive development of computers, semiconductors, and software nurtured the spread of digital information, which in turn led into the revolution. In 1947 John Bardeen, Walter Brattain, and William Shockley at Bell Laboratories patented a semiconductor transistor to displace the omnipresent large vacuum tubes in military and civil communication and control devices. Jack Kilby and Robert Noyce in 1959 independently invented an integrated circuit that combined resistors, capacitors, and transistors on a single silicon wafer. Another major breakthrough came in 1971 when Marcian (Ted) Hoff invented a microprocessor, which incorporated hundreds of thousands of circuit components, thus becoming a computer on a chip.

Integrated circuits and microprocessors interacted with the development of computers and software. Microprocessors allowed designers in the 1970s to introduce personal, or desktop, computers that were much smaller than the giant computer mainframes that opened the computer era. In the 1970s, production of transistors, integrated circuits, microprocessors, and computers skyrocketed in California's Silicon Valley. In 1977 Steve Jobs and Steve Wozniak introduced the Apple II computer and later a Macintosh computer with icons, a mouse, and a pull-down menu, which were first developed at the Xerox Corporation's Palo Alto Research Center and by Douglas Engelbart.

Initially mistakenly considered peripheral to computer or hardware development, software soon became a major component in the evolving information revolution. Using William Gates's MS-DOS operating system, IBM introduced a personal computer with spreadsheet and word processing software. Gates's Microsoft Company soon became the world's leading software producer.

There were other interactions at the technical core. The Advanced Research Projects Agency (ARPA) of the U.S. Defense Department funded the interconnection of computers in a system named the ARPAnet, which, after 1971, became the basis for the Internet. Its use expanded dramatically after Tim Berners-Lee, a scientist at CERN, the European particle physics laboratory, in 1991 wrote the prototype for what has become known as the World Wide Web. The usefulness of the Web increased in 1992 when Marc Andreesen, a student at the University of Illinois, and Eric Bina composed a program for a web browser that allowed Web users to search the Internet effectively.

INVENTOR-ENTREPRENEURS

During the late nineteenth and early twentieth centuries, independent, Edisonian-type, inventor-entrepreneurs presided over the invention and development of power, transportation, communication, and other large technological systems of the second industrial revolution. As explained in the earlier edition of *American Genesis*, however, scientists and engineers in industrial research laboratories displaced the independents during the twentieth century. As a result, inventive activity tended to focus upon improvement in existing systems, such as those cultivated by the industrial corporations sponsoring the industrial laboratories, rather than upon breakthrough inventions and the inauguration of new systems.

Because of the demise of the independent inventors and the rise of industrial laboratory scientists, we would not expect the former to be playing a major role in the origins of the information revolution, but they have. Instead of finding them in laboratories of their own design, as we found Edisonian-type inventors, we will find them in universities, where faculty, students, and researchers have spent the information revolution presiding over breakthrough inventions, often supported by government funds.

During World War II, for instance, J. Presper Eckert and John Mauchly at the University of Pennsylvania designed and built the ENIAC, which many historians consider the first general purpose, mainframe computer. A few years later, John von Neumann at the Princeton Institute for Advanced Study introduced a basic computer design known as "von Neumann Architecture" that influenced the design of computer components and circuitry for years. In the early 1950s, Jay Forrester and his MIT associates built the Whirlwind computer, which calculated and controlled in real time.

In 1967 Wesley Clark, a Washington University researcher, suggested that mainframe, or host, computers of different characteristics could be interconnected in a single system. Acting upon his insights, Joseph Carl Robnett Licklider, who had long associations with MIT, used his position as a section director at ARPA to fund the development of the ARPAnet by researchers mostly in universities. The list of university-based information age pioneers could be greatly extended.

Stanford University faculty and students have played a disproportionate role in inventing computer and Internet hardware and software. John Hennessey of Stanford started MIPS, the company that developed RISC chips upon which Silicon Graphics machines are based. Jim Clark, a Stanford engineering professor, founded Silicon Graphics, and then with Marc Andreessen, a student at the University of Illinois, founded Netscape. Scott McNealy, who has a degree from Stanford, and Andy Bechtolsheim, who did graduate studies there, founded Sun Microsystems along with Vinod Koshla, a Stanford graduate, and Bill Joy of the University of California, Berkeley. Jerry Yang and David Filo, Stanford graduate students, developed Yahoo! as a search engine and portal for the Internet.

What do the Edisonian independents and university information revolution inventors have in common? All were relatively free, as compared to industrial research laboratory personnel, to choose their problems. As a result, the university inventor-entrepreneurs have been responsible for a disproportionate share of the breakthrough, as contrasted with cumulative inventions and developments. Like the Edisonian independents, the university-based, inventor-entrepreneurs often started their own companies in order to manufacture their inventions. During the information revolution these companies have become commonly know as "startups," but they differ little in their origins and development from the companies that the Edisonian types commonly established.

THE EVOLUTION OF STARTUPS[2]

Like the Edisonian companies, Silicon Valley startups display the following trajectory: eureka moment; simultancity of invention; formation of a development team; acquisition of funding; acquisition of momentum; and transfer of management from inventor-entrepreneurs to management entreprencurs.

A eureka moment is a flash of insight that promises a solution to a problem. In February 1987 Jerry Kaplan, an Internet pioneer, had a eurcka moment. While typing his scribbled notes into a computer, his companion on a cross-country flight, Mitch Kapor, the inventor of the Lotus spread sheet, asked if there might not be an easier way of entering information into a computer directly. After a brief nap, Kaplan awoke with an idea for a pen-input device rather than a keyboard. He and Kapor realized that the insight combined simple, familiar elements into a radically new thing. Thus was born the idea for an early, albeit premature, version of the widely used Palm pocket computer. Kaplan then formed a startup company to exploit his eureka moment.

The problems that inventor-entrepreneurs choose are often reverse salients, or lagging components, in developing systems. For instance, during the early years of the Internet a number of inventive persons realized at about the same time that a search engine was needed to find web pages. When Filo and Yang were developing their Yahoo! search engine, a number of other persons were simultaneously doing the same. Because inventor-entrepreneurs often cluster at reverse salient sites, simultaneity of invention and of startup establishment are common.

Once embarked upon a project, inventor-entrepreneurs, whether of the late nineteenth century or of the late twentieth century, organized a development team. One characteristic of the teams is their informality. A memorable image shows Edison sitting in his Menlo Park Laboratory surrounded by a dozen or so of his mechanics, chemists, and model builders. One of them is playing a pipe organ as food is brought in for a late evening break.

Silicon Valley and Cambridge, Massachusetts, startup teams favor a flat rather than a hierarchical relationship. Team members are known for their informal dress and their casual but idea-laden exchanges in hallways and over coffee. Frank Heart, an engineer who headed a Cambridge team that developed a major ARPAnet hardware component, recalls the project as "a labor of love"; members of the team called the work "fun." He encouraged the team to reach decisions by consensus.

Obtaining funding for a project presents a difficult hurdle to inventor-entrepreneurs. Edisonian and Internet startups appealed to capitalists, but the character of the capitalists and the approach to them has changed. Internet startups raise funds from venture capitalists who amass funds from investors and turn them over to startups, whose business plans appeal to the venture capitalists, many of whom locate in Palo Alto. Inventor-entrepreneurs, especially those in universities, also seek funds from government, including the Advanced Research Projects Agency. Because of the technical literacy of the venture capitalists and government project managers, the approach taken by startup entrepreneurs has to be far more technically and managerially detailed than that taken by the Edisonian types.

Momentum can work for or against inventor-entrepreneurs. After a young company has acquired resources and market share, its momentum can carry it forward over obstacles. On the other hand, the momentum of long-standing corporations with a competing product can thwart inventor-entrepreneurs. In the early 1960s Paul Baran proposed a digital packet-switching communications network of the kind later adopted for the ARPAnet. Engineers and managers at AT&T, which then dominated the long distance telephone system, talked politely to Baran, but he found them resolutely committed to their analog system. Presiding over a totally integrated system, AT&T insisted that new technology had to fit into its system. One AT&T engineer, after an exasperating session with Baran, told him that a digital system probably wouldn't work, but if it did AT&T would prevent the launching of a competing system.

Inventor-entrepreneurs of the late nineteenth century presided over a project from invention to innovation, or entry into the market. As their companies grew large, they often turned over management to manager-entrepreneurs. Similarly, the inventor-entrepreneurs of startup companies have given way to managers, often at the insistence of the venture capital firms backing them. The young founders of Excite, a search engine provider, chose a seasoned manager to head their growing firm. The *Wall Street Journal* praised their decision to bring in "adult supervision" as a critical component of Excite's successful transition from a startup to a public company.

THE EVOLUTION OF THE INTERNET

If the information revolution shares characteristics with previous industrial revolutions, then it is possible that the Internet will evolve like the electric power network of the second industrial revolution. Both are electrical transmission networks, one conveying energy and the other information.

The evolution of electric light and power systems (hereafter "electric power systems") can be broken down into phases according to the type of entrepreneur dominating a phase and in accord with the geographical spread of the systems. From about 1880 to 1900, independent inventor-entrepreneurs dominated the development of electric power in the United States. They solved the major problems emerging in the systems, which were essentially technical or engineering problems. During this era, the area of electric supply typically extended over an urban district, not an entire urban area.

Inventor-entrepreneurs also dominated the initial phase of Internet evolution, which began in about 1972 with the ARPAnet. Electric power and Internet inventor-entrepreneurs solved similar problems, too. In the late 1880s they found a way to interconnect heterogeneous, independent, direct-current and alternating-current power systems with transformers, motor-general sets, and frequency changers, thus creating an integrated "universal system." Similarly, Wesley Clark, as noted above, suggested that heterogeneous mainframe, or host, computers could be interconnected in a single system.

Between about 1900 and 1920, merger, acquisition, and the rise of the universal system made possible urban electric power systems supplying an entire city, such as New York or Chicago. Manager-entrepreneurs displaced inventor-entrepreneurs as the prime problem solvers in this new phase of electric power history. While technical problems continued to emerge, especially those associated with the introduction of steam turbines, it was chiefly managerial problems that threatened to frustrate the growth of power systems. Besides the obvious problem of providing an organizational structure for numerous white-collar and blue-collar employees, the manager-entrepreneurs endeavored to increase the market for electricity. An even subtler problem for them was the management of load to improve the load factor, or to increase the utilization of the plant capacity.

Analogously, Internet manager-entrepreneurs began to come to the fore as the twentieth century closed. For instance, they dominated the headlines in 2000 when Time Warner, a media company providing content information for web sites and high-speed cables for transmission of information, merged with America Online (AOL), a large Internet service provider or portal. Stephen Case, the chair and chief executive officer of AOL, who became chairman of AOL Time Warner, was a manager-entrepreneur primarily interested in increasing his company's market share. Robert W. Pittman, who became president of AOL, began as a high-profile figure in New York's media scene where he ran the Movie

Channel. Gerald Levin, the chair and chief executive officer of Time Warner, was known as an aggressive manager. AOL's headquarters were in the Washington suburbs of Northern Virginia, not in Silicon Valley. New York investment banks, such as Morgan Stanley Dean Witter, not Palo Alto venture capitalists, played a leading role in the merger.

From 1920 to 1930 financier-entrepreneurs displaced the electric power managers as prime problem solvers. The financier-entrepreneurs responded to the high capital requirements of interconnecting power systems, urban and rural, into giant regional networks that brought various economies. Because large regional systems deeply influenced the economy and quality of life, local and state governments increasingly regulated the utilities. In response, financier-entrepreneurs also cultivated their negotiating skills and political influence. Similarly motivated by economic considerations, project leaders at the Advanced Research Project Agency funded ARPAnet so that interconnected computers could share costly software and hardware. Robert Kahn and Vincent Cerf found a way to interconnect computer networks of different characteristics, the ARPAnet among them, thus creating a global Internet with its attendant economies.

Power system financier-entrepreneurs established large holding companies that by 1924 controlled two-thirds of the generating capacity of the electric supply industry. Holding companies, such as Stone & Webster, not only financed the technical and organizational improvements in the numerous utilities under their control, but also often constructed the new facilities and managed utilities. As the twentieth century closed, holding companies were also spreading in the Internet world. Venture capital and management consulting companies were taking equity in the companies that they nurtured as startups.

Having considered analogies between power and Internet networks, we should strike a skeptical note as well. A major development in the history of electric power was the establishment by the administration of Franklin Delano Roosevelt in the 1930s of the Tennessee Valley Authority, a government entity that became a major power generator in addition to providing flood control, preventing soil erosion, nurturing afforestation, and cultivating agricultural and industrial development. Future government ownership of any part of the civil Internet seems highly unlikely today.

INFORMATION AGE MANAGEMENT

Tightly coupled to technological change, new managerial practices and organizational forms have evolved during the information revolution. Hierarchy, specialization, standardization, centralization, expertise, and bureaucracy became the hallmarks of management during the second industrial revolution. Flatness, interdisciplinarity, heterogeneity, distributed control, meritocracy, and nimble flexibility characterize information age management. The organizational culture of Silicon Valley at the epicenter of the revolution has been described as information sharing, collective in learning, informal in communication, fast moving, flexible in adjustments, entrepreneurial, start-up inclined, and thoroughly networked.[3]

Virtual corporations that focus upon management and that outsource manufacturing and other functions to contractors also display the information revolution organizational style. Because of its rapid and deep interconnectedness, the Internet allows a virtual corporation to preside over a project by scheduling and coordinating subcontractors. Not invested in facilities to manufacture components made by its contractors, a virtual corporation can nimbly shed its contractors and move to another domain.

INFORMATION AGE ARCHITECTURE

Industrial revolutions brought substantial cultural changes, especially in the realms of architecture and art. The second industrial revolution stimulated the unfolding of an International Style of architecture and a richly diverse style of machine art. What might the information revolution introduce?

The architecture of Robert Venturi and Denise Scott Brown and of Frank Gehry suggests one course architecture might take. Husband and wife, Venturi and Scott Brown broke away from the orthogonal, geometrical buildings and urban grid plans of the early twentieth century International Style. Venturi's "Vanna Venturi/Hughes House" (1964) in the Chestnut Hill section of Philadelphia initiated a widespread and sustained reaction against the order and system of technology-driven architecture. Contradicting the renowned International Style architect Mies van der Rohe, whose buildings expressed the subtle, clean, and orderly lines of the engineering world and who said "less is more," Venturi rejoined that "less is a bore." Instead of order and control, the Vanna Venturi/Hughes House expresses complexity and contradiction. Venturi argues in his highly influential manifesto, *Complexity and Contradiction in Architecture* (1966) that early International Style was, like a young person, seeking clarity and simplicity, while his architecture suited a mature culture able to deal with complexity.

Recently Venturi and Scott Brown have designed buildings that they call electronically decorated sheds. In harmony with the information age, they use electronic, rather than machine signs and symbols to carry information on their buildings. They recall, but alter, medieval churches whose scenographic windows and paintings conveyed messages about the faith to the flock, many of whom were illiterate. Venturi reminds us, computer generated, electronic (LED) video displays allow messages to be adjusted to the context and plurality of audience. Electronic messages are not like their medieval predecessors, cast in stone.

Other information age architects now design buildings using graphics software originally intended for airplane fuselage design, automobile styling, and movie animation. Wes Jones, a Los Angeles architect, calls the results "blob" architecture, which refers to designing spaces using the computer's capacity to manipulate three-dimensional forms as if they were made of soft putty.

The noted architect Frank Gehry, who designed the highly praised art museum in Bilbao, Spain, takes a slightly different approach.[4] Gehry and his office have used a CAD (computer aided drafting) software called CATIA (computer aided three-dimensional interactive application) developed by the French aerospace industry. Gehry's partner, James Glymph, first recognized the potential of CATIA as a source of computer models and construction drawings when Gehry was designing a 189-feet-long and 115-feet-tall giant fish made of steel panels to be placed on Barcelona's waterfront for the 1992 Olympics. Gehry first modeled the fish in soft materials and then turned it over to his model builders to scale up. Using a large model, they took numerical coordinates from which CATIA generated a computer model from which fabrication and erection contractors could take detailed specifications for use. Besides the Barcelona fish, Gehry employed the procedures for the cloud-like forms of the Guggenheim Museum Bilbao (1991–1997), the Walt Disney Concert Hall in Los Angeles (1987–2003), the Experience Music Hall in Seattle, Washington (1995–2000), and the Ray and Maria Stata Center at the Massachusetts Institute of Technology (begun 1998).

INFORMATION AGE ART

Artists and animators, too, increasingly use computer hardware and graphics software initially developed by engineers and computer scientists. Ivan Sutherland, an MIT engineer, for instance, in 1963 introduced Sketchpad, a path breaking, interactive graphics system for manipulating two- and three-dimensional objects. Later Sutherland and colleague Dale Evans created a computer science department at the University of Utah that trained graduate students in interactive graphics software. These students then designed, among other software, RenderMan, which was used in making *Toy Story*, the first entirely computer-animated film, as well as the dinosaurs in *Jurassic Park*, and the cyborg in *Terminator 2*. Hundreds of computer-literate animators and artists were soon creating animated, virtual reality special effects as seen in such popular films as *Titanic* and *Saving Private Ryan*.

The U.S. military has become a major incubator for computer graphics and simulations. A positive feedback relationship exists between the designers of military battlefield and air-war simulations and the designers of commercial video games, cinema special effects, and computer animated movies. This military-entertainment cooperation took a quantum leap in 1999 when the U.S. Army awarded a $45 million, five-year contract to establish an Institute for Creative Technologies at the University of Southern California (USC).

Artists of a more traditional kind are drawing upon the software of computer-using graphic designers and animators. This cooperation may influence art in ways comparable to the manner in which linear perspective influenced painting and design during the early European Renaissance. Enthusiasts also contend that computer hardware and software will influence artistic practice as much as photography did in the nineteenth century. Today artists use digital resources to meld text, images, and sound, to alter and animate images, to change the representational into the abstract, and to make three-dimensional sculpture from three-dimensional computer models. As the computer and software costs decline and quality and complexity increase, the number of artists using digital resources expands.

The Whitney Museum in New York City and the San Francisco Museum of Modern Art both had digital art shows in 2001. The art works at the Whitney often recall abstract and minimalist styles that are familiar to the experienced museum visitor—styles done in the past with conventional paint, acrylic, pastels, and watercolors. The show's originality stemmed mostly from the artists' use of computers and software as tools to create art. Some of the artists programmed their own software, but most used available software. As their familiarity with the digital tools increases, they will find new forms that are especially suited for the new tools.

INFORMATION AGE BIOLOGY

The information revolution has also altered the way science is done, and scientific developments have brought changes associated with the revolution. James Watson and Francis Crick provoked a flurry of activity in contemporary molecular biology with their discovery of the DNA double helix in 1953. Emphasizing information and control, they and numerous other biologists sought explanations for the behavior and interactions of chromosomes, genes, amino acids, and large protein molecules. Many focused upon heredity and others on development. Because some molecular biologists, especially those with physics backgrounds, left feedback and context out of their explanations for organic growth and reproduction and assumed that a linear flow of control information moved from DNA genes to the human's protein manufacturing "factory," critics considered their approach reductionist. Evelyn Fox Keller, a historian of science, criticized many molecular biologists and geneticists for taking this reductionist linear approach and avoiding the tough intellectual demands required when analyzing simultaneous interactions within complex biological systems.

In contrast to reductionist molecular biologists, some developmental biologists introduced the interactivity of both feedback and systems to explain growth. They conceived of information sources distributed throughout the human organic network. They took their metaphors from cybernetic systems, such as computer networks, instead of from the simpler mechanical machine metaphor used by many molecular biologists. These and like-minded scientists in essence argued that genes alone do not explain embryogenesis; a self-organizing and self-steering cybernetic system should be the explanatory model.

From an informational point of view, the molecular has been anything but reductionist, claims to the contrary notwithstanding. Timothy Lenoir, a historian who follows the evolution of molecular biology, has argued that, unlike theoretical physics with its elegantly stated postulates and the predictions drawn from them, molecular biology has recently become a messy science with an overwhelming flood of experimental observations. To achieve a degree of coherence, this mass of data needs to be organized and analyzed.[5]

In response, biologists and computer scientists have collaborated to develop "bioinformatics," which can be defined as the deployment of information technologies to investigate biological problems, or the fusion of communication and experimentation. Some bioinformatics enthusiasts go so far as to say that biology has been transformed into an information science.

Not only have computer scientists helped the molecular scientists with heuristics, or problem-solving approaches, that bring some order out of the chaos of data, but they have also encouraged biologists to interact with one another through computer networks. As biologists seek fuller understanding of their subject matter, their field is becoming a data-bound subject dependent upon massive computer power. With justification, some observers of the field's development suggest that the wet lab with its emphasis upon observation is giving way to workstations connected to massively parallel computing giving access to enormous banks of data originating in universities, private enterprise, and federal government facilities. Even experiments are carried out in "silico" rather than in "vitro."[6]

WILL THE INFORMATION REVOLUTION CHANGE EVERYTHING?

Enthusiasts have insisted that the information revolution will change everything.[7] Avoiding hyperbole, we can nevertheless reasonably ask if the information revolution will change our lives as much as the second industrial revolution. The answer is both yes and no. About the time the second industrial revolution began in 1880, a twenty-year-old person would not have seen incandescent lights, airplanes, automobiles, or, most probably, telephones. By the age of forty, she or he would know about all of these things. The physical world was dramatically altered.

A twenty-year-old in 1980 would not be familiar with personal computers, blob architecture, computer-generated animation, or digital art, as she or he would two decades later. These innovations notwithstanding, the physical appearance of the human-built world of things has not changed as dramatically during the information revolution as it did during the second industrial revolution.

The dramatic changes over the past decades are not visibly obvious, but have nevertheless changed our lives. Though mostly invisible, the Internet has drastically changed the tempo of communication and provided an incredible mass of information sources. It has also stimulated the development of distant learning. The Massachusetts Institute of Technology is making hundreds of its courses freely available on the Internet. The computer has changed management practices, but these are only sensed by those directly affected and, then, not physically. Silicon Valley is new under the sun, but its physical presence does not match that of once-smoky Pittsburgh or the automobile factories of Detroit. Bill Gates has captured the American imagination, but has not become a folk hero like Edison or Ford.

The most subtle and unseenand unforeseenconsequences of the information revolution is the greatly increased potential for the control of things and people. MIT professor Norbert Wiener in 1948 called attention to this development in his *Cybernetics, or Control and Communication in the Animal and the Machine*. At about the same time, Claude E. Shannon of Bell Laboratories associated communication and control with information. A critical question today is who uses information to control things and people, and for what ends.

T. P. H.
June, 2003

NOTES

1. For more on the Information revolution, see my *Human-Built World: How to Think about Technology and Culture* (Chicago, Illinois: University of Chicago Press, 2004): pp. 96–103, 147–52.
2. This section is based on my "Nothing New under the Sun? A Comparison of Edisonian and Silicon Valley Startups," *STS Nexus* 2. Spring (2002): pp. 26–34.
3. AnnaLee Saxenian, *Regional Advantage: Culture and Competition in Silicon Valley and Route 128* (Cambridge, Mass.: Harvard University Press, 1996): pp. x–xi, 2–3. Steven Levy, *Hackers: Heroes of the Computer Revolution* (Garden City, N.Y.: Anchor Press/Doubleday, 1984): pp. 20–24.
4. Mitchell, William J., "Roll Over Euclid: How Frank Gehry Designs and Builds," *Frank Gehry, Architect*, ed. J. Fiona Ragheb (New York: Guggenheim Museum, 2001): pp. 353–63.
5. Lenoir, Timothy, "Science and the Academy in the 21st Century: Does Their Past Have a Future in an Era of Computer-Mediated Networks?" Paper presented at a conference entitled "Ideale Akademie: Vergangene Zukunft oder konkrete Utopie?" held at Berlin Akademi der Wissenchaften, 12 May 2000.
6. Although my brief survey of bioinformatics suggests the impact of information revolution on medical science, I must ask my readers to turn to those trained in the history of medicine for an appreciation of the ways in which information has dramatically transformed health care.
7. In chapter 4 of *Human-Built World: How to Think about Technology and Culture* (2004), I discuss overly enthusiastic reactions to the information revolution, pp. 103–9.

ACKNOWLEDGMENTS

I am grateful to many persons and organizations for encouragement and support. For time free from teaching and administrative duties and for research and writing, I wish to thank the John F. Guggenheim Memorial Foundation and Joel F. Conarroe, the SEL Foundation and Gerhard Zeidler, Riksbankens Jubileumsfond and Nils-Eric Svensson, the University of Pennsylvania and Michael Aiken, the Technische Hochschule Darmstadt and Helmut Boehme, the Wissenschaftskolleg zu Berlin and Wolf Lepenies and Peter Wapnewski, the Wissenschaftszentrum Berlin and Meinolf Dierkes and Wolfgang Zapf, and the Andrew W. Mellon Foundation.

I am especially indebted to Joachim Nettelbeck and Georg Thurn, friends and colleagues, for unstinting support and constant encouragement while I was carrying on research and writing in Berlin. Agatha, my wife, and I are grateful to Svante Lindqvist, who supported us in Stockholm when the writing burden was unusually heavy. In Munich Otto Mayr, in Starnberg Charlotte and Johannes Ottow, and in Berlin Gesine and Hans-Werner Schütt assisted us in countless thoughtful ways while we were researching and writing there. Among those who, along with me, were Fellows at the Wissenschaftskolleg zu Berlin (Institute of Advanced Study) in 1983–1984 and who took interest in my work were Yehuda Elkhana, Timothy Lenoir, and Martin Warnke. In Darmstadt Evelies Mayr was a supportive friend and colleague, as was Hanns Seidler, while I was Foundation Research Professor at the Technische Hochschule. Mary Anderson and Everett Mendelsohn of Cambridge, Massachusetts, and Grafton, Vermont, were ever giving of encouragement and wise counsel over the years when this book was in preparation. To them we owe a special debt of gratitude.

At the University of Pennsylvania, the faculty of my department and especially my chairperson, Rosemary Stevens, provided a challenging and stimulating intellectual environment during the decade when this book was germinating. This book reflects, I believe, the intellectual spirit of the department. Graduate and undergraduate students at Penn heard and commented perceptively and helpfully on early drafts of the book. The administrative staff in the department, including Sylvia Dreyfuss, Patricia Johnson, Marthenia Perrin, and Joyce Roselle, cheerfully provided help in numerous ways. Also at Penn, Nancy Bauer was ever-encouraging and supportive. I learned much from the discussions that took place over several years with the participants in the Mellon Seminar in Technology and Society, and I am especially appreciative of the contribution that Alfred Rieber made in organizing and presiding over the seminar. At the University of Pennsylvania Jane Morley and Karl-Eric Michelson served imaginatively and faithfully as research assistants; Julie Johnson not only assisted as a researcher, but she participated in the selection and location of illustrations as well. At the Darmstadt Technische Hochschule, Wiltrud Ankenbrand was an extremely resourceful research assistant. Elliot Sivowitch, a senior curator at the National Museum of American History, Smithsonian Institution, and Robert E. Kollar, Chief Photographer of the Tennessee Valley Authority, proved notably resourceful in helping with the selection of illustrations.

In Berlin at the Wissenschaftskolleg an informal reading group consisting of Philip Fisher, Agatha Hughes, Timothy Lenoir, Elaine Scarry, Fanny Waldman, and Elaine and Norton Wise commented perceptively on several chapters. Students at the Royal Institute of Technology in Stockholm were unusually encouraging when I shared with them selections from the book while it was in the making. Also in Sweden the seminar in the history of science at Uppsala headed by Tore Frangsmyr, the technology and society seminar at Lynköping, and the SCARSS seminar at Uppsala proved attentive and critical audiences. I had the opportunity to present sections of the book as discussion lectures at the Swedish Royal Academy of Engineering Sciences and at the Swedish Royal Academy of Sciences. Stanislaus von Moos not only had me present my ideas at his History of Art and Architecture seminar at the Universität Zürich, but he repeatedly provided information and advice while the book was in process. In the United States, faculty seminars at a number of institutions asked me to present chapter sections for comment, and I am grateful to each of them.

Among the archives and libraries that furthered my research were those at Akademie der Künste in Berlin, American Physical Society, AT&T Archives, Bauhaus Archiv in Berlin, Burndy Library, Deutsches Museum in Munich, Edison National Historic Site, Ford Museum and Archives, Hagley Museum and Library, History Office of the National Aeronautics and Space Administration, History Office of the U.S. Department of Energy, Institute of Electrical and Electronics Engineers, Library of Congress, Loyola University of Chicago, National Air and Space Museum of the Smithsonian Institution, National Museum of American Art, National Museum of American History, Royal Institute of Technology in Stockholm, State Historical Society of North Carolina, Tennessee Valley Authority, University of Pennsylvania University Museum, Van Pelt Library and Rare Book and Manuscript Collection of the University of Pennsylvania, Wissenschaftskolleg zu Berlin, and Wissenschaftszentrum Berlin. Among those associated with these libraries and archives and others who gave advice in the selection of documents, books, and photographs were Joyce Bedi, Inge Böhm, Margareta Bond-Fahlberg, Gesine Bottomley, Mary Bowling, Hildegard Bremer, Alice Buck, Harold Dorn, Magdalena Droste, Monika Fenkohl, Sylvia Fries, Michael Grace, Sabine Hartmann, Jane G. Hartje, Sheldon Hochheiser, Agatha Heritage Hughes, Robert Johnston, Robert Kollar, Ann Kottner, Margarete Lehmann-Haslsteiner, Dorothea Nelhybel, Gerd Paul, Cynthia Read-Miller, Ursula Reich, Gudrun Rein, Lee D. Saegesser, Wolfgang Schatton, Harmer F. Schoch, Marsha Siefert, George Stephenson, Regine Sühring, Petra Thoms, Achim Wendschuh, Sabine Wieczorek, and Jon Williams.

Many of the ideas in the book stemmed from lengthy discussion with participants at international seminars on the social construction of technology, the history of systems, postmodern architecture and technology, and the history of technical education held at University of Edinburgh, Scotland; Twente Technische Universität, The Netherlands; Wissenschaftszentrum Berlin; Max Planck Gesellschaft, Köln; ASSI, Tierni, Italy; Wissenschaftskolleg zu Berlin; and Cité des Sciences et de l'Industrie, Paris. Among the participants who helpfully criticized my emerging themes were Wiebe Bijker, Tom Burns, Michel Callon, Edward Constant II, Denise Scott Brown, David Edge, Brian Elliott, Eric Forbes, Robert Fox, Louis Galambos, Anna Guagnini, Bernward Joerges, Todd R. La Porte, Bruno Latour, John Law, Maurice Lévy-Leboyer, Harry Lintsen, Donald Mackenzie, Renate Mayntz, Charles Perrow, Trevor Pinch, Julius Posner, Arie Rip, and Robert Venturi.

I am indebted to Byron Dobell and Fred Allen of *American Heritage*, who saw the promise of this book early on, and to Hal Bowser, whose interview with me in *American Heritage* brought my work to the attention of Daniel Frank, now executive editor of Viking Penguin. Whenever I tell my colleagues of Dan Frank's cordiality and wisdom and his unwavering support throughout the publishing process, I get reactions tinged with envy. Such editors, I believe, are rare. I also wish to thank others, including Michael Millman, at Viking Penguin for the extremely high quality of their contributions to the book.

Those who have read and commented on various chapters include Mark Adams, David Brownlee, Bernard Carlson, Richard Hewlett, David Hounshell, Lucian Parke Hughes, Stanislaus von Moos, Alice Kimball and Cyril Stanley Smith, John Staudenmaier, Frank Trommler, and Alexander Vucinich.

While I was conceptualizing, planning, researching, writing, and editing this book, Agatha Chipley Hughes was always at my side. This close personal and professional association with her provides the richest and most rewarding experience arising from the making of this book. This book is as much hers as mine, as all of our friends and close associates know.

T. P. H.

Berlin and Philadelphia ▪ September 1988

AMERICAN GENESIS

INTRODUCTION

THE TECHNOLOGICAL TORRENT

This book is about an era of technological enthusiasm in the United States, an era now passing into history. Literary critic and historian Perry Miller provides a marvelous image of Americans exhilarated by the thrill of the technological transformation. They "flung themselves into the technological torrent, how they shouted with glee in the midst of the cataract, and cried to each other as they went headlong down the chute that here was their destiny. . . ."[1] By 1900 they had reached the promised land of the technological world, the world as artifact. In so doing they had acquired traits that have become characteristically American. A nation of machine makers and system builders, they became imbued with a drive for order, system, and control.

Most Americans, however, still see themselves primarily as a democratic people dedicated to the doctrine of free enterprise. They celebrate the founding fathers and argue that the business of America is business. They celebrate technological achievements, too, but they see these as fruits of free enterprise and democratic politics. They commonly assume that Americans are primarily dedicated to money making and business dealing. Americans rarely think of themselves principally as builders, a people whose most notable and character-forming achievement for almost three centuries has been to transform a wilderness into a building site.

A major reason that a nation of builders does not know itself is that most of the history it reads and hears instructs otherwise.

Perceptive foreigners are not so prone to sentimentalize America's founding fathers, frontiersmen, and business moguls. Other peoples have looked to the United States as the land of Thomas Edison, Henry Ford, the Tennessee Valley Authority, and the Manhattan Project. Foreigners have made the second discovery of America, not nature's nation but technology's nation. Foreigners have come to Philadelphia to see Independence Hall, but those who wish to understand the foundations of U.S. power have asked to see Pittsburgh when it was the steel capital of the world, Detroit when most automobiles were made there, the Tennessee Valley Authority when engineering was transforming a poverty-stricken valley into a thriving one, and New York City because its skyscrapers symbolized the technological power of the nation. The Manhattan Project, which produced the atom bomb, reinforced the belief throughout the world that America was the technological giant. Until the space-shuttle disasters and an embarrassing series of launching failures, the National Aeronautics and Space Administration symbolized America's technological creativity.

Americans rightly admire the founding fathers, who displayed extraordinary inventiveness as they conceived the Declaration of Independence and framed the Constitution, but Americans have embodied comparable, if not greater, inventiveness in the material constitution, the technological systems of the nation. Perhaps the myth that they are essentially a political and business people may be emended, if they reflect more on the technological enthusiasm and activity they have displayed throughout their history, but most obviously during the century from about 1870 to 1970. The enthusiasm reached its height during the middle decades of the period, then subsided, especially after World War II. This book, despite its emphasis on invention, development, and technological-system building, is not a history of technology, a work of specialization outside the mainstream of American history. To the contrary, it is mainstream American history, an exploration of the American nation involved in its most characteristic activity. Historians looking back a century from now on the sweep of American history may well decide that the century of technological enthusiasm was the most characteristic and impressively achieving century in the nation's history, an era comparable to the Renaissance in Italian history, the era of Louis XIV in France, or the Victorian period in British history. During the century after 1870, Americans created the modern technological nation; this was the American genesis.[2]

In popular accounts of technology, inventions of the late nineteenth century, such as the incandescent light, the radio, the airplane, and the gasoline-driven automobile, occupy center stage, but these inventions were embedded within technological systems. Such systems involve far more than the so-called hardware, devices, machines and processes, and the transportation, communication, and information networks that interconnect them. Such systems consist also of people and organizations. An electric light-and-power system, for instance, may involve generators, motors, transmission lines, utility companies, manufacturing enterprises, and banks. Even a regulatory body may be co-opted into the system. During the era of technological enthusiasm, the characteristic endeavor was inventing, developing, and organizing large technological systems—production, communication, and military.

The development of massive systems for producing and using automobiles and for generating and utilizing electric power, the making of telephone and wireless networks, and the organization of complex systems for making war reveal the creative drive of inventors, engineers, industrial scientists, managers, and entrepreneurs possessed of the system builder's instincts and mentality. The remarkably prolific inventors of the late nineteenth century, such as Edison, persuaded us that we were involved in a second creation of the world. The system builders, like Ford, led us to believe that we could rationally organize the second creation to serve our ends. Only after World War II did a handful of philosophers and publicists whom we now associate with a counterculture raise doubts about the rationality and controllability of a nation organized into massive military, production, and communication systems. Their doubts increased as the nation's technological pre-eminence waned.

If the nation, then, has been essentially a technological one characterized by a creative spirit manifesting itself in the building of a human-made world patterned by machines, megamachines, and systems,

Americans need to fathom the depths of the technological society, to identify currents running more deeply than those conventionally associated with politics and economics. Indeed, many of the forces that Americans need to understand and control in order to shape their destiny, insofar as that is possible, are now not primarily natural or political but technological. We celebrate Charles Darwin for discerning patterns in the natural world; we do not yet sufficiently appreciate the importance of finding patterns in the human-made, or technological, world.[3] The purpose of the understanding is not simply to comprehend the impressively ordered, systematized, and controlled, but to exercise the civic responsibility of shaping those forces that in turn shape our lives so intimately, deeply, and lastingly.

A history stressing the technology of an era of technological enthusiasm should be no more celebratory than a history stressing the politics and business of a gilded age. The tendency of popular histories and of museum exhibits of technology uncritically to unfold a story of problem-free achievement unfortunately leaves readers and viewers naïve about the nature of technological change. When more histories of technology that take the critical stance of the best histories of politics are written, Americans will realize that not only their remarkable achievements but many of their deep and persistent problems arise, in the name of order, system, and control, from the mechanization and systematization of life and from the sacrifice of the organic and the spontaneous.

This history, then, argues that inventors, industrial scientists, engineers, and system builders have been the makers of modern America. The values of order, system, and control that they embedded in machines, devices, processes, and systems have become the values of modern technological culture. These values are embedded in the artifacts, or hardware. Modern inventors, engineers, industrial scientists, and system builders, those who flourished in the century of technological enthusiasm, concerned themselves with the production of goods and services and with preparations for, and the waging of, war. Their influence, however, did not end with these activities. Their numerous and enthusiastic supporters from many levels of society believed their methods and values applicable and beneficial when applied to such other realms of social activity as politics, business, architecture, and art.

This history, however, does not argue technological determinism. The creators of modern technology and the makers of the modern world expressed long-held human values and aspirations. Although the inventors, engineers, industrial scientists, and system builders created order, control, and system, in so doing they responded to a fundamental human longing for a world in which these characteristics prevail. They became the instruments of all those, including themselves, who were uneasy in a seemingly chaotic and purposeless world and who searched for compensatory order. In this sense, technology was, and is, socially constructed. As historian and social critic Lewis Mumford so eloquently insisted decades ago, technology is both a shaper of, and is shaped by, values.[4] It is value-laden.

Despite the drive for order, system, and control among the practitioners and enthusiasts of technology, the history of technology, like the history of politics, is complex and contradictory. Framers of constitutions have also tried to establish timeless, all-embracing systems of checks and balances. Neither they nor the designers of machines, devices, and processes have found the one best solution that pleases everyone and resists change. Contrary to popular myth, technology does not result from a series of searches for the "one best solution" to a problem. This book does not present technology as engineers are taught even today to think about it: as an absolutely one-best-way solution to problems. Instead, it presents practitioners of technology confronting insolvable issues, making mistakes, and causing controversies and failures. It shows the practitioners creating new problems as they solve old ones. This book intends to present the history of modern technology and society in all its vital, messy complexity.

Technology in the age of technological enthusiasm meant then, as now, different things to different people. The efforts of textbook writers notwithstanding, technology can be defined no more easily than politics. Rarely do we ask for a definition of politics. To ask for *the* definition of technology is to be equally innocent of complex reality. For many people, technology is goods and services to be consumed by the affluent, to be longed for by the poor. Others, such as inventors and engineers, see technology as the creation of the means of production for these goods and services. Further up the ladder of power and control, the great system builders, people like Ford, find consumingly interesting the organizing of the material world into great systems of production. Still others analyzing modern technology find rational method, efficiency, order, control, and system to be its essence. Taking into consideration the infinite aspects of technology, the best that I can do is to fall back on a general definition that covers much of the activity described in this book. Technology is the *effort* to organize the world for problem solving so that goods and services can be invented, developed, produced, and used.[5] The reader, however, can accept instead of a definition the historian's traditional approach of naming a subject and defining it by examples of his or her choice.

This book centers less on ideas and more on people, especially American inventors, engineers, system builders, architects, artists, and social critics. The organizations and movements of a modern culture, the institutional frameworks and symbolic structures in which inventors, system builders, and others acted are not, however, neglected. Among the organizations considered are the inventor's workshop, the industrial research laboratory, the business corporation, the government agency, and the military-industrial complex. Among the movements included are the international style in architecture; the Futurists, Constructivists, Dadaists, and Precisionists in art; scientific management and progressivism in production and politics; and the conservationist and counterculture advocates among the social critics. Throughout, references to modern culture refer to the devices, machines, processes, values, organizations, symbols, and forms expressing the order, system, and control of modern technology, and to the thought and behavior mediated by these and their expression.[6]

There is a pattern to this book analogous to that of the growth of the large technological systems about which it is written. The early chapters treat the invention of systems; the middle section deals with the spread of large systems; and the final chapters recount the emergence of a technological culture, of mammoth government systems, and counterculture reaction to systems. The remarkable achievements of independent inventors and industrial research opened and greatly shaped the age of technological enthusiasm. Philosopher Alfred North Whitehead believed that the invention of a method of invention was the greatest invention of the era.[7] Men and women assumed, as never before, that they had the power to create a world of their own design. Independent inventors experienced their heyday during a gilded era after the United States had emerged from the Civil War, and they forged a massive productive enterprise that ended up dominated by giant corporations. The historians Charles and Mary Beard called this the era of "the Second American Revolution," referring to the momentous technological, economic, political, and social changes.[8] Mumford saw it as the beginning of the modern, or neotechnic, era in the history of technology and society.[9] The inventions of the independents provided the foundations for the rise of the industrial giants, especially the newly burgeoning electrical industry. Edison, the Wizard of Menlo Park, became the heroic figure of the era, but there were other independent inventors, such as Elmer Sperry, who were impressively creative and more professional. The inventors continued to flourish as their country competed successfully with the great European powers for industrial supremacy. As World War I approached, the inventors became involved in inventing for the military. The military establishment funded their inventive activity and used their creations to develop new weapons, strategy, and tactics.

By the beginning of World War I, American inventors had helped to establish the United States as the most inventive of all nations. Only Germany, recently united, seemed a competitor for the title. Inspired by German achievements, leading American corporations such as General Electric, Du Pont, General Motors, and Bell Telephone also established industrial research laboratories. Industrial scientists widely criticized the haphazard methods of the independent inventors and claimed that the mantle of creativity had fallen onto their own shoulders. Yet there flowed from the industrial laboratories inventions with a conservative cast improvements rather than dramatic innovations. During World War I in the United States, the scientists, especially those with graduate training in physics, effectively challenged the role of the independent inventors as the source of improvements in military systems. The war-waging nations, dependent on their inventors and scientists, innovated and counterinnovated with the submarine, airplane, tank, and poison gas much as large corporations in peacetime contended for market advantages with innovations. Technology was capable of creating not only a new life-supporting world, but also a deadly environment.

The inventions and discoveries of the inventors and the industrial scientists became part of large systems of production that expanded impressively during the interwar years. These systems were the work of the system builders, whose creative drive surpassed in scope and magnitude that of the inventors. Designing a machine or a power-and-light system that functioned in an orderly, controllable, and predictable way delighted Edison the inventor; designing a technological system made up of machines, chemical and metallurgical processes, mines, manufacturing plants, railway lines, and sales organizations to function rationally and efficiently exhilarated Ford the system builder. The achievements of the system builders help us understand why their contemporaries believed not only that they could create a new world, but that they also knew how to order and control it. Frederick W. Taylor, father of scientific management, became famous, or notorious, throughout the industrial world for his techniques of order and control. Several chapters of this book focus on the American system builders.

American technology, especially its systems of production, fascinated European industrial managers, bureaucrats, social scientists, and social critics. Fordism and Taylorism for them symbolized the essence of the modern American achievement. Fordism and Taylorism spread throughout Europe, much as Japanese managerial techniques would into the United States after World War II. Lenin and other leaders of the Soviet Union displayed even greater enthusiasm for Fordism and Taylorism than the Americans had. When the Soviet Union embarked on a Five-Year Plan that specified mammoth regional systems of technology based on hydroelectric power and prodigiously rich stores of Siberian natural resources, it turned to American consulting engineers and industrial corporations for advice and equipment. The Soviets constructed entire industrial systems modeled on the steel works in Gary, Indiana, and hydroelectric projects on the Mississippi. In Weimar Germany after World War I, many persons believed that Taylor and Ford had the answer not only to production problems but to labor and social unrest as well. They labeled Ford's ideas white socialism, believing this to be an answer to Marxism. Many Europeans, especially Weimar Germans, decided that democracy, American technology, and a new European and modern culture could restore war-devastated Europe and create a good society. In the Soviet Union, Lenin predicted that Soviet politics, Prussian railway management, American technology, and the organizational forms of the trust-building entrepreneurs would bring the new socialist society.[10]

Modern technology was made in America. Even the Germans who developed it so well acknowledged the United States as the prime source. During the interwar years, the industrial world recognized the United States as the pre-eminent technological nation, and the era of technological enthusiasm reached its apogee. Modern technological culture, however, was defined in Europe. The Europeans held up a mirror in which the Americans could see themselves as the raw materials of modernity which the Europeans wanted to fashion into modern culture. European engineers, industrialists, artists, and architects came to America to admire its "plumbing and its bridges"[11] and made, as we have observed, the second discovery of America—the great systems of production.

From the turn of the century on, avant-garde European architects and industrial designers searched for ways to combine American modes of mass production and the principles of quality design. In so doing they were inventing the forms and symbols for a modern technological culture. In the 1920s, at the Bauhaus in Dessau, Walter Gropius and his architect and artist associates brought the movement to a climax by contributing greatly to the establishment of the modern or international style of architecture and design. This style expressed in construction methods and in formal design the principles of modern American technology. The dire housing shortage following World War I spurred Gropius and other avant-garde architects to apply the mass-production methods attributed to Ford and the scientific management methods of Taylor. A description of the construction of great housing settlements in Dessau and Berlin makes this clear. In France, Le Corbusier fervently and eloquently articulated the technological age. In his journal *L'Esprit nouveau*, published in the 1920s, he sought to define verbally and visually the modern in art, architecture, and interior and industrial design. He believed that American engineers had found the heart of modern design by joining, in their bridges, ocean steamers, grain silos, and automobiles, a mathematical exactness with rational methods of production and rational design. The architects who adopted the engineers' techniques and infused them with the aesthetics of the artist were, he was convinced, creating the modern style. Le Corbusier was more enamored of order and system than the engineers themselves.

Painters, too, became self-consciously modern. The Italian Futurists, around the turn of the century, saw modern technology as a way of destroying traditional culture in Italy. Social and artistic radicals, they found Italy backward and oppressive. Modern motor cars, not Renaissance museums, held the key to the future for Italians. The Futurists celebrated the dramatic and dynamic artifacts of modern technology—"adventurous steamers that sniff the horizon . . . deep-chested locomotives whose wheels paw the tracks . . . the sleek flight of planes. . . ."[12] After the Russian Revolution of 1917, the Soviet artists of the Constructivist movement, several of whom were graduate engineers, also envisioned art as a means of radically transforming culture, of bringing into being the new Soviet society. Vladimir Tatlin conceived of "machine art" and El Lissitzky of new elements of style from which a modern art and architecture could be created that would influence the character of the new man in the modern social system. In Germany after the war, the artifacts and order of the technological world fascinated the artists of the Neue Sachlichkeit school. Their visual vocabulary included "order," "clarity," and "harmony." They thought these to be the principles of technological rationality and the governing principles of the human-made world.

In 1915 Marcel Duchamp and Francis Picabia came to New York and emboldened a few American artists to look to technological America rather than Europe for the subject matter, forms, and symbols of the modern. The American Precisionists Charles Sheeler and Charles Demuth, and the Russian American Louis Lozowick painted technological landscapes and objects inspired by the development of modern systems of production. Their work was exemplified by Sheeler's series of paintings and photographs of Ford's River Rouge plant.

Leading American architects did not adopt a formal vocabulary characterized by a technological or machine aesthetic until the 1930s, when Gropius, Ludwig Mies van der Rohe, and other avant-garde architects, emigrating from Nazi Germany to the United States, brought with them the international style. The paradox remains that, although modern technology originated in America, modern painting and architecture inspired by it germinated and took root first in Europe.

The Great Depression and the violence and destruction made possible by modern technology during World War II dampened technological enthusiasm, but technological systems entered a new stage in the United States when the government became involved in their cultivation. Franklin Delano Roosevelt inaugurated the Tennessee Valley Authority, a government-funded, -designed, -constructed, and -operated project that systematically developed the resources of an extensive river valley. Once again the United States provided the world a model of modern technology. During World War II, the United States poured unprecedented resources into the Manhattan Project, a technological system of unprecedented size. When President Dwight Eisenhower later warned his nation about the increasing momentum of the military-industrial complex, he referred to the rise of great systems of armament production modeled on the Manhattan Project. The Strategic Defense Initiative, or Starwars, exemplifies the most recent military-industrial (and university) complex.

The dropping of the bombs on Hiroshima and Nagasaki starkly revealed for many the threat of uncontrolled, destructive, technological creativity and the massive size of technological projects and systems in which the government was involved. Subsequent and largely unsuccessful efforts to bring about control of the nuclear arsenal heightened these anxieties. Rachel Carson in *Silent Spring* and others who followed her lead stimulated an increased concern about environmental costs of large-scale production technology. The wasting of Vietnam by military technology brought the growing reaction to a head. A counterculture erupted. Reflective radicals of the 1960s, both in America and abroad, attacked modern technology and the order, system, and control associated with it. The counterculture called for the organic instead of the mechanical; small and beautiful technology, not centralized systems; spontaneity instead of order; and compassion, not efficiency. Paul Goodman, Herbert Marcuse, and other intellectual leaders of the counterculture unerringly aimed their attacks at technological rationality and system. Mumford, whose critical concern about technology and society antedated that of the counterculture, also wrote of megamachines. Jacques Ellul also criticized the technological systems that he and Mumford feared were determining the course of history.

Time has dampened the bitterness and vision of the counterculture. Today technological enthusiasm, although much muted as compared with the 1920s, survives among engineers, managers, system builders, and others with vested interests in technological systems. The systems spawned by that enthusiasm, however, have acquired a momentum—almost a life—of their own. They involve the surviving technological enthusiasts, persons whose income derives from the systems, large corporations, government agencies, and politicians beholden to those with vested interests in the systems. A multitude of persons persuaded that armaments and the producers of them are critical for the nation's defense and survival adds to the momentum of military-industrial systems. The age of technological enthusiasm has passed, but it has left behind a burden of history. Those who know the history and the burden may be able to rid themselves of it or turn it to their ends.[13]

1
A GIGANTIC TIDAL WAVE OF HUMAN INGENUITY

No other nation has displayed such inventive power and produced such brilliantly original inventors as the United States during the half-century beginning around 1870. Periclean dramatists, Renaissance artists, British engineers during the Industrial Revolution, late-nineteenth-century Berlin physicists, and Weimar architects in the 1920s all stimulate memories of similarly remarkable creative eras. During such times as these Sophocles and Euripides wrote *Oedipus Rex* and *Medea;* architect engineers, including Francesco di Giorgio Martini, Leonardo da Vinci, and Michelangelo left notebooks, bridges, canals, fortifications, and secular and religious buildings of surpassing ingenuity and beauty; George Stephenson and Isambard Kingdom Brunel laid the railroads and built the bridges that transformed the face of Britain; Hermann von Helmholtz and Max Planck revealed the conceptual power and elegance of modern physics; and a few avant-garde architects, among them Walter Gropius, Le Corbusier, and Ludwig Mies van der Rohe, established the international school of modern architecture. As yet, however, we have not realized the remarkable quality of a comparable era in American history when the independent inventors, Thomas Alva Edison and Orville and Wilbur Wright among them, introduced electric lighting, airplane flight,

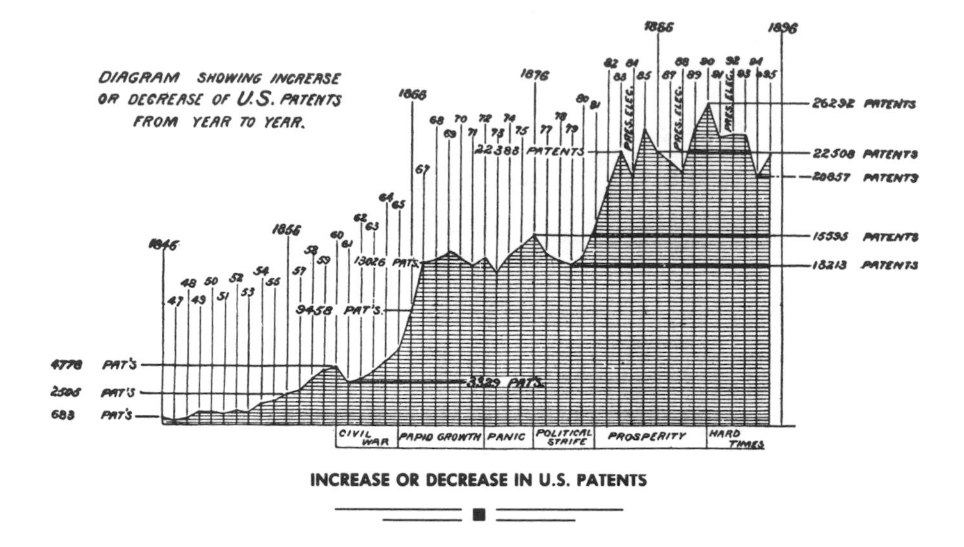

INCREASE OR DECREASE IN U.S. PATENTS

wireless transmission, and a multitude of inventions that shape the modern world.

In 1896 a writer in the *Scientific American* referred to the remarkable outpouring of U.S. patents since the Civil War, exuberantly insisting that his was "an epoch of invention and progress unique in the history of the world. . . . It has been," he observed, "a gigantic tidal wave of human ingenuity and resource, so stupendous in its magnitude, so complex in its diversity, so profound in its thought, so fruitful in its wealth, so beneficent in its results, that the mind is strained and embarrassed in its effort to expand to a full appreciation of it."[1] The number of patents issued annually more than doubled between 1866 and 1896, and the number for each person increased more than 1.75 times. The historian Daniel Boorstin has observed that "all the resources which had been used to lay tracks across the continent, to develop an American System of Manufacturing in its several versions, now went into American Systems of Inventing."[2] Not only were tens of thousands of Americans inventing at the grass-roots level, but a singular band of independent inventors was also flourishing during the decades extending from about 1870 to 1920.

Before the rise, about 1900, of the industrial research laboratory, and long before that of the large government-funded national laboratories that originated in World War II with the militarization of nuclear power, the nation's technical inventiveness was concentrated in the independent inventors. The role and characteristics of the large laboratories are fairly well understood today because they are well publicized and still with us, whereas the role of the inventors has been sentimentalized, trivialized, or forgotten. Yet, if we wish to understand the nation's rise to industrial and technological pre-eminence, we ought to fathom the complex character and manifold activities of the independent inventors. Instead of accumulating more biographical sketches of a heroic cast, we need to discover and understand the characteristics the inventors shared.

The era of the independent inventors began about the time Alexander Graham Bell invented his telephone and Edison opened his Menlo Park laboratory in 1876. The failure of the U.S. Navy board of inventors, headed by Edison during World War I to fulfill its expectations, in addition to the success of a group of physicists called to solve wartime problems, signaled the end of the golden era of the independent inventors. After World War I, industrial scientists displaced the independents as the principal locus of "research and development activity," the new name for invention. During the intervening decades, as the independents flourished, the United States became not only the most inventive nation, but also the world's industrial leader, thereby surpassing the United Kingdom, whose leaders had long belittled the industry and technology of its former colony. Between 1895 and 1900, U.S. coal production overtook the British—by 1915 the young nation's production had doubled that of her rival; between 1885 and 1890 the United States also moved ahead of the British in pig-iron and steel production; and leadership in the production of heavy chemicals passed from British to American hands between 1900 and 1913. The United States moved ahead in the well-established fields of heavy industry and took a clear and substantial lead in electric light and power, a new industry at the front edge of high technology.[3] These, then, were epochal decades, and ones to whose technological and industrial achievements the independent inventors substantially contributed.

The list is long of outstanding inventions attributable to the U.S. independent inventors during their era of preeminence. These inventions include Bell's telephone, Edison's incandescent lamp, phonograph, and motion-picture system, and William Stanley's, Nikola Tesla's, and Elihu Thomson's contributions to the development of electric light-and-power transmission. The Wright brothers introduced the internal-combustion-engine airplane, and Reginald Fessenden, Lee de Forest, and Edwin Armstrong pioneered in wireless telegraphy and telephony, or radio. Elmer Sperry and Hiram Stevens Maxim spread their inventive activities across a number of fields, including electric lighting, but Sperry is best remembered for his gyrocompass and automatic control devices for the navy and Maxim for his machine gun. The list reveals the prominent role that the American independents played in the rise of the electrical industry and the application of invention to military purposes during the armaments race preceding World War I. They were major contributors to the era of rapid industrialization sometimes called the "second industrial revolution."

In our exploration of the nature of invention and the style of the independents, we shall refer to the activities of several of those who flourished in the half-century after 1870. Listed chronologically, they are Hiram Stevens Maxim (1840–1916), Alexander Graham Bell (1847–1922), Thomas A. Edison (1847–1931), Elihu Thomson (1853–1937), Nikola Tesla (1857–1943), William Stanley (1858–1916), Elmer Sperry (1860–1930), Reginald Fessenden (1866–1932), Wilbur Wright (1867–1912) and Orville Wright (1871–1948), Lee de Forest (1873–1961), and Edwin H. Armstrong (1890–1954). They shared some similar experiences and characteristics. Maxim, Edison, Tesla, Sperry, and de Forest grew up in rural communities; neither Maxim, Edison, Thomson, Stanley, Sperry, Fessenden, nor the Wright brothers obtained college or university degrees. All of the independents at early ages displayed precocious interest in mechanical and electrical devices. Maxim's early jobs included one in Boston in the shop of a scientific-instrument maker. Edison began as a telegraph operator. Thomson taught chemistry at Central High School in Philadelphia, but became a full-time inventor within a few years. Young Stanley worked as an assistant for the inventors Maxim and Edward Weston. Tesla, after university, found employment as an engineer in the telegraph and electric-lighting field; Sperry began as an inventor of electric lighting. Fessenden worked as engineer and inventor for both Edison and

George Westinghouse. Later he became a professor of electrical engineering at Western University of Pennsylvania (Pittsburgh). The Wright brothers operated a bicycle shop in Dayton, Ohio. After a Ph.D. at Yale University, de Forest became a technical assistant with Western Electric, the Chicago telephone manufacturer. Armstrong started as an assistant in the electrical-engineering department at Columbia. Bell was a professor of elocution at Boston University when he invented and developed his telephone. All began work on major inventions before they were thirty years old.

Several characteristics of each prove memorable. Edison is the Nestor of the independents and history's best-known inventor. Only Leonardo da Vinci evokes the inventive spirit as impressively but, unlike Edison, Leonardo actually constructed only a few of his brilliant conceptions. Edison acquired more than a thousand patents and brought countless inventions into use, a process commonly known as innovation, to distinguish it from the prior, conceptual, stage of invention. Other contemporary independent inventors so often told anecdotes about themselves that resemble episodes in Edison's life that we have reason to suspect that he, who was so well publicized, was often their conscious model. Less well known than Edison, Sperry provides a better model for an aspiring inventor today than Edison. Sperry's involvement in high, or complex, technology was deeper than Edison's. Sperry should be remembered as the father of cybernetic, or feedback-control, engineering. Whereas Edison veered off in later life from invention into the organization and management of industrial processes and research, Sperry rarely strayed from the craft of invention. The U.S. National Academy of Sciences perceptively elected Sperry to membership earlier than Edison.

Bell was the archetypal amateur inventor of genius. Possessed by the idea of voice transmission by electricity, he enthusiastically pursued his vision during the hours free from his professorial responsibilities. After successfully launching his invention and reaping a handsome income from it, he did not continue to pursue invention in a professional way, but followed his interests wherever they led him in science and technology. He is not remembered for any other major commercial invention. Tesla best fills the popular image of inventor as genius. He did nothing to lessen the impression. Darkly handsome, he dressed elegantly, lived alone in exclusive New York hotels, moved with New York's social and financial elite, built laboratories of spectacular architecture, and gave demonstrations of high-voltage phenomena that rivaled nature's displays of lightning. Today he is remembered for the introduction of modern electric-power transmission. Fervid admirers insist that he also envisioned inventive possibilities then not fully comprehended, possibilities that they felt he could have brought to fruition if the funding had been at his disposal. The fame of the Wright brothers has increased and spread as aeronautics and astronautics have increasingly absorbed the resources of our era. Not until recently have historians of science and technology taken over the cultivation and preservation of the story of the Wrights from the popularizers. As a result, the Wrights now emerge from history not simply as dogged tinkerers, but, like so many other contemporary independent inventors then, as methodical, even scientific, in solving technical problems.

De Forest and Fessenden were the foremost American inventors and early developers of wireless transmission. De Forest possessed the added distinction of having invented the three-element vacuum tube that launched modern electronics. Fessenden, inordinately proud of the scientific nature of his inventions, led the way in transforming wireless from telegraphy (transmission by code) to telephony (transmission by voice). De Forest's career is instructive, for it reminds us that independent inventor-entrepreneurs had often to be ingenious promoters of publicity and resourceful fund-raisers. He also revealed how dependent inventors were on one another for stimulating ideas. His critics, who included Fessenden, insisted that de Forest was so dependent that his inventions were derivative. As a result, de Forest knew patent lawyers and the courts as well as workshops and laboratories. Armstrong's disputes with de Forest about priority are legendary. Armstrong's technical and scientific peers sided with him in the dispute about who invented the vacuum tube used for radio reception and transmission, but the courts found the legal nuances favoring de Forest. Endless litigation with de Forest and others appears to have broken the brilliant Armstrong's spirit. He committed suicide after he had become involved in particularly harassing, lengthy litigation.

Thomson ranks high among the independent inventors because he, like Sperry, displayed methodological characteristics in the workshop and the laboratory as inventor and in the business world as entrepreneur that recommended him for emulation then and now. He also chose to solve problems in the rapidly expanding field of electric light and power which allowed him to ride the crest of a wave to become a founder of the General Electric Company. Stanley merits our attention because he possessed the essential instinct of the fruitful independent to withdraw prudently whenever possible from the hurly-burly world of promotion and finance to the isolated and sustaining spaces of the workshop and the laboratory. Maxim is included among those whose lives we shall use to characterize independent inventors because he possessed the essential quality of choosing problems to solve that kept him in the mainstream of technological change, no matter how tortured the course.

Despite their major role in American history, the independent inventors remain poorly understood. A host of older popular biographies and children's books have reduced the independents, in the public's imagination, to one-dimensional heroes, thereby mystifying their creativity. Fixed in the public mind is the image of Edison in a Napoleon-like pose brooding over an incandescent lamp that, we are told, was invented by a hunt-and-try technique assisted by a stroke of genius. In the lore of invention, the lamp sprang without precedent from his massive brow. Also part of the currency of invention is the Wright brothers' tinkering with bicycles in a Dayton shop and with a flimsy airplane at Kitty Hawk, North Carolina. Dependency of pioneers on the prior experiences and publications of others and painstaking experimentation are foreign to popular stories of heroic creativity. Whether writing of Edison, the Wrights, Tesla, or other independents, the early biographers who established the heroic-inventor myths rarely asked whether the inventors had a method of invention, whether they depended on assistants and laboratories, whether their problem choices were as critical for their successes as their problem solutions, whether money raising took an inordinate amount of their time, whether military support spurred them on, and whether the mysterious creative process proved to be much like everyday problem solving punctuated by rare eureka moments.

Perhaps Sperry, the most professional of the independent inventors of their golden era, can best establish the tone that we shall adopt in our interpretation of them. In later life Sperry, founder of the field of practical cybernetics, author of more than 350 patents in a wide range of fields, the inventor described as contributing more to the modernization of the U.S. Navy than any other, and the first among the independents to be elected to the National Academy of Sciences, said:

> Think as I may, I cannot discover any time in which I have felt in the course of my work that I was performing any of the acts usually attributed to the inventor. So far as I can see, I have come up against situations that seemed to me to call for assistance. I was not usually at all sure that I could aid in improving the state of affairs in any way, but was fascinated by the challenge. So I would study the matter over; I would have my assistants bring before me everything that had been published about it, including the patent literature dealing with attempts to better the situation. When I had the facts before me I simply did the obvious thing. I tried to discern the weakest point and strengthen it; often this involved alterations with many ramifications which immediately revealed the scope of the entire project. Almost never have I hit upon the right solution at first. I have brought up in my imagination one remedy after another and must confess that I have many times rejected them all, not yet perceiving the one that looked simple, practical and hard-headed. Sometimes it is days and even months later that I am brought face to face with something that suggests the simple solution that I am looking for.[4]

INDEPENDENT INVENTORS

A key to understanding the independent inventors lies in realizing that their freedom from organizational entanglements left them free to choose their own problems. Contrary to popular myth, however, they were not heroic poverty-stricken loners working in dilapidated garrets, using only a hunt-and-try approach. The successful independent inventors customarily worked with a few assistants, mostly craftsmen, and in small laboratories or workshops that they designed and owned. They used an empirical approach when science and scientific method failed, but they also relied on organized information and experimental techniques like those used by experimenting scientists. Edison declared that he could apply scientific law and follow logical method when he was experimenting in the electrical field, but that the state of chemical science in his areas of interest drove him to hunt-and-try. As inventor-entrepreneurs supported by advanced facilities engaged in complex creative work, they helped establish a tradition that finds its counterpart today in the inventor-entrepreneurs of Route 128, near Boston; in Silicon Valley, California; and in inventive scientists still finding a measure of independence in university environments, despite the omnipresent mission-oriented grants.

The ranks of the independents in the later nineteenth and early twentieth centuries included both professionals, like Edison, and nonprofessionals, like the Wright brothers. Nonprofessionals and professionals cannot be distinguished by the quality or complexity of their creations. The difference between them is in the means of livelihood. The professionals spent most of their lives inventing various devices, with income from these providing their means of support. Characteristically, the nonprofessionals concentrated on a single invention and depended on regular employment, often unrelated to their inventive interests, until income from their major invention made them wealthy. As we have noted, Bell typifies the nonprofessional in his enthusiastic concentration on, and singular success with, the telephone.[5] Edison the professional, by contrast, accumulated more than a thousand patents spread over a bewildering variety of technological fields. The Wright brothers were also nonprofessionals in their single-minded concentration on the airplane. Bell and the Wrights depended on income from other work while pursuing their inventive activity. Only after the telephone and airplane became commercial successes could they depend on the income from their several patents or from the companies founded on the basis of these. On the other hand, Sperry—with his more than 350 patents ranging from streetcars to automatic airplane pilots—was a professional. He derived sufficient income from these to support himself, his family, and his inventive activity.

In order properly to appreciate the independents, both nonprofessional and professional, we must see them as more than inventors. As inventor-entrepreneurs they also presided over the introduction of their inventions into use. Certainly their enthusiasm lay in creation, but they realized that, unless nurtured, the brainchild might not survive. In alliance with capitalists, but not employed by them, these inventor-entrepreneurs often established companies to manufacture and sell their patented inventions, but at the same time they managed to distance themselves from the routine managerial responsibilities of the companies. Edison's first love was invention, not the money raising and organizing of companies so often necessary in order to bring his lighting system into being. About 1880 he founded an entire family of interrelated manufacturing firms that became Edison General Electric. His contemporary Thomson, also an inventor of electric-light systems, helped found the Thomson-Houston Electric Company to manufacture and market his inventions, but he, too, realized within a few years that management and finance were best left to others. In 1893 the managers and financiers of the two companies, led by J. Pierpont Morgan, epitome of the financier-entrepreneur, merged them to form the General Electric Company. The engineering departments, and later the industrial research laboratories of these companies, with their far greater material and personnel resources, took over responsibility for the inventive activity that made possible the further expansion of electric light-and-power systems. When other independents, like Sperry, found their particular innovative talents no longer needed, they, too, withdrew—or were forced out—from the companies they founded.

The activities in 1906 and 1907 of Fessenden, an inventor of wireless telegraphy and wireless telephony, impress us with how thoroughly invention and entrepreneurship were mixed. Fessenden was supervising tests of a U.S. Navy wireless installation in Puerto Rico at the same time that he was trying to get a bill through Congress to allow him and his

REGINALD FESSENDEN AND HIS GANG OF WIRELESS OPERATORS AND EXPERIMENTERS

company to sue the government because the navy was making and using without payment a wireless receiver on which he had a patent. He was experimenting and developing a wireless telephone, the success of which soon brought him widespread recognition. In May 1906 he had a short interview with President Theodore Roosevelt to present his objections to the navy's using the Fessenden receiver without compensating him. About the same time that he was seeing the president, he was supervising trans-atlantic transmission of wireless signals with apparatus he had invented and was developing. In January 1906 he prepared to counter the possibility that Guglielmo Marconi would take the occasion of a lecture in New York City to announce that his company was regularly transmitting two-way transatlantic messages. Fessenden planned to have one of his representatives at the lecture to respond that Fessenden's National Electric Signaling Company had already been regularly transmitting between the U.S. mainland at Brant Rock, Massachusetts, and Britain, and that it was ready to transmit any message that the chairman of the proceedings might have.[6]

Even though the independent inventors were entrepreneurs presiding over change and innovation, they tended to count patents as well as — perhaps more than—money as symbols of success. As their memoirs and biographical sketches show, the number of patents acquired gave them status among their inventor peers. A major reason the independent inventors have not been—and are not—understood today is that we confuse their drives and goals with those of the financial or business entrepreneur. The independents reserved their enthusiasm and primary creative thrust for the act of invention; they performed the entrepreneurial function of establishing companies because they wanted to bring their inventions into use. They had to establish companies because they found that firms busily presiding over well-established technologies were usually not interested in nurturing radically new technologies with which their employees had no experience and for the manufacture of which their machines and processes were not suited. The independent inventors also found that others did not nurture their inventions with the tender care and undivided attention they themselves lavished on their brainchildren.

MACHINE SHOPS AND LABORATORIES

Characteristically, independent inventors withdrew to spaces of their own choice or design. In their withdrawal, the inventors were like avant-garde artists resorting to the atelier or the alternative life-style of a historic Montmartre, Schwabing, or Greenwich Village. Aware of the unorthodoxy of their ideas, inventors and artists intensified their feelings of being outsiders by their physical withdrawal. Working in their retreats, intellectual and physical, they created a new way, even a new world, to displace the existing one. The inventors created machines and processes among which they felt at home, and the artists invented pure and ordered spaces filled with music, painting, and sculpture. In their childhood or adolescence, many inventors and artists had encountered a world that disappointed them. Thus their withdrawal to spaces of their own making, filled with the devices of their imagination, became a response, an attempt to change or make anew the world.

The withdrawal to isolated spaces of their own choosing and design not only removed inventors and artists from the constraining influences of the status quo but also sheltered them from the hostility or ridicule of those whose established views and institutions the inventors' new ideas would undermine. Sperry's wife, Zula, said he was never happier than when he was working in the company of his enthusiastic assistants. Tesla treasured those hours in his Manhattan laboratory pursuing his insights into the mysterious electrical and magnetic fields of force.[7] Tesla, thin and more than six feet tall, in his dress, manner, and mind personifying dramatically the creative genius, said:

> It is providential that the youth or man of inventive mind is not "blessed" with a million dollars. The mind is sharper and keener in seclusion and uninterrupted solitude. Originality thrives in seclusion free of outside influences beating upon us to cripple the creative mind. Be alone—that is the secret of invention: be alone, that is when ideas are born.[8]

Edison was more representative of the independents. He, too, wished to be free of distractions, to work in a shop or laboratory with loyal assistants at hand to carry out his ideas. At Menlo Park, New Jersey, in 1876 he established an invention factory that was a cross between Camelot and a monastic cloister. At that time the place of withdrawal for inventors was the machine shop or the chemical laboratory. Edison provides a memorable example of the inventor withdrawing to the atelier space with the assistants, tools, and apparatus able to extend his creative powers. From 1869, when he resigned his job as a Western Union operator in order to devote himself full-time to invention and various telegraphic enterprises, until 1876, when he moved into his famous Menlo Park invention factory, Edison centered his inventive activities in a series of machine shops and made use of the talents of a number of model builders. Like Bell a few years later, he frequented the model-building shop of Charles Williams. There, with the help of George Anders, one of Williams's men, he worked out his first patented invention, the vote recorder of 1868.[9] On establishing himself with space in the Williams shop in January 1869, Edison gave notice in a telegraph journal that he would

ARTIST'S RENDITION OF EDISON'S MENLO PARK COMPOUND CONVEYS ATMOSPHERE OF SECLUSION AND CREATIVITY.

devote his full time to bringing out his inventions.[10] After he moved to New York in 1869, he joined electrical engineer Frank Pope to form a consulting firm whose advertising stressed the designing of instruments, the making of experimental apparatus from rough sketches, and the conducting of experiments.[11] Once he had received the backing of a number of investors wishing to acquire interests in his improvements in telegraphic apparatus, and begun the manufacture of some of his telegraphic instruments, he rented or purchased several machine shops and factory buildings in Newark, New Jersey. In 1870 he rented a large building and bought the finest machines and best tools, spending more than $30,000 during the first week of equipping the establishment. Edison, however, did not care to supervise production, so he used the best machinists and machines to help him in his ceaseless inventing and model building.[12] At another of his factory-and-machine shops in Newark, he set aside a room on the top floor as a place of invention and development and, in 1871, began to keep a notebook of his daily experimentation and his reflections. "This will be a daily record containing ideas previously formed, some of which have been tried, some that have been sketched and described. . . ."[13] Trying, or experimenting, meant model building. During the half-decade in Newark when he staffed this and his other machine shops, he sought out skilled craftsmen who had "light fingers" and brought in men trained mostly as clockmakers and machinists.[14] Here he hired Charles Batchelor, a machinist from England, and John Kruesi, a Swiss clockmaker and machinist, who translated his ideas into precise drawings and models, not only in Newark but during the marvelously inventive Menlo Park period.

Working closely with the Western Union Telegraph Company and financiers in New York who supported his inventive activity and who bought the patented devices he manufactured, Edison nonetheless felt constrained by the goals they defined for him and the managerial chores that manufacturing imposed. Once he amassed sufficient money, he withdrew—but not too far—to Menlo Park, a rural railroad stop about halfway between New York City and Philadelphia, to establish in the spring of 1876 his "invention factory." From Menlo Park he could call on the manufacturing and financial resources of New York and Philadelphia, but also escape their suffocating influences. At the start of one of his notebooks, Edison penned the remark that he was not inventing for "damn capitalists." According to Matthew Josephson's biography of Edison, the combined responsibility of manufacturing and inventing had "taxed even his superhuman energies."[15] Edison longed for the exhilaration of exploring undiscovered intellectual ground in the company of a few loyal and kindred spirits, of experiencing those eureka moments of discovery, of solving problems of his own choosing, and of transforming ideas into devices with the aid of the best craftsmen, tools, and machines available. In an industrial and urban environment like Newark or New York, the clutter and clamor of the existing world preoccupied him and diverted the energies of his craftsmen.

Before Edison and his band of sturdy craftsmen arrived, Menlo Park had about a half-dozen farm dwellings. Edison's father, Sam, supervised the construction of the first two-story, barnlike, rectangular building. Originally the building had a small office partitioned off for Edison, a little library, a drafting room, and the machine shop on the lower floor,

THE WIZARD OF MENLO PARK

and a chemical laboratory on the second. Edison lavishly equipped Menlo Park, bringing in, by great horse-drawn trucks, instruments, chemicals, books, a steam engine, and machine tools. No other combination of special machine shop and chemical laboratory in America was probably so well equipped, not even in the universities. Years later, Henry Ford the industrialist, a close friend and great admirer of Edison the inventor, restored the entire Menlo Park complex at the Ford-museum site in Dearborn, Michigan. Thus Ford erected an appropriate monument to the greatest invention of the nineteenth century—a method of invention.

In time, Edison added other buildings as his projects required. In the cozy invention village, he lived in a farmhouse with his wife and children, Marion Estelle ("Dot") and Thomas Jr. ("Dash"). Later, William, a second son, was born. Two of Edison's closest associates, Batchelor and Kruesi, and their families lived in two of the other six houses in the village. Mrs. Jordan's boarding house provided a comfortable if austere environment for others. In the spring of 1876, there were about twenty Edison craftsmen in the village. Supported by his people and his things, happy and expansive with his new freedom and the capacity to pursue his vision, Edison sent a patent-lawyer acquaintance his new address and an invitation: "brand-new laboratory . . . at Menlo Park, Western Div., Globe, Planet Earth, Middlesex County, four miles from Rahway, the

THOMAS A. EDISON (SIXTH FROM LEFT) WITH HIS MECHANICS, CHEMISTS, AND MODEL BUILDERS AT MENLO PARK. PIPE ORGAN IN REAR FACILITATED MARATHON NIGHTS OF EXPERIMENTATION.

prettiest spot in New Jersey, on the Penna. Railway, on a High Hill, Will show you around, go strawberrying."[16]

Before he left Newark, he declared "in all seriousness" that he proposed at Menlo Park "a minor invention every ten days and a big thing every six months or so."[17] Because his recent rate of patent acquisition had reached about forty a year, his aspirations were reasonable. As the newspapers drew public attention to the invention compound and to the minor and major inventions that did emerge, many of which touched everyday life directly, in the public mind Menlo Park acquired a quality of enchantment. Edison became known as the Wizard of Menlo Park. Some people imagined Edison, when the last rays of the evening sun shrouded the laboratory in a mysterious glow, as a Faustian figure brewing exotic substances and mastering powerful forces. The artist who made the widely circulated painting of a snow-covered Menlo Park suggested a Saint Nicholas at the North Pole working with his elves to make new gadgets for an eagerly waiting world. Today, less infatuated with the technological cornucopia and more persuaded that technology is in league with science than with craft and cunning, we can see Edison and his full-bearded craftsmen in their isolated compound at Menlo Park as precursors of J. Robert Oppenheimer and his nuclear scientists at their mountain fastness in Los Alamos, New Mexico, when they unleashed the terrible destructive power of the atom. Like Edison before him, Oppenheimer was exhilarated by the withdrawal to a citadel of creation, a location that he had known earlier as a youth in a private New Mexico school.

Edison fulfilled his desire to withdraw in order to create, not only when he built Menlo Park, but again when he planned and constructed a much larger invention factory at West Orange, New Jersey. In his plans for this place of invention, he once more showed his need for independence. In the new facility he had the private workroom of which he had long dreamed, and which he did not have at Menlo Park. He spoke of it as "a special or secret" place for "special things I want sub rasa [sic]."[18] In the private room he conducted his experiments and tests. He wanted only a few chairs, a worktable, and materials related to the project on which he was working at the time; but nearby, on call in the same building, he needed his craftsmen and their tools.[19]

Edison left the invention factory at Menlo Park in 1881, first moving to New York in order to be on hand for the construction of the pioneer central lighting station on Pearl Street and to preside over the expanding manufacture and sale of electrical lighting. He pursued his first love, experimentation, on the top floor of an associate's factory at Seventeenth Street and Avenue B. Having accumulated substantial money from his inventions and business, Edison in 1886 decided to invest at least $100,000 in the new complex. He then chose a site in West Orange, at the foot of the hill where he had purchased an estate, Glenmount, for Mina Miller, his socially prominent second wife, and for the family of his first wife, who had died in 1884. In this rural setting, he was again away from the tumult and distractions of New York, but close enough to exploit its finance, manufacture, and ferment of new engineering ideas and activities. His initial sketches show that he envisaged a monumental and prestige-enhancing building in the French mansard style with an ingeniously concealed smokestack. Encircling an interior court, the complex conveyed prestige and privacy, even secrecy. Edison, however, abandoned these plans and turned to several architects for the design

GENERAL VIEW OF MENLO PARK AND EDISON'S LABORATORY.

INTERIOR OF THE LABORATORY.

EDISON'S PERFECTED ELECTRIC LIGHT.

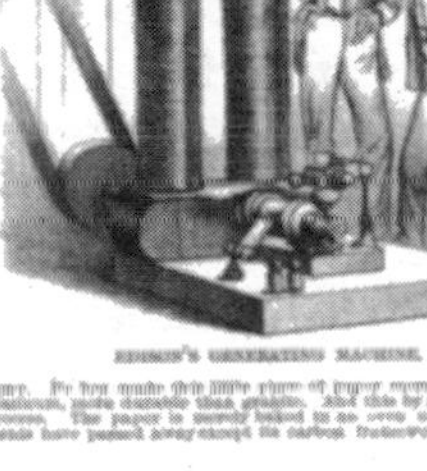

EDISON'S GENERATING MACHINE.

MAKING LAMPS FOR ELECTRIC LIGHT.

EXHAUSTING AIR FROM GLASS "LAMPS."

NEW JERSEY.—THE WIZARD OF ELECTRICITY—THOMAS A. EDISON'S SYSTEM OF ELECTRIC ILLUMINATION.

THE MANY FACETS INVOLVED IN INVENTING AND DEVELOPING THE EDISON SYSTEM OF ELECTRIC LIGHTING

EDISON "INSOMNIA SQUAD" AT THE WEST ORANGE, NEW JERSEY, LABORATORY, 1917 (EDISON, FAR RIGHT)

of a complex of less embellished factorylike buildings that facilitated flexible expansion so that he could pursue projects besides electric lighting.

Those who then portrayed Edison, the American hero, as a plain and pragmatic hunt-and-try inventor unencumbered by science and organized knowledge would have been surprised to learn of the emphasis he gave to a library. Handsomely paneled in dark-stained pine and graced by a large clock given to him by his employees, his library had alcoves and balconies stocking technical and scientific journals, a wide selection of books, and volumes of patents. He also displayed a collection of minerals and ores that were cared for by a chemist curator. Edison had a roll-top desk and a conference table in the library, and at the center of the room a large potted plant adorned displays of new inventions and products. Tradition has it that Mina Edison insisted that a cot for him be placed in one of the alcoves so that her husband would not catnap on the floor.[20] The presence of the cot also enhanced the myth of an inventor, indifferent to the regular sleep needed by mere mortal men, who unremittingly pursued his goals.

Edison said, as we have noted, that he could depend more on science for the solution of mechanical and electrical problems, but that he had to fall back on hunt-and-try in chemical matters. In West Orange he proved the point. He had "8 loads of experimental stuff" brought in, for he believed that "an experimenter never knows five minutes ahead what he does want." He boasted that he had "everything from an elephant's hide to the eyeballs of a United States Senator."[21] Fessenden, the wireless pioneer who worked for Edison for a time, always found what he needed in the storerooms, including an emergency snack. An entire building given over to chemistry housed several chemists with German doctorates. There they worked at their benches, saving another bench for Edison.

Edison, as always, depended heavily on the skilled mechanics and craftsmen to transform his ideas and sketches into mechanical and electrical models with which he could experiment and then test. These men were the "muckers" among a staff consisting, as one of them recalled, of "learned men, cranks, enthusiasts, plain 'muckers' and absolutely insane men."[22] Among them since the Menlo Park days were Batchelor, who assisted Edison in planning and outfitting the new facilities, and William K. L. Dickson, who helped him develop motion pictures. Be-

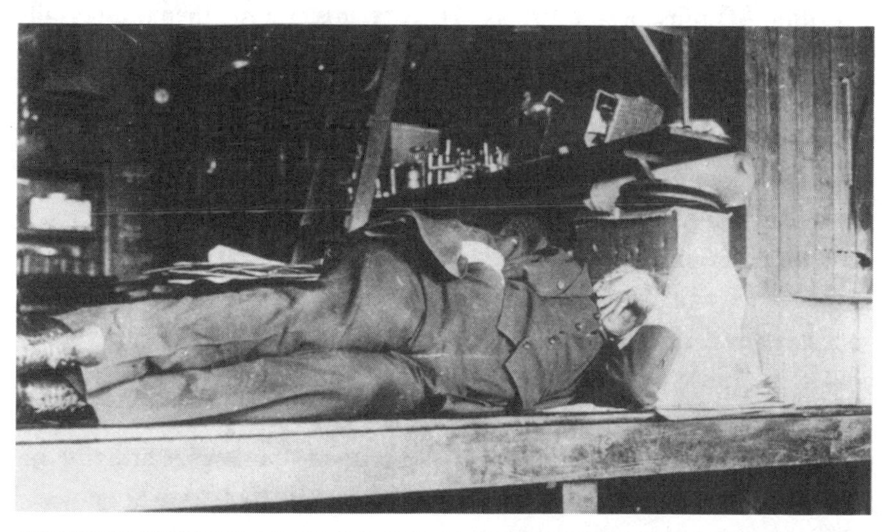

PUBLICITY ABOUT EDISON'S CATNAPS AFTER LONG INVENTIVE SESSIONS CONTRIBUTED TO THE LEGEND.

sides the chemists, Arthur E. Kennelly, a physicist who later became a professor at both the Massachusetts Institute of Technology and Harvard University, was at Edison's side in West Orange.

It would be a mistake, however, to see Edison at West Orange as creating an environment to attract the bright young scientists later lured, early in this century, to the major industrial research laboratories, such as that at General Electric. Willis R. Whitney, who headed General Electric Research Laboratory and had a German Ph.D. in science, came to industry from an academic post at MIT. He did what he could to establish an academic atmosphere at the GE laboratory. Edison could not play the role. De Forest, the radio pioneer who visited Edison's lab in 1912, described him in the midst of concentrated inventive activity as having slept little and having just had his first "wash" in three days, insisting that his helpers join the "Insomnia Squad." Coatless, beltless, in a grimy white boiled shirt with a bow tie off and under his left ear, and "pants almost open," Edison to the admiring de Forest was "an inspiring, almost pathetic sight."[23]

Edison's laboratory established a pattern for the laboratories of other inventors. Edward Weston, another independent inventor, had a laboratory remarkably like that at Menlo Park. Weston was a prolific inventor, the number of his U.S. patents in the 1890s falling behind only those of Edison, Thomson, and Francis H. Richards. Remembered especially for his electrical inventions, among them generators, incandescent lamps, and measuring instruments, Weston in 1886 built a well-equipped laboratory behind his residence in Newark, New Jersey. Like Edison's, his laboratory complex facilitated a wide range of experimentation and the patenting of a striking variety of inventions. The Weston complex also had steam-driven machine tools, a chemical lab, a physics lab with an array of electrical testing instruments, and a library with ten thousand volumes, including rare history-of-science editions. He regularly employed five assistants and often more. An admiring *Scientific American* reported that in its "main rooms and in its offices and smaller apartments everything is contained that can aid the inventor in quickly bringing his ideas into concrete form and determining their value when so presented."

EDWARD WESTON'S CHEMICAL LABORATORY IN NEWARK, NEW JERSEY

Weston's secretary recalled that the inventor worked fifteen to twenty hours a day in the lab, often throughout the night.[24]

In March 1895, when fire destroyed Tesla's laboratory at 33–35 South Fifth Avenue in New York City, Charles A. Dana of the New York *Sun* wrote that the event was more than a private misfortune, it was a calamity for the entire world. Dana believed that Tesla's research and inventions were more important for the human race than the activities of all but a handful of men alive, perhaps the number that could be counted on one thumb.[25] Edison offered the facilities in his lab to restore Tesla's fire-destroyed equipment. In the spring of 1899, Tesla established in Colorado Springs his most remarkable laboratory. Situated in a great meadow and silhouetted against a background of towering Pikes Peak, the hulking laboratory building was topped by a 145-foot mast carrying a thirty-inch metal sphere. Filled with expensive equipment, the laboratory enabled Tesla to explore the possibilities of the wireless transmission of messages and of appreciable amounts of power through the atmosphere and the earth. He predicted that he would be able to transmit power throughout the world from the Niagara Falls hydroelectric plant. An excited reporter wrote that Tesla would electrify the whole earth. Like many other scientists and inventors of the decade, Tesla was aware of James Clerk Maxwell's 1873 treatise on the propagation of electromagnetic waves through space, and of Heinrich Rudolph Hertz's demonstration of them between 1886 and 1888. Others had also shown that energy could be transmitted by electromagnetic induction.

Tesla, however, was unique in the scale and drama with which he performed his experiments. His laboratory was equipped, he announced, to send messages to Paris; this was about a year before Marconi sent transatlantic wireless message. Tesla did not demonstrate this claim, but he did display unforgettable visual and aural effects. In his laboratory, he had large high-frequency transformers, and a "magnifying transmitter" that he considered his greatest invention. With these and electricity from the local utility, he built up large charges of electricity and discharged the energy in powerful waves. When operating, the transmitter stimulated fiery arcs in lightning rods spread over a radius of about ten miles. Nearby horses bolted as they felt the shocks from it in their shoes. On occasion, Tesla and a few assistants caused fabulous displays of lightning within the laboratory. Witnesses reported that at night the electrical display filled the sky with sound and color.

Edison once commented that Tesla was always about to discover something. Tesla believed that through the Colorado experiments he had set the earth in electrical resonance, produced the electrical potential of about twelve million volts—greater than that of any other experimenter—passed a current around the globe sufficient to light two hundred incandescent lamps, and demonstrated on a small scale the feasibility of wireless transmission of power. He also said he received messages from space, most likely from Venus or Mars. Besides all this, he had found the isolation, the rarefied atmosphere, and the spectacular mountains spiritually exhilarating. Menlo Park paled in comparison.[26]

For his elaborate laboratories and experiments, Tesla ceaselessly searched for funds. Like the other independents, he instinctively feared the constraining ties that subsidies entail. One of his biographers, Mar-

NIKOLA TESLA

TESLA DISCHARGES HIS ARTIFICIAL LIGHTNING.

garet Cheney, observes that Tesla, a loner by preference who resented any form of control, found corporate involvements distasteful. Slow-thinking, plodding engineers as associates drove him mad with impatience. So, if he had to deal with a corporation, he preferred to transcend the bureaucracy and deal one-to-one with the president or chairman of the board. Choosing to approach powerful barons of finance or system builders like John Jacob Astor, Morgan, and Samuel Insull, he nevertheless feared that he would be "Astored," "Melloned," or "Insulled."[27]

The far less flamboyant Sperry also withdrew as far as possible from the harassing entanglements of business and manufacture. Early in his career as an inventor, Sperry saddled himself with day-to-day managerial responsibilities for a company founded in 1883 to manufacture his patented arc light and generator. For a person of managerial aspirations, these could have been taken as challenges and learning experiences but, when the company was unable to meet its payroll, twenty-three-year-old Sperry feared that he would "break down . . . between work and worry."[28] Assisted by only two or three business associates, Sperry was occasionally so exhausted and depressed that "I didn't care whether school kept or not. . . ."[29] His yearly output of patents fell sharply. Discouraged but resourceful, he wrote, "do you suppose for one moment that I am going to fail in the great Chicago where one can turn a hundred ways—no sir." He found the way out of the morass in 1888, when he established what today is called a research-and-development company and laboratory to bring his and others' patents to fruition. The exhilaration and creative drive he felt when freed of routine appears in a letter to his fiancée, Zula Goodman: "I worked nearly all last night on an invention and got it this morning, too. I knew it was to come out correctly if I but could find the time and could have the heart to work upon it. . . . It is one of the most valuable I have yet been able—under His guidance, Dear—to bring out. . . . *Just* won't you and I turn out the inventions, though?"[30] For the rest of his life, Sperry stayed mostly free of administrative entanglements.

Stanley, the American inventor and developer of electrical transmission by transformers, also distanced himself from harassing detail. A Westinghouse Electric Company executive recalled that Stanley, who worked for several years as a consultant for the company, did his best work "away from contact with the every-day never-ending mental work, discipline and industry, of either the main office, or the shops, or the working laboratory at Pittsburgh."[31] Commissioned by Westinghouse to work on the transformer, Stanley and his wife packed their few belongings, "shook the dirt of dreadful Pittsburgh from us and hastened to the green hills of Berkshire [Massachusetts] to build a laboratory, and succeed or perish in our work."[32] A laboratory was a place of retreat, experimentation, and contemplation for him as for other independent inventors. Several years later, when he established an electrical-manufacturing company based on his patents, he took care to establish an independent research laboratory that he headed and that developed his inventions for the manufacturing company.

After becoming a director and chief electrician for the company he established in 1880 to manufacture his arc-lighting inventions, Thomson installed a machine shop and a private workshop that he named the "model room" rather than "laboratory." In it he placed the men and installed the equipment for building the physical models needed to experiment with his inventions as they passed through the various stages of development. He also had an adjoining patent library, for he, like the other inventors, needed to know how others had tried to solve problems on which he was working. The model room became his personal domain, his place for invention. As the company changed hands, expanded, and became the Thomson-Houston Electric Company in 1883, he resisted suggestions that he enlarge the model room by hiring other inventors and additional machinists. He wanted only five or six machinists, whom he could personally supervise, and one or two clerks to handle correspondence, especially that dealing with patents. In 1888 he was discomfited when company mergers brought in several other inventors.[33] Thomson, like Sperry, shied away from management and established his niche in the company as the inventor and developer of new products. The model room was off-limits to factory employees and visitors, and not simply to prevent industrial espionage. Working there, Thomson "perfected a method of invention which brought together his skills in analyzing problems, visualizing and sketching solutions, and building models."[34] From 1880 to 1885 Thomson averaged twenty-one patent

YOUNG ELIHU THOMSON'S IN-HOME LABORATORY

applications annually; from 1885 to 1890 the average doubled. He not only distanced himself physically from activities other than invention, but in 1890 he also secured a company agreement that he would not be limited to inventive problems presented by the marketing and production departments, but would also be able to choose his own. When his company merged with Edison General Electric in 1892 to form the General Electric Company, Thomson chose to withdraw even further from business and manufacturing and to establish a new laboratory near Boston, far from GE's headquarters and under his personal supervision.[35] Asked to become a director of GE, he declined, protesting that his value was in the realm of ideas, not management.[36]

MODEL BUILDERS AND CRAFTSMEN

With the rise of factory production and the displacement of human skills by machine, numerous social critics lamented the passing of the era of the craftsman. In England, William Morris celebrated the joy of work and called for the recovery of medieval crafts. In the United States, the iconoclastic economist and public intellectual Thorstein Veblen, writing in 1914 on the instinct of workmanship, argued that "Chief among those instinctive dispositions that conduce directly to the material well-being of the race, and therefore to its biological success, is perhaps the instinctive bias here spoken of as the sense of workmanship."[37] But in the industrial era tool users were giving way to machine tenders. Still, in the model rooms, laboratories, and machine shops used by independent inventors, craftsmen of conspicuous skill thrived. Kruesi, with an intuitive grasp of Edison's three-dimensional concepts, presided over the machine shop at Menlo Park. He transformed a quick sketch of Edison's into the first phonograph. Always at Edison's right hand, Batchelor displayed marvelous control in his fingers.[38] Thomson brought Edwin Wilbur Rice, Jr., one of his former students at Philadelphia's Central High School, to work as his assistant in the model room and on the factory floor. Rice did sketches of new inventions and supervised the machining of these. When GE was formed, Rice became technical director; Kruesi became manager of the plant at Schenectady. Sperry attributed his company's success in manufacturing the precision gyroscopic devices to the skill of his machinists, many of them Swiss. Edison, always seeking to generate the spirit of serendipity, considered those model makers his "muckers."

When the independent inventors could not establish their own model rooms and machine shops, they turned to model-building shops that served a number of inventors. In 1868 a wood engraving on the cover of *Harper's Weekly* magazine portrayed "The Model Maker."[39] Not only Edison but also Bell, in their invention and development activities, resorted to the skilled services of Charles Williams and his machine shop at 109 Court Street in Boston.[40] Williams advertised the manufacture of telegraph instruments, galvanic batteries, and the sale of telegraphic sup-

MENLO PARK: THE CREATORS OF THE EDISON SYSTEM OF ELECTRIC LIGHTING (EDISON, THIRD ROW BACK, CENTER)

plies of all kinds.[41] Other inventors interested in electrical devices also congregated at the Williams machine shop, for telegraphy was high technology then. In business for twenty years, he made, on order and in small quantities, telegraphic and electrical signaling devices, such as hotel annunciators and fire-alarm systems, but he also catered to the needs of a procession of inventors. Williams's shop was "one of the largest and best fitted in the country," and a place where "individual initiative was the rule." "Wild-eyed inventors, with big ideas in their heads, and little money in their pockets," found kindred spirits and stimulation there.[42]

When Bell used Williams's shop, there were about twenty-five employees working on the third floor and attic of the Court Street building.[43] A dozen or more hand lathes, a steam engine, and several small steam-powered lathes, as well as a forge, created spaces filled with grease, grime, steel filings, dust, and noise—so unlike the conventional image of a

MACHINISTS AND MODEL BUILDERS AT MENLO PARK

research-and-development laboratory. For Bell, the Williams shop spelled Boston as much as, or more than, Harvard University, the Massachusetts Institute of Technology, or Boston University. Later, living in Washington when he was famous, he wrote, "Washington is no place in which to carry out inventions. If we only lived in the neighborhood of a large city, I could have apparatus made in a large workshop as I did at Williams' in Boston."[44] To understand Bell's plight there, we need to recall that invention is rarely if ever an act, but usually a process involving the conceptualization, probably visualization, of various means to an end or of solutions to a problem; the embodiment of these in models; and subsequent experimentation with the models to discover how well the means fulfill the end in mind.

Sperry depended on at least two model builders in New York when he was developing his gyrostabilizer and gyrocompass for ships. A few scientists, the Frenchman Jean Bernard Léon Foucault among them, had explicated gyroscopic principles. From them he learned of the remarkable property of a spinning wheel mounted on gimbals so that the wheel's axis of rotation could move freely to maintain its orientation in space despite the movements of the earth or any other body, such as a ship, on which the spinning wheel, or gyro, was mounted. He knew that, if the gyro's axis of rotation was aligned parallel to the earth's axis of rotation, then the gyro axis, like the axis of the earth, would point "north" and "south." Sperry sometimes demonstrated the property to a lay audience by holding a gyro while he spun around in place. The gyro axis would align with his spin axis. Also known was the "precession" property of the gyro. If a force were directed against a gyro with the intent of realigning the gyro's axis of rotation, the remarkable gyro responded with a counterforce of equal magnitude. Sperry intended to use this precessional reaction of a massive gyro wheel spinning rapidly to absorb the force of the waves and to maintain a ship on an even keel. Because the gyro axis reacted with a motion at right angles to the external force imposed on it, he called this precession the moving of the force "around the corner."

Sperry, however, could not rely on scientific treatises or on engineering texts for practical information about the behavior of gyros aboard ships, for the science was generalized, and only a handful of inventors and engineers had ventured into the field of application, and their results were tentative. Not owning ships and employing crews, he could not afford to learn by full-scale installations. So Sperry turned to the model builder, who could preliminarily test ideas in simple ways on a small scale. Knowing that every ship had an unvarying period of roll—a battleship in his day had a period of about sixteen seconds—Sperry used a pendulum with the same period to represent a ship, and a small electrically driven gyro to represent the gyrostabilizer. He turned to the model-building firm of Charles E. Dressler & Brother of 143–45 East Twenty-third Street in New York to build the model. On completion it had two tripods carrying a horizontal hinged member that was free to rotate, and from which was suspended a length of gas pipe weighted by a bench vise. A Dressler-built gyro was mounted on the rotating crosspiece of the pendulum. With strings and pulleys, Sperry could precess, or operate, the gyro and cause a reactive force. He carefully timed the period of the pendulum and measured the reactive, stabilizing force of the precessing gyro. With this device, Sperry demonstrated for the first time the principle of active gyroscopic precession, for he dampened the oscillation of the heavy pendulum in fifteen to twenty seconds.[45]

Dressler & Brother advertised that it made instruments and apparatus for "medical doctors, chemists, physicians, dentists," and "special scientific apparatus for schools and technical institutions." The shop also engaged in "developing of inventions." Dressler published a special brochure about his small electrically driven gyroscope that sold for $125. (Sperry did not use this.) The brochure ventured that "the person who discovers the formula governing the rotation of gyroscopes . . . may reveal the long sought goal: one fundamental physical law governing the universe, and may receive the Alfred Nobel $40,000 cash prize." The gyro was called "a physicochemical atom model in motion, or a kinet," and "a psychological machine which shows affections."[46] In 1909, when he was developing both his gyrostabilizer and his gyrocompass, Sperry rented space in the Fred K. Pearce Co., another model shop, at 18–20 Rose Street in the shadow of the Brooklyn Bridge. In a ten-by-fifteen-foot enclosure, Sperry and his two assistants prepared drawings that they handed over to Pearce mechanics to use in making models.

Throughout his inventive career, Sperry helped establish companies based on his patents, then withdrew and left the running of the companies to others. In 1914, however, at the age of fifty-four, he broke his own rule to found the Sperry Gyroscope Company, an invention, development, and manufacturing firm. As we would expect, he made a model-building and machine shop an important part of the company. The Sperry company manufactured gyrocompasses, naval fire-control systems, and other complex instruments and devices for which he acquired the finest machine tools for precision engineering and the most skilled master machinists, some of whom had been Pearce employees. Sperry, as we have noted, especially appreciated the Swiss immigrants among his craftsmen. He attributed to their formidable skills the success of his firm in discouraging the rise of American competition in the field of gyro technology. Like Sperry's craftsmen, many of Edison's had brought their skills from the Old World. New York probably had a disproportionate share

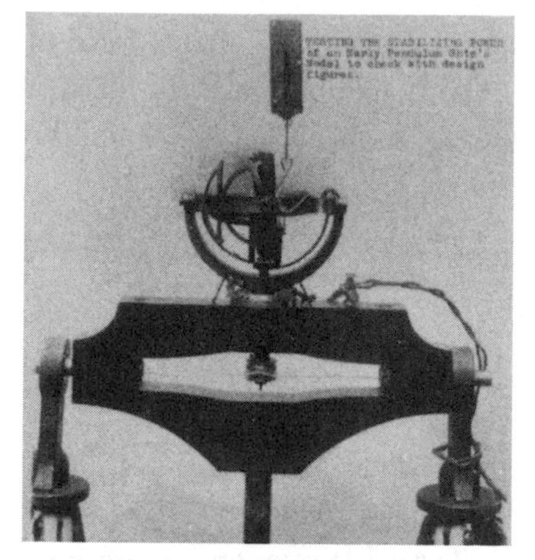

MODEL TO INSTALLATION: SPERRY GYRO AND PENDULUM SIMULATION OF A ROLLING SHIP

MODEL TO INSTALLATION: EXPERIMENTING WITH MODEL GYROSTABILIZER ON ROWBOAT

MODEL TO INSTALLATION: SPERRY GYROSTABILIZER PLACED ABOARD NAVAL SHIP U.S.S. *WORDEN* FOR TESTING

of foreign-born and -trained craftsmen who had entered the country with the waves of late-nineteenth-century immigrants.

THEORY AND EXPERIMENTATION

At the same time that the independents in the electrical and mechanical fields relied heavily on skilled mechanics and model builders, they also needed men trained in science and chemistry. The inventors used science in the form of organized information and also in the form of theory, if this was available. Often it was not.

Because of outrageous, off-the-cuff, sometimes teasing remarks to newspaper reporters innocent of technology and science, Edison has left an impression that he had no use for science and scientists. Even though he roguishly dismissed long-haired scientists, however, he counted them among his friends and numbered them among his staff. Young Francis Upton, a Princeton graduate in science with postgraduate education at the University of Berlin, who joined the inventor's staff in the fall of 1878, contributed to the complexity and effectiveness of the Edison approach. Edison openly relied on him. Francis Jehl, another Edison assistant at Menlo Park, years later recalled that Upton coached Edison in science and provided him with theoretical insights into electric circuits and systems.[47] Yet Upton insisted, "I can answer questions very easily after they are asked, but find great trouble in framing any to answer."[48] Edison, he continued, the laboratory director with remarkable powers of concentration and a single-minded pursuit of an objective, chose the problems.

Sperry carefully chose as assistants from the leading engineering schools young engineers educated in science and technology. During the development of the gyrostabilizer, Sperry relied heavily on Carl Norden, a graduate of the advanced engineering-and-science program at the world-famous Polytechnic Institute in Zurich. Thomson had lectured on electrical science before establishing himself as an inventor. Tesla studied engineering at Graz Polytechnic in Austria and completed his education at Prague University. In the field of radio, Fessenden was a professor of electrical engineering at Purdue University and at Western University of

Pennsylvania (Pittsburgh); de Forest had a doctorate in physics from Yale University; and Armstrong took his engineering degree at Columbia University and became, as we have noted, a professor there. On the other hand, we must take care not to see the independents simply as appliers of science; often their experiments forged ahead of theory.

There is also a widespread assumption that persons who excel in invention, research, and development are well grounded in mathematics. Yet most independent inventors were mathematically unsophisticated. Edison depended on Upton for mathematics. He once said he did not need math because he could hire all the mathematicians he needed. Sperry turned to the gifted naval officer D. W. Taylor for a mathematical analysis of gyroscopic behavior. De Forest seems never to have had a firm grasp on the theory of his own triode tube. Tesla brilliantly conceptualized rotating magnetic fields, but his concepts were visual, not abstract. The Wright brothers experimented imaginatively, but their aerodynamical theories were mathematically simple, low-level abstractions.

His biographer insists that Armstrong, inventor of frequency modulation (FM), had a fundamental understanding—comprehended in physical terms—of his regenerative and superheterodyne radio circuits, but that he had a strong aversion to mathematical abstractions and a lifelong quarrel with mathematicians.[49] When John Carson of AT&T, in a carefully reasoned and mathematically grounded paper of 1922, dismissed Armstrong's new system of frequency modulation for radio, the inventor's distrust in purely mathematical arguments and their authors only increased. After he had proved the efficacy of FM, he took this as one more demonstration that invention depended on experimentation and physical reasoning. Armstrong "never allowed Carson to forget . . . [his] blooper" and never lost the opportunity to "rub it in" that AT&T had fumbled its chance with frequency modulation.[50] Armstrong summed up his feelings in a seething paper entitled "Mathematical Theory vs. Physical Concept."[51]

Independents could not depend on science and abstract theory as guides into the future, because they were exploring beyond the front edge of technology and of knowledge. They probed beyond the realm of theory and the organized information that makes up packed-down science. Theory available to the independents usually explained the state of the art, not what was beyond it. Academic scientists working on their own frontiers did not customarily oblige the inventors by obtaining information or conceiving theories related to the areas in which the independents were working. As we shall see, the rise of the industrial research laboratories staffed by scientists seeking a fundamental, or theoretical, understanding of technology tended to change this. They did fundamental research in order to find explanations and theory for the devices, processes, and machines that industrial corporations were developing or manufacturing. Frequently, however, their theories were not complex enough to encompass the complexities of technology. As Armstrong insisted, mathematicians often forced the messy world of invention and engineering into an overly ordered and reductionist mold that precluded a thorough and usable explanation of how physical things worked. The same was true of scientists.

Scientists, unfamiliar with the details of new technology such as that being introduced by independents, often exasperated the inventors by insisting that they apply theory that the inventors knew was outmoded. Some scientists arrogantly ridiculed the empirical approach of the so-called Edison hunt-and-try method at the same time that they reasoned from anachronistic theory. Edison was impatient with stiff-necked, academic scientists who argued that the theory of electric circuitry, developed for arc lights, was valid for the newer incandescent lighting. Similarly, in the field of bridge building, Robert Maillart, the pioneer of reinforced-concrete construction, had to suffer unsolicited and erroneous suggestions from theoreticians who believed that the elegant theory worked out for older stone-and-iron construction was applicable.[52]

Deprived of adequate theory, the inventors resorted to experimentation, empirical but often insightful. In 1870 one of Edison's business associates complained that the inventor carried out too many experiments, some useless. Edison replied, "No experiments are useless." The inexperienced were simply too impatient and shortsighted, he believed, to appreciate the nature of experimentation. "Galieo [sic] discovered," Edison added, "the principle of accurate . . . [horology?] in the swinging Lamp of Pisa. It wouldn't be a very sage remark to say—why damn it that lamp ain't a

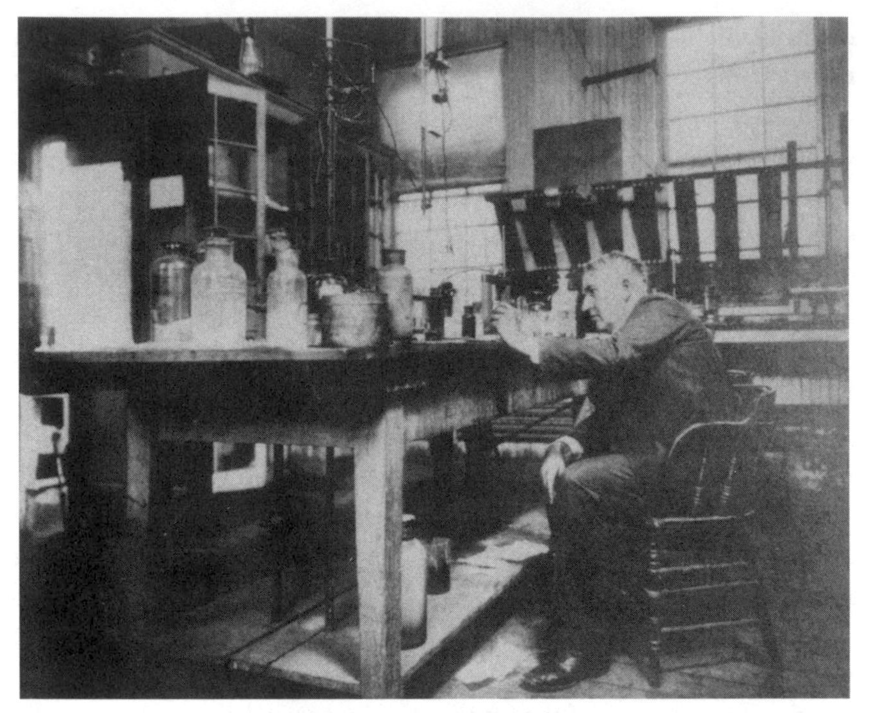

EDISON POSING AS AN EXPERIMENTER

clock."[53] Consideration of several sets of Edison's experiments provides some insight into the way in which he used experiments throughout the process of invention and development. Always close at hand and often making the experimental observations and notes were the indispensable Batchelor and Kruesi. Jehl recalled the case of the telegraphic stock ticker. Progressive changes and refinements appeared in the succession of models:

> He displays cunning in the way he neutralizes or intensifies electromagnets, in applying strong or weak currents or in using polarized magnets; and with aggressive certainty commands either negative or positive directional currents to do his work. You can see his natural talent shaping the resistance of the electromagnets to a certain ratio as regards that of the line.[54]

Here Edison was testing the validity of insightful concepts and refinements based on them. Like a scientist experimenting, he used quantitative as well as physical models. Reasoning allowed him to try out his concepts with quantitative models; the physical models allowed, at least for him, more complex testing. His progressive conceptualizations, experiments, and refinements brought him forty-six patents on the stock ticker.

Edison's invention and development of a carbon telephone transmitter designed to compete with the transmitter of Bell provides another example of experimentation and model building. About 1873 he had conceived of a rheostat for varying the resistance of an electric circuit and had constructed a measuring instrument based on the concept. The rheostat consisted of fifty or more silken discs filled with fine particles of graphite. He could obtain any degree of resistance from four hundred to six thousand ohms by increasing or decreasing the pressure on the stack of discs, for the pressure varied the conductivity of this semiconductor device. In 1877 he transformed the rheostat into a telephone transmitter or speaker. Pressure from the voice's sound waves varied the resistance and flow of current in the electric circuit that included the telephone receiver. For the telephone transmitter he found he needed a semiconductor of lower resistance than the silken discs filled with graphite. In 1877 he experimented with a model transmitter using plumbago as a semiconductor button between two platinum discs. The pressure of sound waves vibrating a diaphragm was mechanically transferred to the button, and this varying pressure changed the conductivity of the plumbago button. Since the button was in the telephone circuit, the current flow in the circuit mirrored the pattern of the sound waves. Edison patented this model in December 1877. Finding, however, that the volume of sound was insufficient, he experimented with other semiconductors, including oxides and sulphides, in the rheostat-become-telephone-transmitter, until a button of lampblack was found to give excellent results. In February 1878 he applied for a patent covering the use of lampblack in the transmitter. Up to this point in his experimentation, he had used a rubber component to carry the vibrations of a transmitter diaphragm to the carbon button. More experimentation and model variation revealed that vibrations need not be transmitted, but only diaphragm pressure changes, so a metal spring that did not wear like the rubber was substituted and a heavier diaphragm used. More tests showed that a spring was not needed, but simply a solid substance. The excellence of articulation and the volume of sound served as criteria of improvement as experimentation proceeded.[55]

Scientists unfamiliar with invention and development often denigrated this empirical approach, not realizing that to hunt-and-try was to hypothesize and experiment in the absence of theory. Thomas Midgley, the chemist and inventor responsible for the tetraethyl-lead additive for gasoline, remarked that the trick was to change a wild-goose chase into a fox hunt. Thomson, Sperry, and Edison, like other independent inventors, treasured their model builders, their chemists, their scientists, and their laboratories, because these facilitated experimentation, the lifeblood of invention. The inventor needed an environment in which to test new ideas and early models of the inventions that embodied them. That environment needed to be less complex and more ordered and controlled than the world in which the inventions would be used. In their initial conceptualization of a device, process, or machine, the inventors were rarely able to encompass all of the complexities that the developed invention needed for survival. The development of inventions in the laboratory through experimentation typically involved a gradual increase in the complexity of the invention and in the environment in which it was being tested. Laboratory experimentation continued until the invention was functioning in a laboratory environment almost as complex as that in which it would be used. Then it was launched onto the market. After this innovation, the inventor could no longer order, systematize, and control his or her brainchild. Perhaps this is why inventors flourished in the supportive spaces of the laboratory and model shop and tended to avoid the world of business and administration.

CHOOSING AND SOLVING PROBLEMS

Invention can be seen as the process of solving new problems. Accounts of successful inventors often focus on their problem-solving techniques. Yet their choice of problems provides as much insight—and perhaps more—into the character of the independent inventors and an explanation of their successes—or failures—as does their problem solving. Their independence, or freedom from organizational constraints, allowed them to choose problems that when solved created the nuclei of new technological systems. As we shall see when we consider the process of inventing in industrial laboratories, industrial scientists were often constrained to choose problems to solve that would improve and spur the growth of existing systems in which the corporations were heavily invested. The system-originating inventions can be labeled *radical*, the system-improving ones *conservative*.

SALIENT CHARACTERISTICS

Independents invented a disproportionate share of the radical inventions.[1] This is perhaps the characteristic that most obviously distinguished them from inventors and scientists working in industrial or government laboratories, and that best explains their notable successes during their golden era. Their radical successes also sustain the arguments of those who today lament the constraints that organization places on creativity, whether the organization is a university, a government, or a private enterprise. The independents preferred to strike for the breakthroughs or improvements in nascent systems rather than for the incremental improvements in well-established technological ones. Elmer Sperry said, "If I spend a life-time on a dynamo I can probably make my little contribution toward increasing the efficiency of that machine six or seven percent. Now then, there are a whole lot of arts that need electricity, about four or five hundred per cent, let me tackle one of those."[2] Organizations did not support the radical inventions of the detached independent inventors because, like radical ideas in general, theirs upset the old, or introduced a new, status quo. Radical inventions did not fulfill the needs or solve the problems of existing organizations. Radical inventions required new nurturing institutions. Such inventions often de-skilled workers, engineers, and managers, wiped out financial investments, and generally stimulated anxiety in large organizations. Large organizations, private or government, sometimes rejected the inventive proposals of the radicals as technically crude and economically risky, but in so doing they were simply acknowledging the character of the new and radical. In the late nineteenth century, gas-lighting, railroad, and telegraph companies did not preside over the invention and development of electric lighting, the automobile, or the radio. Independent inventors brought them into being, along with the new companies and utilities needed to nurture them. Calling such inventions *radical* associates them with the traditional definition of political ideas that challenge established political institutions. Much as a large number of related political ideas of radical cast may cause a destructive political revolution, interrelated radical inventions often cause a technological revolution.

Characteristically, independents preferred to create systems, rather than to improve the systems of others. They realized that they could not deploy the facilities and personnel that corporations and government agencies could commit to improving the well-established systems over which they presided. The systems invented by the independents were sometimes striking breakthroughs solving salient problems, such as the controls for the Wright brothers' airplane; at other times independents invented systems that provided alternatives to existing ones, but they would venture this only if the existing ones were new, unrefined, and not yet presided over by a large corporation with massive resources. Lee de Forest and Reginald Fessenden, for instance, invented wireless telegraph systems that competed with the earlier one of Guglielmo Marconi, but the latter's company was not especially large and was handicapped in competing in an American market because it was British. Edison, Thomson, and Sperry invented entire systems of arc or incandescent lighting in the early days of the industry. Joseph Swan, the British inventor whom his countrymen celebrate as the inventor of the incandescent lamp, erred in trying to integrate his incandescent lamp into a system with a generator designed by others for other purposes. As a result, the components did not harmoniously and optimally interact. Unlike Edison, Swan did not have the substantial funding that allowed the invention and development of an entire system. A representative of a company selling the Swan lamp told an Edison representative that Edison's great advantage was in having a coherent system.

BELL AND THE WRIGHTS

The various independents differed from one another in the particulars of their problem-choosing styles. In order to survive, most of the independents had to choose problems that would lead to patentable and commercially successful inventions. But there were some, like Alexander Graham Bell and Orville and Wilbur Wright, who were free even of this constraint. They had occupations that supported them so they could explore more freely as inventors. The case of Bell and Elisha Gray demonstrates the differences in problem choice between a thoroughly independent inventor and one who, though nominally independent, felt constrained by the need for commercial results and was willing to improve the existing system of a manufacturer. Bell, a professor of speech and vocal physiology at Boston University, did not feel the compunction of the professionals Thomson, Edison, and Sperry to choose a problem the solution of which was likely to bring a financial return in the short run.

Before Bell concentrated on the invention of the telephone, he had, for several years after 1872, worked in an unhurried fashion on a multiple telegraph, a device for which the telegraph companies saw an immediate and pressing need. The device would reduce the number of wires needed to transmit a given number of messages. Copper for wires was a major expense then for the telegraph companies. Because the need was so obvious, a number of inventors in the early 1870s patented various versions of a multiple telegraph. Gray, an independent inventor remembered for a number of telegraph inventions, was prominent among them. In 1874 both Bell and Gray saw the possibility, while working on multiple telegraphy, of transmitting musical and vocal sounds electrically by wire. Gray, however, dependent on invention for income, put the possibility aside as he vigorously pursued the goal of a practical multiple telegraph of obvious economic importance. Bell, to the consternation of his father-in-law, who was backing his multiple-telegraph work because of its obvious practicality, put aside the practical device to pursue what seemed to many the will-o'-the-wisp of voice transmission, or telephone.[3] Bell's knowledge of speech, or voice, as a teacher of elocution, and his concern about his wife's deafness probably helped provoke his decision. The result, after months of experimentation, was a striking, breakthrough invention, but one for which no one else saw a need for several years. Bell the nonprofessional had let enthusiasm direct him, while the more professional and practical Gray gave up the chance to develop what eventually proved to be one of the most profitable inventions in history.

The problem choice of the Wright brothers, part-time inventors, also culminated in a striking, breakthrough invention. Enthusiasm and a vague longing for fame and fortune explain more about their choice of inventing an airplane than do rational economic considerations. In 1920 Orville Wright offered an explanation for the commitment.[4] He insisted that the origins of their greatest invention extended back to childhood. Orville recalled that their interest in flight began when their father brought them a toy rubber-band-driven helicopter. Later, in 1896, the publicity surrounding the tragic death in flight of the glider pioneer, Otto Lilienthal, rekindled their interest in flight. Thus, when they became young bicycle manufacturers, they began looking for books on the subject. In

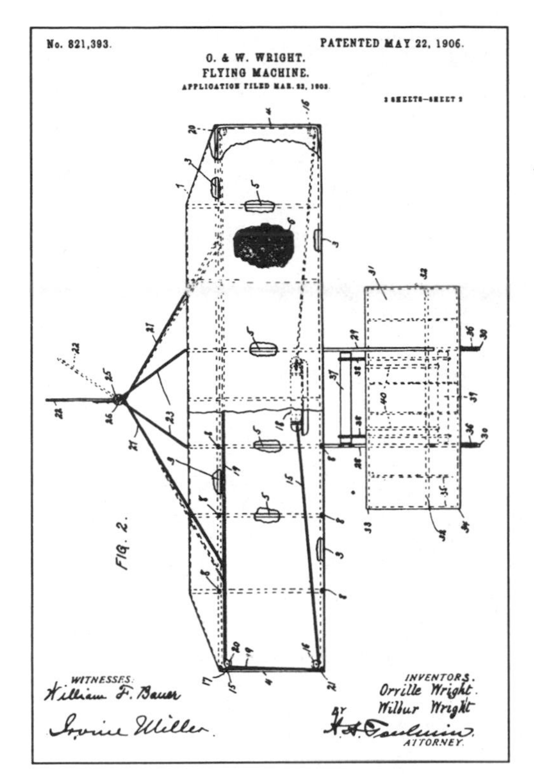

ORVILLE AND WILBUR WRIGHT PATENT ON FLYING MACHINE

the spring of 1899 a book on ornithology suggested to them that birds could be the model for human flight, as Leonardo da Vinci and countless others had imagined many centuries earlier. After the analogy of birds and gliding man spurred their enthusiasm for gliding, they wrote to the Smithsonian Institution for references to books and articles on the problems. A public-serving and informed Smithsonian responded with reprints and references to works by Lilienthal and Samuel Pierpont Langley, plus Octave Chanute's *Progress in Flying Machines* (1894), a detailed history and analysis. The Wright brothers then struck up a long-lasting, fruitful correspondence with Chanute, a distinguished civil engineer, railroad-and-bridge builder, and enthusiastic student of flight.

The response of the brothers to the references provides one of many indications that they were not the simple, empirical experimenters they were often portrayed as being in the popular and juvenile literature that has, unfortunately, shaped the public image. After reading the various pertinent books and articles in their field, they resorted to a procedure

WRIGHTS' CYCLE SHOP

common among experienced inventors: they carefully analyzed the history of others who had attacked the common problem—gliding or powered flight—to seek explanations for their failures. This approach was similar to Sperry's close reading of the patent literature and to de Forest's and Fessenden's searches for the weak links in Marconi's wireless system. With considerable perspicacity they decided that maintaining equilibrium in flight was the critical and unsolved problem, the stumbling block for the earlier inventors and experimenters. "We at once set to work to devise a more efficient means of maintaining the equilibrium," Orville Wright confidently reported.[5] Their willingness to take on the challenge after remarking the failures of notables who had worked on the problem of flight, including Lilienthal; Sir George Cayley, internal-combustion-engine inventor; Hiram Stevens Maxim, the independent inventor; Charles Parsons, the turbine inventor; Bell; and Langley, scientist and secretary of the Smithsonian Institution, suggests the cool self-assurance—or innocence—of the young bicycle manufacturers. Similarly, instead of being discouraged by the problem encountered by Johann Philipp Reis, German inventor of a telephone, Bell confidently forged ahead in developing a practical telephone, believing he could take the steps—not giant leaps—needed to solve the problem. Edison was notable for his confidence that he could succeed precisely where others had failed.

Chanute, their friend and adviser, later described the Wright brothers as improvers, not inventors, of the airplane. Their improvements, however, were breakthrough inventions that spelled the difference between an impractical system and a practical one. He listed as their contributions wing warping, placing the elevator in front, employing the prone position for the pilot, making wind-tunnel experiments, using propellers and transmission systems, and utilizing reliable motors. The first three contributed to maintaining the equilibrium of the airplane, a problem the Wrights stressed in describing their contribution, but Chanute gave the Wrights sole credit for only the last three improvements. He said that Paul Renard, a French glider-constructor, had placed the elevator in front, that he himself had suggested a prone pilot, and that Louis-Pierre Mouillard, an expert on ornithology and a designer of gliders, had patented a version of wing warping. So Chanute had the Wrights looking more like diligent mechanics than inspired aerodynamicists. During the last five or six years before his death in 1910, relations between their mentor and the brothers cooled considerably, because the press played up differences among them with regard to Chanute's contributions to their work.[6]

As they approached their goal, the Wrights unremittingly pursued the solution, but even then financial reward was not driving them. In 1901 Wilbur Wright confided to Chanute that they felt that time spent on their aeronautical experiments "was a dead loss in a financial sense."[7] Even after their successful powered flight in December 1903, the brothers seemed to have thought that the possibility of making even a modest amount of money from their invention could come true only if they entered aeronautical contests, such as that planned for the Saint Louis

ORVILLE WRIGHT AND WILBUR WRIGHT (LEFT TO RIGHT) IN CYCLE SHOP, 1897, DAYTON, OHIO

EXPERIMENTING AT KITTY HAWK, NORTH CAROLINA

World's Exposition of 1904.[8] When Chanute advised them in 1902 to patent their invention, he added that this was not because money was to be made from it but only to avoid unpleasant disputes as to priority.[9]

DE FOREST AND FESSENDEN

De Forest and Fessenden, pioneers in wireless communication, chose problems that when solved would result in an invention that would become the nucleus of a new system to improve on existing ones. When de Forest began his inventive activity in 1899, he concentrated on improving wireless detectors, or receivers. "I began," he recalled, "a serious systemized search through *Science Abstracts*, *Wiedemann's Annalen*, *Comptes Rendus*, and other physics journals, seeking to find some hint or suggestion that might possibly be a clue to the development of a new device which could be used as a detector for the reception of wireless signals."[10] He probably also scanned the technical journals, such as *Electrical World & Engineer*, where the number of articles on detectors sharply increased after 1900.[11] Unlike some other inventors, de Forest candidly acknowledged that he was seeking ways to improve on others' inventions. Later he did not hesitate to patent his improvements, and to claim originality. Fessenden and de Forest separately entered the market with wireless telegraphy systems after Marconi had introduced his. Each had analyzed the Marconi system and found the weak link or component.

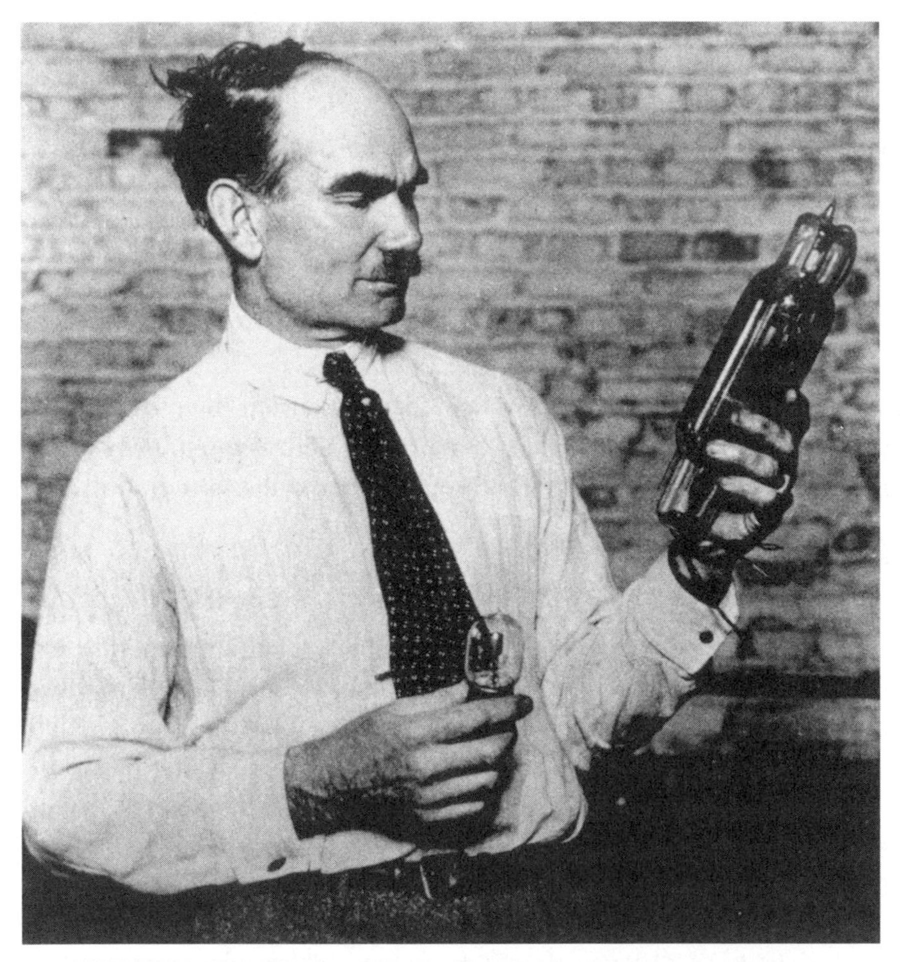

LEE DE FOREST, WITH EARLY (SMALLER) AND LATE (LARGER) THREE-ELEMENT VACUUM TUBES

REGINALD FESSENDEN

Having singled out the detector, or receiver of wireless waves, as the weak component in the advancing wireless front, each then patented another kind of detector and embodied it in a complete wireless telegraphy system involving antennae, transmitter, and other components.

Fessenden's system was the more original of the two. He based his detector on the principle of continuous waves rather than on the discontinuous ones employed by Marconi.[12] By 1906 Fessenden's variation culminated in voice and music transmission, an achievement not possible with the Marconi technology, which transmitted coded dot-dash telegraph signals. De Forest, on the other hand, gave "the impression of a man scrambling for a foothold in a rapidly shifting technology, trying now one device and now another, inventing what he could, borrowing (to use a neutral term) what he could not."[13] In 1903 Fessenden had de Forest as a guest in his home and showed him his laboratory at Fort Monroe, Virginia. De Forest found Fessenden using a Wollaston-wire detector, or liquid receiver, of improved design. Dr. Frederick Vreeland, Fessenden's assistant, surreptitiously told de Forest that he had invented the device. Later Vreeland wrote to Thomas H. Given and Hay Walker, Fessenden's business partners, telling them that the detector, the salient component in the Fessenden system, was his invention and asking them to consider "his claim."[14] Walker decided that Vreeland "is evidently out for blackmail." He cautioned Fessenden, "Remember it is not the facts of the case that counts [sic], but the proof that he sets up that counts."[15]

After seeing the liquid receiver, de Forest later wrote, "we ourselves [thereupon] resolved to use a Wollaston wire-rectifier detector or its equivalent."[16] Perhaps he anticipated that Vreeland's claim could be used to thwart any infringement action by Fessenden. De Forest also tells us that he found that the physicist and electrical engineer Michael Pupin had disclosed the essence of the invention before Fessenden had patented it. The courts eventually found de Forest infringing. His "American system of wireless" seems to have been anticipated in its fundamental components by the disclosures, or patents, of Marconi and of Oliver Lodge, as well as of Fessenden.[17] De Forest, however, was only one of the many independent inventors who improved on, or invented around, the prior systems of others, as the plethora of patent interferences and infringements of the era bear witness.

TESLA'S CHOICES

Nikola Tesla's choices of problems are more difficult to explain than those of other independents. His choices over the span of his career have the quality of "pure" invention, a concept analogous to pure science. Like a pure scientist unconcerned about commercial advantage, Tesla usually made choices that followed the will-o'-the-wisp of his interest and vivid imagination. This can be explained in part by his becoming famous in his thirties for the invention of a system of electric, polyphase, power transmission. For a time, the substantial income from the sale of the patents on this system to the Westinghouse Company sustained his inventive activity. Afterward, his reputation and his dramatic experimental demonstrations of his discoveries and inventions in the field of high-voltage electricity attracted the support of such persons as J. Pierpont

Morgan. His appeal to the public and to Morgan was only enhanced by his elegant style of dressing and living. For those in awe of him, he exuded an air of mystery arising, it seemed, from his communion with cosmic creative forces.

The original idea that culminated as polyphase power transmission came to him in 1877 under circumstances suggesting pure invention and anticipating the way in which he made problem choices later in his career. His professor at Graz Polytechnic in Austria influenced him to think along lines that culminated in his polyphase power system. Academics have stimulated other inventors besides Tesla. Professor William Anthony, a Cornell professor, pointed out problems to Sperry. Professor Carl von Linde alerted Rudolf Diesel when he was a student in Munich to search for an efficient heat engine; and Frank Fanning Jewett, professor of chemistry at Oberlin College in Ohio, stimulated Charles M. Hall, the American inventor of a process for making aluminum commercially, to begin his quest. When Jewett told his students that anyone who invented a commercial process for making aluminum would "be able to lay up a great fortune," Hall turned to a classmate and whispered, "I'll be that man."[18] The professors, well read in the technical and scientific periodicals and in touch with the technical community, knew the critical problems in developing technological systems. Tesla's professor alerted him and the other students to the problem of destructive sparking occurring at the brushes of an electric motor. Then, Tesla recalled, he saw the problem as one of designing a motor without brushes. Yet it was not until five years later that he experienced a eureka moment that brought him the solution in a flash. And not until four years after that did he apply for patents to cover the invention, by which time an Italian professor, Galileo Ferraris, had claimed priority. Several years earlier Ferraris had described, but not patented, the rotating electromagnetic field—the principle Tesla had embodied in his brushless motor and polyphase system of power transmission.[19]

In the mid-1880s, at a time when the knowledgeable inventive community was well aware of the need, Tesla began sustained development of a polyphase electric-power transmission system. The nearly simultaneous patenting of polyphase motors and generators by Tesla, Friedrich

August Haselwander in Germany, C. S. Bradley in America, Jonas Wenström in Sweden, and Michael Dolivo-Dobrowolsky in Germany provides evidence of the clustering, or concentration, of inventive activity.[20] Clustering has often occurred during the history of invention, because, as we have seen, inventors keep abreast of one another's activities. Tesla did not use his patents to establish his own company to manufacture the system. Instead, he sold the patents to the Westinghouse Company, which assigned a team of particularly gifted design and development engineers to work with him on the problem of transforming his ideas into a practical system. Tesla was retained as a consultant for a year, though the system was not ready for the market for several years.[21]

Supported by income from Westinghouse royalties, he remained an independent inventor. His inventive concepts grew grand—some thought unrealistic—but, unlike Edison and other professionals, he was not engaged by entrepreneurship—the founding of companies. Tesla's style remained that of the inventor, an inventor gripped by abstract concepts of energy and the application of these. His vision of rotating magnetic fields was one of swirling energy. Later, as we have seen, he envisioned energy transmission over great distances without wires, and he also invented devices for wireless communication and control. He often discussed his concept of universal energy, which, he believed, permeated space, and which man could harness wherever he wished. Men, he said, would eventually attach their machinery to "the very wheelwork of nature," and "I expect to live to be able to set a machine in the middle of this room and move it by no other agency than the energy of the medium in motion around us."[22] Historian Arnold Toynbee, in generalizing about the development of technology, as he did about so much of history in his multivolume A *Study of History* (1951–61), described its progress as "etherealization," a term aptly descriptive of Tesla's inventive style.

Idiosyncratic in many ways, Tesla acted like the other independents in keeping his distance from bureaucratic organizations. He resembled Edison and Sperry, the professionals, in ranging widely in his problem choices. He worked briefly with the Edison company in Paris, for Edison himself in New York, as a consultant for Westinghouse, and for several years for a small electrical-manufacturing firm bearing his name. But his ultimate commitment, after achieving some financial independence about 1890, was to the life of the professional with a small research laboratory. This gave him a wide latitude of problem choice. Among the fields into which he ventured as an inventor were electric lighting, electric power, wireless communications, automatic controls, wireless power transmission, turbines, air conditioning, and vertical-takeoff airplanes. He acquired more than one hundred U.S. patents.

EDISON AND SPERRY: THE PROFESSIONALS

Bell and the Wright brothers can be styled amateur or part-time inventors if we do not take "amateur" to suggest that their inventions were any less imaginative or subtle than those of the full-time, professional inventors. Fessenden and de Forest were full-time inventors notable for their single-minded concentration on wireless. Tesla is difficult to classify. Edison, Sperry, and Maxim, however, can be characterized as representative professional inventor-entrepreneurs in their years of full-time dedication to invention and their establishment of companies to exploit their inventions. Their style of problem choice was characterized by its wide-ranging nature. Maxim invented, among other things, a machine gun, an incandescent lamp, smokeless gunpowder, and a steam-powered aircraft.[23] Among Sperry's more than 350 patents were major contributions to the technology of electric light and power; mining machinery; electric railways; electric automobiles; batteries; electrochemistry; gyroscopic guidance, control, and stabilization; gunfire control; aviation instruments; and others. The range of Edison's inventive activity is also impressive. The more than one thousand patents taken out in his name include those covering the telegraph, phonograph, telephone, electric light and power, magnetic ore separation, storage batteries, concrete construction, and motion pictures. Such a range would have been unlikely in an industrial research laboratory tied to product lines. Thomson, an Edison competitor in the electrical field, took out 696 patents, including major ones on arc lighting, incandescent lighting, electric-generator controls, alternating-current transformers, electric motors, electric welding, electric meters, and X-ray devices.

Prudent drawing on their prior experience also characterized the Edison and Sperry style of problem choice. Unlike the naïve inventors, they did not dream that anything was possible for them. They chose problems most likely to be solved by the particular inventive characteristics they had acquired. Edison learned about the behavior of electrical currents as an experimenting telegraph operator. This experience propelled him through a sequence of related problem choices. It gave him a tacit understanding of the conservation of energy, for he observed firsthand not only the transformation of electricity into magnetism and magnetism into motion, but the fact that motion in the presence of magnetism produced electricity. Edison's insistence that he had thoroughly read and digested the works of Michael Faraday, the British discoverer of electromagnetic induction, seems plausible. He also knew from the work of early-nineteenth-century scientists and inventors that electricity could produce heat and light. The relationship between the resistance of metals and the flow of electrical energy must also have engaged his interest. As a telegraph operator and experimenter, he learned much about electrochemistry from batteries. His experiments with the telephone called his attention to the convertibility of sound waves into mechanical motion and electrical waves. So his sequential inventions of duplex telegraphy, telephone transmitter, phonograph, incandescent-lighting system, and electric railways become comprehensible, for they involve ingenious transformations of energy. Even his venture into iron-ore separation involved electricity and magnetism, and his work on the storage battery recalled the electrochemistry of the telegraph system. In this sense he was an applier of electrical science, but his intimate knowledge of the behavior of electrical devices and his ingenious application of accumulated tacit knowledge distinguished him from most scientists, who tended to be more verbal and theoretical. Edison's involvement with motion pictures and his effort to obtain rubber from plant substances seem, however, out of character for the master of energy transformations.

Sperry can aptly be characterized as a solution looking for feedback problems. A superficial reading of the titles of his more than three hundred patents leads one to assume that there was genius, but little order or pattern, in his inventive activity. A closer reading proves the contrary. The titles of his patents were broad—as is customary—but scrutiny of them reveals that the important ones were mostly in the field of feedback controls. His patents on electric light pertained to the automatic control of arc-light carbons; the patents on electric generators had to do with the control of their output; streetcar patents dealt with the control of these; his numerous and seminal patents for ship-and-airplane stabilization focused on feedback-control devices; and his famous gyrocompass had feedback mechanisms. In short, Sperry's style was characterized by an expertise and a remarkable range of applications of the principle of feedback control. As the father of modern feedback controls, he explored a field now described variously as cybernetics, automatic controls, and automation.

Another salient characteristic of Sperry was his determination to ferret out the difficult, or pretty, technical problems. After his unhappy youthful experience in trying to compete with larger and more experienced firms in manufacture and sales, he decided that he could compete most effectively in the domain of invention and development. He made up his mind always to choose the difficult technical problems that mediocre, ruthless, driving men without technical and inventive talent would avoid. The characteristic Sperry invention then became a highly complex device, such as a gyrocompass, incorporating feedback control and demanding precision manufacture. Heavily financed, large-scale manufacturing enterprises found that they could not compete with Sperry in the area of these pretty problems.

Analysis of his patents also shows that Sperry typically entered a field of industrial activity, such as electric light or electric streetcars, when it was new and developing rapidly as measured by the rapid tide of capital investment. He remained in the field about five years, then left it for another as industrial corporations, grown large, assigned a staff, or department of inventor-engineers, to solve the problems of the expanding technological systems over which the corporation presided.[24] Sperry knew that the problems attacked by industrial inventors and engineers were usually ones of refinement, especially suited to collective responses by well-equipped research teams, and ones especially amenable to solution by those immersed in the particulars of the technological system in which the problems were emerging. Such responses by Sperry characterize the attitude of this independent, who repeatedly turned down long-term associations and positions with large corporations of which he was not a major owner. He had an innate sense that these commitments would restrain his problem choice and the resultant exhilaration of the ninety-five-percent breakthroughs. He probably agreed with Charles Kettering, another major inventor and entrepreneur of the early twentieth century, who said, when he heard Charles Lindbergh had flown the Atlantic alone, that it certainly could not have been done with a committee. (Paradoxically, Kettering became head of General Motors Research Corporation.) But Sperry, too, eventually established his own industrial enterprise, the Sperry Gyroscope Company, with its own small staff of inventor-engineers. He delayed this transition, however, until he was fifty years old and had approximately two hundred patents. And even after establishing the company, he left the routine problems to his staff, preserving freedom of problem choice for himself. Several of the major inventions of his later years were not in the gyroscope field, to which his company was committed.

In order to identify problems, Sperry and the independents followed closely the patent activity of other inventors. As a young man, he waited eagerly for the latest issue of the *Official Gazette* of the U.S. Patent Office. In it he found abstracts of the most recently issued patents. Digests of patents could also be found in the technical journals appearing in increasing numbers as the century ended. For $1.00 the *Scientific American* would supply a copy of any patent issued since 1867. Inventors could also learn of the inventions of peers at the meetings of the newly proliferating technical societies. By following the news of patenting, the independent inventors could discover where the others, including the independents, were concentrating and, therefore, where problems likely to be solved by invention were located. Sperry said, "I was a constant student of electrical inventions, as patented in the U.S. and English patent offices . . . [and] took scientific and electrical papers by means of all of which I was enabled to keep posted as to the advances made in the art. . . ."[25] Sperry kept himself posted in order to participate in those "advances." Edison did, too—both of his laboratories had extensive libraries housing the latest technical and scientific periodicals, as we have seen.

Sperry's close scrutiny of patents helps explain his choosing to concentrate, as a neophyte inventor, on arc lighting. At that time, the number of arc-lighting patents issued by the U.S. Patent Office was increasing dramatically—from eight in 1878 to sixty-two in 1882.[26] Sperry had identified a cluster that resulted from other inventors' problem choices. From reading the patent claims, he could pinpoint the problems precisely and try to find a solution (invention) that improved upon but did not interfere with the prior inventions of others. Also Professor Anthony informed young Sperry about the need for better regulation of arc-lighting systems when he made a special trip to Ithaca to ask the professor about technical problems needing solution.

REVERSE SALIENTS AND CRITICAL PROBLEMS

After choosing a particular problem on which to work and conceiving of sketchy solutions, often at the instant of a eureka moment, the independent inventor faced a sequence of development problems. These arose in the course of transforming the bright seminal idea into models and seeing them through stages of increasing complexity. At each stage the evolving models had to be tested by experiment, as we have seen, to learn whether they could function in increasingly complex environments. Ultimately the inventor would test the invention in the intended-use environment. During this process of developing the invention, the inventor frequently resorted to a problem-identification technique that suggests the image, or metaphor, of a reverse salient in an expanding military front. One of the most horrendous reverse salients in history developed near the French town of Verdun during World War I. The Germans believed that, before a general advance could continue, they had to eliminate this reverse bulge, or reverse salient, in their front lines extending along the Western Front. The French were equally determined to hold this projection, or, from their point of view, salient, into German-held territory. A military front line has salients and reentrants (reverse salients) all along its length.

The reverse salient in an advancing military front proves an apt metaphor for a technological system, because the system, like a military advance, develops unevenly. Some components in a technological system, like some units in the military front, fall behind others. In the case of the military, *ahead* and *behind* can be determined by physical distance. Some components in technological systems can be said to be behind others if the former function less efficiently and act as a drag on the system. A telephone relay that distorts a message in a long-distance system is an example. Components that malfunction and cause breakdowns of a system can also be seen as being behind. Insulators in high-voltage transmission lines that fail and cause electrical shorts provide another example of a reverse salient in a technological system. In other cases, a reverse salient in a technological system can be one that adds disproportionately to the cost of the system. Rare materials in system components are a case in point. Inventors must correct reverse salients in systems they are developing. A perspicacious inventor often is particularly adept in discovering the reverse salients. Realizing that each component in a system affects, and is affected by, the characteristics of other components in it, the inventor looks for the component that has figuratively fallen behind the others. Having decided, for instance, to improve on Marconi's detector, Fessenden and de Forest had to design the other components in the wireless system to harmonize with it. As we shall see, Bell Telephone Company engineers and scientists concentrated on inventing improved relays for an expanding transcontinental telephone transmission. General Electric Company engineers and scientists focused on solving problems involving insulation in high-voltage transmission lines. As with military lines of advance and retreat, the appearance of reverse salients and salients along the expanding front of a technological system is continuous, because the front is in a state of constant flux. As a result, inventors and generals are seldom without problems to solve. "Reverse salient" suggests the fluidity of the course of technological-system development; other metaphors suggesting rigidity and simplicity, such as "bottleneck," do not work so well.

We might consider also what Edison wrote about his invention of a system of electric lighting:

> It was not only necessary that the lamps should give light and the dynamos generate current, but the lamps must be adapted to the current of the dynamos, and the dynamos must be constructed to give the character of current required by the lamps, and likewise all parts of the system must be constructed with reference to all other parts, since, in one sense, all the parts form one machine, and the connections between the parts being electrical instead of mechanical. Like any other machine the failure of one part to cooperate properly with the other part disorganizes the whole and renders it inoperative for the purpose intended.
>
> The problem then that I undertook to solve was stated generally, the production of the multifarious apparatus, methods, and devices, each adapted for use with every other, and all forming a comprehensive system.[27]

Once independents embarked on the voyage of inventing a system, there were beacons all along the way. They concentrated on the sequence of reverse salients as the appropriate problem choice.

Because Edison's extensive notebooks and his papers have been preserved, we have a better understanding of his method of finding reverse salients and solving the critical problems associated with them than we have for other inventors. In the case of the electric-lighting system, he used the reverse-salient technique not only to launch the incandescent-lighting-system project, but also to plot the course of developing it. Edison chose to invent and develop an electric light system in 1878, when he observed how close a number of inventors had come to introducing a practical incandescent bulb, a lamp for which there was a market because the arc lights then in use as streetlights burned too brilliantly for small, enclosed spaces. During the nineteenth century and before Edison began his work, at least twenty different types of incandescent lamp had already been invented.[28] Like other independents, Edison scanned the inventions of earlier inventors seeking systems that were not quite technically or commercially successful, then proceeded to correct by invention the reverse salients and bring the systems onto the market. Like the Wright brothers, he possessed the confidence to take the step that made the difference between failure and commercial success. Larger resources, longer concentration on the problem, and superior knowledge partly explain their successes.

In 1878 he decided that the principal reverse salient in the failed electric-lighting systems was the short-lived filaments in the incandescent bulbs. By autumn he believed he had found the critical problem—to design a rheostat that would intermittently and imperceptibly shut off current to a platinum filament to allow it to cool before melting. Some months later his analysis became more complex and fruitful. Having identified a reverse salient, Edison framed the critical problems that had to be solved to correct the salient. Taking a holistic view of the electric-lighting system, he realized that, even if he solved the technical problem of filament durability, an economic reverse salient remained—the cost of the copper wire needed to transmit energy to the numerous lamps of a large-area system supplied from a central generating station. He foresaw that, if he were unable to reduce the amount of copper wire needed, his system could not compete in cost with urban gas-lighting systems supplied from central stations.

Edison's response to the copper-cost problem testifies to his brilliance as a holistic conceptualizer. Resorting to known laws describing interacting behavior of components in electric circuits, such as Ohm's law, which relates current, voltage, and resistance, he experienced a eureka insight. He foresaw that, by increasing the resistance of the lamp filament, he could reduce the amount of copper in the distribution wires without decreasing the flow of energy. From then on, he and his laboratory associates and assistants pragmatically searched for the right filament material and proceeded to design other system components, such as the generator, to harmonize with the characteristics of the high-resistance incandescent lamps.[29]

EUREKA MOMENTS AND METAPHORS

Eureka moments and instantaneous insights are part of the lore of invention and discovery. Biographers of inventors and analyzers of the act of invention, however, after reverential asides about the mystery of creative genius, tend to pass quickly over such events. On 8 September 1878, Edison experienced a eureka moment when he discussed with the inventor and industrialist William Wallace the flawed incandescent lamp system of Wallace's collaborator, the inventor Moses Farmer. The durability of Farmer's incandescent bulbs, like those of the other early inventors, was inadequate, but he had found two ways of approaching the problem. After reflection, Edison found these and related insights of his own so promising that he telegraphed an associate, "Have struck a bonanza in electric light. . . ."[30]

He also wrote shortly afterward:

> I have the right principle and am on the right track, but time, hard work and some good luck are necessary too. It has been just so in all of my inventions. The first step is an intuition, and comes with a burst, then difficulties arise—this thing gives out and [it is] then that "Bugs"—as such little faults and difficulties are called—show themselves and months of intense watching, study and labor are requisite before commercial success or failure is certainly reached.[31]

Inventors' imaginations and eureka moments often involve the use of metaphor. Independent inventors' frequent resort to both verbal and visual metaphor offers us the most suggestive key to understanding the moment of inventive insight. This is not surprising, for metaphor has often been associated with creativity. Metaphor, being far more than a decorative literary device, is one of our most often used and effective ways of knowing.[32] It has been defined as "the use of a word in some new sense in order to remedy a gap in the vocabulary."[33] The word used in a new sense is the principal subject, and the word to which it is compared, and which is used in a literal and conventional sense, is the subsidiary subject.

Unable to find a word to designate a sunset sky, someone in the distant past may have said, "the color of the sky is orange." Orange as sky color is the principal subject; orange as color of fruit is the unexpressed subsidiary subject. The principal and subsidiary subjects of a metaphor interact. This means that the reader or hearer of the metaphor will, if the metaphor works, selectively project commonplace characteristics associated with the subsidiary subject onto the principal subject. "A mighty fortress is our God" is an example of an "interaction" metaphor. The metaphor assumes analogies of some features of God and of fortresses. The reaction is to project selectively onto God some fortress characteristics, such as sheltering, power, and endurance. If the creator and the recipient of the metaphor do not share commonplaces about the subsidiary subject, and if they do not select similarly from the array of commonplaces, the metaphor will mislead. (A hearer would be puzzled by the association of the Christian God with the spewing of hot oil and the hurling of projectiles.)

Aristotle wrote, in *De Poetica*, "The greatest thing by far is to be a master of metaphor; it is the one thing that cannot be learnt from others; and it is also a sign of genius, since a good metaphor implies an intuitive perception of the similarity in the dissimilar." In the history of science, Isaac Newton's projection of the characteristics of a falling apple on the "fall" of the planets around the sun illustrates metaphorical thinking of the highest order. Poetry, like scientific discovery and technological invention, depends heavily on metaphor. William Blake's poem "The Sick Rose," ostensibly about a beautiful flower destined to be destroyed by an ugly worm ("O Rose, thou art sick! The invisible worm . . . Has found out thy bed"), suggests, among other possible readings, a beautiful woman about to be destroyed by an insidious disease.[34] Not only poets but schizophrenics also make such metaphors. From observing numerous patients, Silvano Arieti, a psychiatrist and an investigator of schizophrenia, believes that the schizophrenic seeing similarities between the predicate characteristics of two dissimilar persons or things will lose the figurative sense of a metaphorical insight and take them to be identical. A patient who longed to be virtuous and who was a virgin identified herself with the

Mary who was also a virgin. A reductionist, interaction metaphor, "I am the Virgin Mary," became her unreal identity.[35]

To invent machines, devices, and processes by metaphorical thinking is similar to the process of word creation. The inventor needs the intuition of the metaphor maker, some of the insight of Newton, the imagination of the poet, and perhaps a touch of the irrational obsession of the schizophrenic. The myth of inventor as mad genius is not without substance. Metaphor provides for the inventor a bridge from the discovered or invented into the realm of the undiscovered. Edison used metaphors extensively when he resorted to analogy, an explicit statement about the similarities juxtaposed in a metaphor.[36] He worked out the quadruplex telegraph, perhaps the most elegant and complex of his inventions, "almost entirely on the basis of an analogy with a water system including pumps, pipes, valves, and water wheels."[37] The metaphor (analogy) for Edison was "A Quadruplex telegraph will be [like] a water system." The inventor must use the future tense when referring to the primary subject, for it has yet to be invented. Edison had in mind the particular characteristics of a water system that could be projected onto the quadruplex telegraph to be invented. Later, thinking metaphorically, he conceived of the interaction between existing illuminating gas-distribution systems and the illuminating incandescent-light system he intended to invent. The analogy stimulated him to invent a system, rather than only an incandescent lamp.[38]

Edison and the other independent inventors who contributed greatly to the rise of the electrical industry in the late nineteenth century may have been so prolific because they had absorbed the essence of Joule's law, a rich stimulus for invention by analogy. In 1843 James Prescott Joule completed and published the results of a series of experiments demonstrating the equivalence, or convertibility, of electrical, mechanical, and heat energy. He generated electricity with chemical batteries, used it to drive an electric motor, then measured the heat given off by the armature of the revolving motor. He also employed mechanical force to drive an electric generator, then measured the amount of electricity generated and the thermal heat given off by the electricity flowing through a circuit. And Joule showed the quantitative relationships among electric current, resistance of a circuit, and heat given off by the circuit. Soon Joule's law became a commonplace of science, one that suggested analogies to fertile imaginations like Edison's. Hence Edison's drawing an analogy between, or conceiving a metaphor involving, mechanical pumps and electrical telegraphs. He and other inventors also saw that various forms of energy, including acoustic, were convertible in practical ways. Edison's telephone transmitter, his electric-lighting system, and his phonograph depended on the convertibility of energy. Similarly, Sperry saw an electromagnet as a metaphor involving mechanical and electrical action. He used electromagnets as automatic electromechanical controls for electric circuits.

Sperry often conceptualized metaphorically. His most intriguing usages were his frequent references to machines as beasts. Working on one of the first airplane stabilizers, he observed in 1923 that "of all vehicles on earth, under the earth and above the earth, the airplane is that particular beast of burden which is obsessed with motions, side pressure, skidding, acceleration pressures, and strong centrifugal moments, . . . all in endless variety and endless combination."[39] He characterized the early ship stabilizer of another inventor as an "English blood ugly . . . a brute of a machine."[40] Throughout his inventive career, Sperry was gaining control, as he put it, of the beasts. He spoke of harnessing "that brute," and then "putting the little fellow to work" after the brute had been brought to heel.[41] We can only ask what deep psychological drive so impelled him to use the beast metaphor in speaking of his inventions, or "these queer dreams of mine."[42]

Like innumerable other inventors, he proceeded metaphorically as he simulated machines and structures for testing. As we have noted, he assumed that the commonplace characteristics of a rolling ship and a swinging pendulum were similar despite their dissimilar appearances; so, when he was developing his ship stabilizer, he used a small pendulum as a ship and mounted a laboratory gyro on it. A full-scale ship would have permitted him to avoid the imprecision of a metaphor, but the cost would have been prohibitive—a pragmatic argument for metaphor.[43]

De Forest also inclined to metaphor. His most renowned invention,

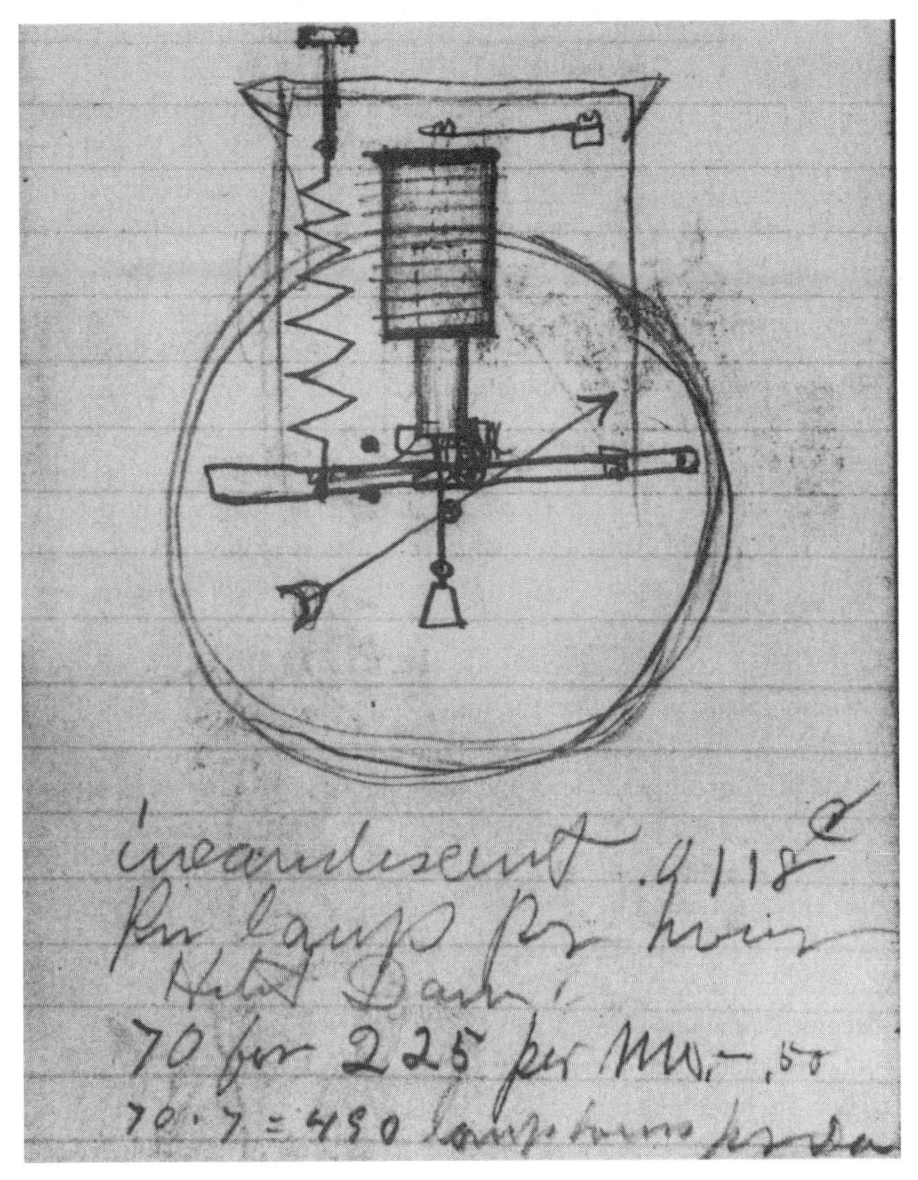

ELMER SPERRY JOTTED DOWN IN HIS POCKET NOTEBOOK INSIGHTS, IDEAS, AND RANDOM ITEMS THAT OFTEN CULMINATED IN A PATENT AND AN INVENTION.

the triode vacuum tube, began its long conceptual history as a chance observation and a durable analogy. Working in Chicago in 1900 and experimenting in free hours with wireless spark-transmitters, he observed that a Welsbach gas burner brightened when the sparks discharged from the transmitter. At the time, de Forest assumed the incandescent particles or the hot gases of the flame were responding to the electromagnetic radiation—Hertzian waves—emitted from the spark transmitter.[44] His earlier observation, as he prepared his dissertation at Yale University, that ionized gas detected wireless waves, probably predisposed him to this interpretation. He was profoundly elated, for, if this were the case with the gas flame, he would have the possibility of developing a wireless detector, or receiver, on the basis of the observed phenomenon. Soon, however, he shut a wooden door between the transmitter and the detector and found that the flame was reacting to the noise of the spark discharge, not to electromagnetism. "I had discovered," he recalled, "simply a novel form of 'sensitive flame'!"[45]

His initial reaction was erroneous, but, if we can rely on his memoirs, a highly important event in the history of technology. He recalled that the illusion that the flame was responding to electromagnetic waves "had persisted in my mind so long and I had cogitated so intently in seeking some explanation for the supposed effect . . . that, notwithstanding this shocking disappointment I remained convinced that the supposed action and effect did nevertheless exist."[46] We must ask if the memoirs are reliable, because inventors are so conditioned by acrid argument over priority, inside and outside the patent courts, that legal briefs tend to become confused with events as they actually transpired. Our skepticism notwithstanding, de Forest's recollections remain important testimony about the inventor's mind and its proneness to metaphor and analogy, for he wrote in his laboratory notebook: "Several have mentioned the weak acoustic action of a coherer [wireless detector]; from analogy then might I not expect an electromagnetic action lurking somewhere in the sensitive flame, since that responds so to acoustic vibration?"[47] A highly ingenious and intricately reasoning mind, in hypothesizing, created a labyrinthine analogy: if coherer receivers respond to sound and to electromagnetic waves, then a flame of hot gases responding to sound may similarly respond to electromagnetic waves. He also fantasized, "Is there not here at least an *analogy* between this effect of elect.-mag. waves on heated gases and the intimate connection between *sun spot* and the magnetic storms that accompany them?"[48]

De Forest's love of analogy also appears in the poetry he frequently wrote, a selection from which he included in his autobiography.[49] His use of metaphor seems to have emerged unconsciously and spontaneously. For instance, when observing under a microscope the flow of minute particles between electrodes in his wireless detector, he imagined:

> Tiny ferryboats they were, each laden with its little electric charge, unloading their etheric cargo at the opposite electrode and retracing their journeyings or, caught by a cohesive force, building up little bridges, or trees with branches of quaint and beautiful patterns.[50]

Spurred on by analogous thinking, he resolved to invent a flaming-hot gas (ionized), or incandescent particle, receiver: a search that culminated in his invention of a gas-filled, three-element electronic tube, the fundamental early invention in electronics.[51] In 1907, when de Forest applied for a patent on the electronic tube, or valve, he still believed that the essential phenomenon in the device was the activity of heated gas permeated by electromagnetic waves. His inventive analogy bore fruit, but again illogically, for the fundamental phenomenon in his tube was electron discharge, a fact that he failed to comprehend.[52]

Tesla, a master of dramatic presentation, whether in the staging of high-voltage electrical experiments or in the recounting of experiences for an admiring public, had one of history's most graphic eureka moments. In 1882, after having pictured in his mind for several years a problem of electric-motor design, he was walking one day in a Budapest park with a friend and reciting long passages of Goethe's *Faust* from memory when he had the eureka flash. As he quoted, "Oh, that spiritual wings soaring so easily / Had companions to lift me bodily from earth,"[53] his imagination indeed soared: "In an instant I saw it all, and I drew with a stick on the sand the diagrams which were illustrated in my fundamental patents of May, 1888."[54] Inasmuch as five fundamental, related patents were issued to Tesla in May 1888, the detail of his insight must have been remarkable.

The history of early wireless is strewn with other instances of invention aided by metaphor. The inventors and scientists often drew analogies among acoustic, electrical, and mechanical phenomena. Seeking to increase the range of wireless transmitters, Columbia University scientist and engineer Michael Pupin drew analogies with the tuning fork. Pupin also had the highly fruitful idea that the as yet not well-understood wireless waves could be better comprehended if they were seen as analogous to the better-understood electrical waves of alternating-current electricity widely used by 1900 for lighting and power. Recognizing underlying similarities between a telephone receiver and a wireless detector helped Fessenden invent the heterodyne principle, which today remains fundamental to radio technology.[55]

Although they are articulated verbally, the metaphors of inventors have often been visual or spatial. Inventors, like many scientists, including Albert Einstein, Niels Bohr, and Werner Heisenberg, show themselves adept at manipulating visual, or nonverbal, images.[56] We can imagine inventors committed to the nonverbal mode as manipulating a visual or spatial alphabet consisting of mechanical, electrical-circuit, and chemical-process symbols, such as gears, induction coils, and the pressure chamber. The eighteenth-century Swedish engineer Christopher Polhem constructed an alphabet of wooden models of fundamental machine components such as ratchets, cams, and drive shafts with which he instructed his students so that their imaginations would be filled with visual symbols that they could creatively combine into new machines.[57] Independent inventors, most of whom were avid consumers of patent and technical literature, also had their heads filled with visual concepts by these profusely and precisely illustrated sources. Some of them had also studied descriptive geometry, a rich stimulus to generating spatial forms and relationships. Decades earlier, Robert Fulton, the artist and steamboat inventor, had urged that the mechanical craftsman and inventor "should sit down among levers, screws, wedges, wheels, etc. like a poet among the letters of the alphabet, considering them as the exhibition of his thoughts, in which a new arrangement transmits a new Idea to the world."[58] It would be a mistake to conclude that to visualize is to simplify, for many of the complex symbols that inventors think about "cannot be reduced to unambiguous verbal descriptions."[59]

Sperry also thought visually—or, as some persons believe, drew on the right side of the brain.[60] An associate recalled that the inventor, while seemingly staring absent-mindedly into space, would suddenly grab a pad, hold it in the air at arm's length, and draw on it. When asked about his behavior, he replied, "It's there! Don't you see it! Just draw a line around what you see." Sperry also had the remarkable ability to visualize a device or machine and see it operate with his mind's eye. With heightened sensitivity, he could locate the points of friction.[61] Teen-aged Sperry's avid reading of the Patent Office's *Official Gazette* not only helped him identify technical problems to solve but probably also helped develop his ability to visualize and operate the devices shown in the profuse and handsomely rendered patent drawings. Perhaps Einstein's reading of countless patent applications when he worked as a patent examiner also reinforced his tendency to think visually.

FINDING FUNDING

Because they avoided salaried positions and long-term associations with large-scale enterprises, the independent inventors faced challenging problems in finding and securing funding. Until the onset of the naval armaments race among the great powers around 1900, the independents had little recourse but to turn for support to private investors. These entrepreneurs, knowing of prior fortunes made in inventions, hoped to discover and invest in another Edison or Bell. After becoming famous as the Wizard of Menlo Park and inaugurating his electric-lighting project, Edison found himself in the unusual position of being able to call on the prodigious resources of Morgan. Established manufacturers, however, rarely funded the radical invention of an independent inventor, because they sought inventions that improved on the existing machines, devices, and processes that they manufactured. Radical inventors who tried for the breakthrough inventions did not do this.

If an independent were fortunate enough to secure backing by a private investor or business entrepreneur, the inventor typically joined his backer to form a company to manufacture and market the independent's invention. Usually the inventor took a large share of the stock issue in return for the patents he assigned to the new company, and the business entrepreneur received capital stock in return for development funds. Often additional shares were sold to raise additional capital. The early career of the inventor Thomson exemplifies the process. Two Philadelphia businessmen, one a dealer in photographic supplies and another a farmer and salesman for the Brush arc-lighting system, agreed to underwrite the manufacture and sale of a dynamo patented by young Thomson and his partner, Edwin Houston. Later, in 1880, Frederick H. Churchill, a lawyer from New Britain, Connecticut, enthusiastically committed to the new science and technology of electric lighting and to profitable investments, organized with Thomson a company to manufacture a complete electric-lighting system based on Thomson's patents. Churchill raised $87,500 as capital for the American Electric Company. Not long afterward, Charles R. Flint, a highly imaginative business entrepreneur who also found electric lighting an attractive new field for investment, became interested not only in Thomson's patents, but in Thomson's future as inventor. Independent inventors often assigned to a new company both existing and future patents in a field. A New York businessman who headed the shipping and trading firm of W. R. Grace and Company, Flint in 1882 came close to acquiring a stable of inventors. Intending to combine their inventive talents into a single corporation, he brought Charles Brush, the leading U.S. arc-light inventor; Edward Weston, who had valuable patents in incandescent lighting; and Thomson together in his office. But they could not reach an agreement.

During his long independent career, Sperry inspired a number of entrepreneurs and investors to establish companies on the basis of his patents. When he was only twenty years old, his knowledge of electrical technology and enthusiasm for invention were so obvious and contagious that the managers of the wagon works in his hometown of Cortland, New York, funded the invention and development of his first arc-lighting system. Widespread publicity surrounding Edison probably persuaded the Cortland investors that a bright, self-taught, and ambitious young Baptist reared in a hard-working small town had the makings of another Edison. Encouraged by his Cortland backers to try out his arc-lighting system in bustling Chicago, and introduced by them to their network of Baptist businessmen and civic leaders there, Sperry in 1883 helped establish the manufacturing company bearing his name. The Reverend Galusha Anderson, first president of the University of Chicago, was one backer of the company.

In 1888 Sperry took the imaginative step of organizing the Elmer A. Sperry Company of Chicago to develop his and others' inventions. This early invention-and-development company, in effect, incorporated his inventiveness. Joseph Medill, mayor of Chicago and editor of the *Tribune*, helped fund Sperry's effort to develop a gas engine. When Sperry invented improvements for streetcars, a small group of investors in Cleveland, Ohio, aware of his previous successes, formed the Sperry Syndicate to acquire his patents and to begin manufacture of the improved devices. Among the investors were officers of a company making arc-light carbons, several bankers, and the son of President Rutherford Hayes, who had his father's financial backing. Sperry usually received cash and stock for the "capital" (his patents) that he signed over to the companies. As companies based on his patents expanded and his stock increased in value, he financed more of his own inventive activities.

Tesla also incorporated his inventive genius by forming an invention-and-development company. Edward Dean Adams, financier of the Niagara Falls hydroelectric project in 1895, helped him about the same time to form the Nikola Tesla Company by subscribing $100,000 in stock with an immediate cash payment of $40,000. When Tesla had exhausted this amount in equipping his New York laboratory, he persuaded the financier John Jacob Astor, in whose Waldorf-Astoria Hotel Tesla lived, to contribute $30,000 toward the new laboratory to be built in Colorado Springs. In 1900 Tesla approached Morgan, telling him that he had already transmitted wireless over seven hundred miles and promising that he had patents and could construct plants for wireless transmission across the Atlantic and the Pacific. Morgan in return promised up to $150,000 to finance a project to bring the inventions into use and to support further research. Tesla said that his benefactor was a great and generous man.

Morgan asked for fifty-one percent of Tesla's electric-lighting and wireless patents, existing and future, in return for his investment.[62] Funded by Morgan, Tesla built a transmission station on Long Island, which he named Wardenclyffe. The famous architect Stanford White designed the main building. By 1903, after large expenditures for equipment, Tesla was again desperate for money. Even though he now held out to Morgan the additional promise of wireless transmission of substantial amounts of power, not only messages, the financier, disenchanted by the absence of results, refused the request for more money. Pursued by his creditors and challenged by Marconi, who transmitted across the Atlantic in December 1901 with relatively simple and inexpensive apparatus, Tesla closed down Wardenclyffe, his "dream plant," in 1906. He said of his failure to transmit power by wireless, "It is a simple feat of electrical engineering—only expensive."[63] After Morgan died in 1913, Tesla turned to Morgan's son for money, holding out the promise that the Tesla ship had almost arrived. Tesla, a single man who preferred to live in fine New York hotels, Astor's Waldorf-Astoria among them, and mix there in the afternoons with the important men of Wall Street, found these men far less responsive as his reach appeared more often to exceed his grasp.[64] After his mother died in 1892, he lived an increasingly solitary life, with pigeons, which he kept in his hotel room, as companions. In his declining years—he died in 1943—Tesla lived alone in a run-down Times Square hotel. Still filled with inventive schemes and still seeking money for them, he wrote to the chairman of the Westinghouse Company, which had bought his electric-power transmission patents a half-century earlier. He offered, this time, a patent on a technique for artificially raising chickens with a feed he named Factor Auctus (Creator of Growth). "You will be grateful to me," he wrote, "when you get the delicious eggs and meat obtained by this revolutionary process."[65]

In his prime and during the development of his electric-lighting system, Edison also depended for funds on great financiers and investment bankers, including the Vanderbilts and Drexel, Morgan & Company. Grosvenor Lowrey, a prominent New York lawyer and general counsel for the Western Union Telegraph Company, a position that brought him into close contact with the highest echelons of finance, promoted Edison's

ARCHITECTURE OF NIKOLA TESLA LABORATORY AT WARDENCLYFFE, NEW YORK, ENHANCED THE AURA OF GENIUS SURROUNDING THE INVENTOR.

projects in these circles. Lowrey's admiration and support for Edison transcended the rationality of a legal and financial adviser; he became a champion of Edison's, deeply committed to helping him fulfill his aspirations. Lowrey promised Edison in 1878 that the income from a successful electric-lighting system would "set [Edison] up forever . . . [and] enable [Edison] . . . to build and formally endow a working laboratory such as the world needs and has never seen."[66]

Even though Edison could depend for substantial support on backers reasonably well informed about technology and science, he, like other independents, also needed to attract the support of a wider range of persons, whose information about technology was superficial and uncritical. Unlike many inventor-entrepreneurs today, Edison could not turn to universities, the military, and the government for research-and-development grants; therefore, he did not cast his appeal in language sufficiently sophisticated for these institutions. Instead, he and his advisers used newspaper publicity to generate enthusiasm and additional financial support. His interview with the New York *Sun* published on 20 October 1878 attracted wide attention, for he announced in easily comprehensible, general language that he had solved the major problems of an electric-lighting system—which, in fact, he had not done and would not do for at least another year. He promised that he would soon light up the entire downtown area of New York with five hundred thousand bulbs; four years later his system supplied only 12,843 bulbs in a few blocks of the Wall Street district.[67] A less enthusiastic interview might have discouraged investors ignorant of the problems of long-term, painstaking development and subscribers who believed in the myth of dramatic bursts of genius and quick financial rewards. Edison and Lowrey maintained enthusiasm and funding during the long development period by staging dramatic demonstrations such as that held on New Year's Eve 1879, when they hired special trains to bring the public—and the financiers—to see a small but impressive lighting system at Menlo Park. Again, the newspapers widely publicized the event. Edison was the center of attraction during the exhibit, for he and his advisers realized that public confidence in him provided a major reason for success in raising funds for the long-term project. The cultivation of Edison as inventive hero might also explain why his name alone is found on many patents and in much publicity, though additional credit was undoubtedly owed to his highly competent laboratory staff.

De Forest and his business associates were masters of demonstration and publicity. His ingenuity along these lines may at times have exceeded that in the technical realm. He acknowledged that he found publicity sweet. In 1901 he persuaded a press association and several financiers to fund installation of his wireless system on a tugboat, so that he could report at sea the progress of the international yacht races. Marconi also covered the event for the Associated Press and the New York *Herald* with apparatus installed on Jordan Bennett's palatial yacht. Despite the radio-wave interference between the competing systems that frustrated satisfactory reception, de Forest heralded the achievement and tried to raise money from twenty-five "capitalists" in New York, some of them Yale classmates. The results were disappointing: "Two or three months of hard hustling among capitalists had shown me a little of the disparity between the inventor's and the investor's points of view."[68] Several months later, however, he met Abraham White, a financial speculator who had netted a fortune in government bonds. De Forest found White "gifted with the optimistic vision that J. Pierpont Morgan and other tycoons whom I had solicited, totally lacked."[69] White and de Forest organized the American De Forest Wireless Telegraph Company with $3 million of capital stock. Shortly thereafter, the company began demonstrating and winning more publicity with impressive transmissions across New York Bay. De Forest recalled that "a gratifying amount of public recognition resulted from this work"—and, he might have added, stock sales.[70] Other highly publicized installations financed by the company followed, and de Forest soon had thirteen patents pending. White equipped an automobile with a wireless transmitter and parked it near stockbrokers' offices to flash American De Forest stock quotations. Soon White and his salesmen, more interested in generating stock sales than in transmitting radio messages, pushed de Forest to build wireless stations wherever the market—for securities—was promising. De Forest, however, in building the over-

DE FOREST AND "HONEST ABE" WHITE INVOLVED IN PUBLICITY AND MONEY RAISING

land wireless stations that attracted investors, learned "how vicious and brain-defying God-made static interference could be."[71] In 1906 the bubble burst: the salesmen sold more stock than the company had issued, excessive expenditures exhausted the treasury, and Fessenden brought an infringement suit against de Forest for using his wireless detector. The company bankrupt, de Forest found himself once again walking the streets of New York, but "with experience, confidence, an international reputation in wireless," if not in business affairs. He ultimately recovered and introduced, among other wireless devices, the modern three-element vacuum tube.[72]

Dramatic demonstrations of electric lighting, wireless, and other new devices opened the purses of investors, great and small, but the fame of the inventors was effective, too. Hence the cult of the inventor. An Edison anecdote about an interview with a cub reporter reveals his awareness of his public image:

> While the reporter was being ushered in, the Old Man disguised himself to resemble the heroic image of "The Great Inventor, Thomas A. Edison." . . . Suddenly gone were his natural boyishness of manner, his happy hooliganism. His features froze into immobility, he became statuesque in the armchair, and his unblinking eyes assumed a faraway look like a circus lion thinking of the Nubian desert. He did not stir until the reporter tiptoed right up to him, then he slowly turned his head, as if reluctant to lose the vision of the Nubian desert. The interview itself [about the storage battery] was insignificant. . . .[73]

Francis Jehl, a laboratory assistant of Edison's at Menlo Park who publicly celebrated Edison the inventor, privately recalled Edison's self-promotion less sympathetically:

> When an abnormal man can find such abnormal ways and means to make his name known all over the world with such rocket-like swiftness, and accumulate such wealth with such little real knowledge, a man that cannot solve a simple equation, I say, such a man is a genius—or let us use the more popular word—a wizard. So was Barnum![74]

Edison, de Forest, and the others might have insisted that they tolerated adulatory publicity as a means to the end of funding, but in time they probably began to believe the publicity-generated image of themselves as inventive geniuses. This might explain the difficulty they had in accepting direction and criticism, especially from those who constrained them by means of purse strings. Edison, de Forest, Thomson, Stanley, Maxim, and Fessenden all disagreed and argued with financial backers.

Fessenden's troubled relations with his two wealthy Pittsburgh businessmen backers, Thomas H. Given and Hay Walker, Jr., are memorable. The climax came in 1910, when Walker's agent tried to remove Fessenden's files from his office along with Fessenden's wife, who had thrown her arms around the files to prevent their removal. Fessenden then received an injunction enjoining him from further participation in the affairs of the company that he had helped found and that was based on his patents.[75] Both William Stanley, whose development of the electrical transformer was funded by George Westinghouse, and Tesla, whose development of the system of alternating-current power transmission was also funded by Westinghouse, came to believe after the event that Westinghouse had treated them unfairly in their contractual arrangements. The manager of the U.S. Electric Lighting Company, formed in 1878 to market Maxim's inventions, sent Maxim to Europe, ostensibly to buy patents but primarily to get rid of the temperamental inventor and his debilitating fixation against Edison, with whom he competed in the development of the incandescent lamp.[76]

Publicity and success may have turned Edison's head and caused him to lose his way as he grew older. In his early years as an inventor, he concentrated on small, precise—even elegant—electromechanical devices, such as the stock ticker, the telegraph, and the telephone. The phonograph of the early years was a simple but highly ingenious mechanical-acoustical device. When he turned to the invention and development of an electric-lighting system, he was still applying familiar and congenial electromechanical principles. After moving to the large laboratory at West Orange and choosing to develop a very large-scale ore-separation process, he seems to have lost his sense of identity as an inventor, for he became an innovating industrialist. Instead of working in his customary surroundings—a laboratory surrounded by skilled mechanics, craftsmen, and appliers of science—he was often out in the field, designing and supervising the construction of mammoth iron-ore-crushing rollers, immense electromagnets, and large materials-handling equipment. He invested in ore separation almost the whole of the fortune his prior inventions had brought.[77] After constructing a large ore-separation plant near Ogden, New Jersey, Edison spent almost a decade improving the technical efficiency of the plant and introducing labor-saving machinery. Instead of seeking new applications of Faraday's, Joule's, and Ohm's laws, he was concerned about mass production, labor saving, and unit costs. The principles he was applying were those of the production engineer and the capitalist, not the master inventor and applier of electrical science. The decline in iron-ore prices from Lake Superior ultimately doomed his venture to failure, for, despite efficiency, his pro-

THOMAS A. EDISON: A CRITICAL CARICATURE ENTITLED *THE DECADENCE OF THE WIZARD OF MENLO PARK*

cess could not compete in price. No amount of technical ingenuity and inventiveness was able to overcome that.[78]

His unrestrained ambition may have driven him to compete with the great industrialists, who were capturing the public imagination and amassing immense fortunes at the close of the century. Sperry, by contrast, avoided the temptation, continuing until the end of his life to prefer the technically fine and "sweet" problems. Edison biographer Matthew Josephson associates Edison's change of style with his second marriage, after the death of his first wife, to Mina Miller, the daughter of an Akron philanthropist; the purchase of a baronial estate for her in West Orange; and the establishment of the new laboratory. "Though not yet forty in 1886, he saw himself as one who was rising to the status of America's ruling industrial barons. . . . Henceforth, all that he undertook must be planned on the grand scale."[79]

It is paradoxical that Edison, who had taken pride in his status as an

THE FACTORY-LIKE FAÇADE OF EDISON'S WEST ORANGE LABORATORY, C. 1895, SHOULD BE COMPARED WITH THE INFORMAL LAYOUT OF HIS EARLIER LABORATORY AT MENLO PARK.

independent inventor, became more deeply involved in the development of large-scale manufacturing processes even after the financial failure of ore separation. His new projects required investments and institutional structures that would more severely constrain his freedom of choice and flexibility of response. He ventured into cement manufacture, also a heavy industrial process, so that he could make use of some of the equipment and know-how developed for ore separation. As the technological momentum of his own doing overwhelmed him, Edison was behaving more like the "small-brained capitalists" he once despised.[80] He also committed himself and his resources to the development and manufacture of storage batteries. During World War I, as head of the Naval Consulting Board, he advocated the establishment of a naval research-and-development laboratory to develop prototypes of heavy naval equipment. He proposed laboratory buildings of heavy concrete, like modern manufacturing buildings, and he wanted the laboratory to concentrate on designing and drawing up specifications for the manufacture of airplanes, submarine engines, small guns, "and everything relating to war machinery."[81] After the war, he tried to cultivate new sources for rubber, another large-scale industrial field. The ingenious inventor of subtle devices, such as a quadruplex telegraph, had indeed become a would-be captain of industry.

BRAIN MILL FOR THE MILITARY

After the turn of the century, as the armaments race intensified, fueled by naval competition especially between Great Britain and Germany but also involving the other great powers, the independent inventors began increasingly to invent for the military. The military's support of the independent inventors launched such large-scale technological systems as wireless telegraphy and telephony (radio), the airplane, and feedback guidance-and-control systems for ships and airplanes. Today hydrogen bombs, ballistic missiles, atomic submarines, and the Strategic Defense Initiative (Starwars) signify the military-industrial complex, but this confluence is not just a post–World War II development. Alliance between the military and profit-making manufacturers of armaments has a centuries-old history in the West. Today, however, when we refer to the military-industrial complex, we usually mean specifically the relationship between the military and those who invent and develop new weapons. This relationship also has a long history. As early as the fifteenth century, rulers and inventive architect-engineers collaborated to bring about the rapid development of cannon design. The interaction became far more intense and highly organized, however, after 1880, when the British Admiralty began specifying the performance characteristics it wanted in its engines, guns, ships, and other equipment, and challenged inventors, engineers, and industrialists to develop those designs. The military also began to pay at least part of the costs for testing inventions. Invention of armaments for the major industrial and military powers then became a "command economy," or the supportive direction of innovation by government.[1]

In the closing decades of the nineteenth century, a naval armaments race gathered momentum and stimulated the command economy and the growth of a military-industrial complex. The British and German navies wove a network of collaboration to further armaments expenditure, invention and development, and manufacture. They forged ties of mutual interest with arms-producing firms like Krupp in Germany and Armstrong in the United Kingdom; with politicians seeking support to push through costly naval construction bills that would benefit owners and white- and blue-collar workers in their industrial districts; and with a jingoistic, sensation-seeking popular press fanning the fires of national rivalry. Within the military services a few officers with engineering experience and knowledge of the rapid rate of innovation in the civilian sectors of the economy played the role of entrepreneurs, and saw to it that innovation, not simply expansion, took place in armaments as well.[2]

Lessons learned in Austro-Prussian and Franco-Prussian wars between 1866 and 1871 spread the conviction that new weapons and communications systems were major modes of military competition, the essence of advanced strategy and tactics. In these wars the Prussians coordinated their railroad system for rapid mobilization and troop movement; they used the field telegraph to maintain contact with, and some control over, field officers; and they equipped the infantryman with a breech-loading rifle to make firing possible from a prone position. Changes in naval technology were more dramatic. During the second half of the nineteenth century iron-hulled, steam-propelled vessels with larger and more accurate guns displaced wooden sailing ships. Inventors and engineers systematically integrated advances in metallurgy, machine tools, explosives, steam propulsion, guidance (compasses), and gunfire-control devices, and introduced the pre–World War I dreadnought-class battleship.

Since World War II intercontinental missiles with nuclear warheads have been the high technology of the armaments race. The prime contractors compete for billion-dollar design, development, and manufacturing contracts as launching, missile, and warhead systems are continually being redesigned. Before World War I the dreadnought-class battleship was the super-weapon, the showpiece on the expanding armaments budgets. The British launched the H.M.S. *Dreadnought*, the first battleship of her class, in December 1906. Equipped with ten twelve-inch guns and capable of a speed of twenty-one knots, the super-battleship had greater firing range and speed than any other ship afloat. A committee, chaired and spurred on by the First Sea Lord, Admiral Sir John (Jackie) Fisher, a "volatile, egocentric, overbearing, belligerent, and bellicose" entrepreneur,[3] designed the vessel. H.M.S. *Dreadnought* was not only an impressive man-of-war because of firepower and speed, but it also had a remarkably advanced system of machinery, including the recently developed steam turbine as the main propulsion unit. Engine rooms containing the old reciprocating steam engines were wet and deafeningly noisy. The far more efficient, compact, durable, smoothly rotating turbines transformed the engine room from a "sodden cacophonous hell to a paradise of quiet orderliness."[4] On her trials, the ship steamed seven thousand miles at 17.5 knots, an achievement far beyond the capacity of any other warship. "No greater single step towards efficiency in war was ever made than the introduction of the turbine."[5] Charles A. Parsons, the British inventor who introduced the modern steam turbine, had attracted the navy's attention with a spectacular display in 1897 on the occasion of the Queen Victoria Diamond Jubilee Naval Review. The *Turbinia*, a small ship with a test installation, broke all regulations and darted at high speed in and out of a line of solemnly anchored battle ships. The *Turbinia* was capable of the astounding speed of thirty-four knots.

Because it could fire on enemy ships and remain out of range of their guns, the *Dreadnought* made existing battleships obsolete and greatly escalated the armaments race. The German government suspended building programs until ships comparable to the *Dreadnought* could be designed, and other navies, including the United States', followed suit. President Theodore Roosevelt, a big-navy proponent, obtained authorization for six ships of the dreadnought class: the *Delaware, South Dakota,*

Utah, Florida, Arkansas, and *Wyoming.* As Roosevelt carried out an ambitious program of naval development, the naval budget escalated after 1900, reaching almost $181 million by 1909–10. The president supported the innovative and technologically informed officers, such as William Sims, who as a young officer had reformed gunnery practices.[6] In 1911 Rear Admiral Bradley A. Fiske, inventor of a number of naval devices, made an eloquent plea that the navy assume more of the financial risk of developing inventions, including those of the independents. A year later the *Scientific American* noted further improvement in the navy's attitude toward inventors.[7] The military-inventor complex was gathering momentum.

As the armaments race proceeded, the United States turned to a much-celebrated resource believed to be uniquely American—the creative genius of its independent inventors. Popular stories and myths of inventors and invention rarely recount the major role that the military played in the early development of aviation, wireless, guidance and control, and analogue computers. Several of the leading independents drew heavily on military funding before World War I to develop their inventions. The navy supported Sperry with guidance and control, and de Forest and Fessenden with wireless telegraphy (code) and wireless telephony (voice). The U.S. Army finally bought Wright airplanes. The British military supported Maxim's development of the machine gun. A closer look at the relations between the independents and the military shows how deep the roots of the military-industrial complex, and especially of the military influence on the shape of invention, reach into the American past.

CONTINUOUS GUNFIRE

It would be a mistake, however, to assume that an organization as bureaucratic and respectful of tradition as the military would unreservedly embrace invention and change. For every pre–World War I story of successful innovation in the military, there is another to tell of ingrained reaction to disruptive change. A classic episode concerns the efforts of William Sims as a junior American naval officer to introduce into the fleet a new system of continuous-aim, rapid gunfire. In about 1900, drawing on prior experience with the apparatus and technique of continuous-aim firing of another innovation-minded British officer, Percy Scott, whom he had met on the China Station, Sims prepared detailed and undeniable evidence of the advantages of the new gunfire system. To all of this the higher naval authorities turned a deaf ear. Sims, however, persisted. His spirit of reform was summed up in his declaration:

> I am perfectly willing that those holding views differing from mine should continue to live, but with every fiber of my being I loathe indirection and shiftiness, and where it occurs in high places, and is used to save face at the expense of the vital interest of our great service (in which silly people place such a childlike trust), I want that man's blood and I will have it no matter what it cost me personally.[8]

Unable to countenance the bureaucratic inertia of his superiors, who claimed that the new gunfire scheme was unworkable at the same time that it was being employed in the British navy and informally tried with success in the United States, Sims resorted to a hoary tactic of those bent on organizational change. To break the crust of conservatism, he used the leverage of a superior force outside the organization—in this case, the naval enthusiast and president of the United States, Theodore Roosevelt. Roosevelt recalled the young officer from China and had him named inspector of target practice, a strategic position from which to bring about the changes in gunnery. As a result, the navy had to institute the countless changes in training, equipment, and personnel that innovation in a large organization demands. Awareness among experienced officers of the disruptive and sometimes demoralizing changes required by innovation explains in part the seeming obfuscation and senseless inertia of the navy vis-à-vis the unqualified enthusiasm and often inexperienced commitment to change of the fiery and inventive young officer. Sims, it should be noted, later became an innovation-minded admiral.

THE WRIGHT BROTHERS

Popular accounts of Orville and Wilbur Wright and the airplane rarely recount their determined efforts to have the airplane used first by the military. The Wrights are not usually recognized as early promoters of the military-industrial complex. Until their successful flight in 1903 the Wright brothers seemed indifferent to financial gain from their invention, but afterward they relentlessly tried to interest the U.S.—or any other—military in buying their airplane. Perhaps they chose the military market because these serious sons of a Protestant minister wished to avoid the kind of sensational publicity and trafficking with capitalists in which other inventors had engaged. In 1905, with a thoroughly tested and reliable machine, they proposed to the U.S. War Department that they supply airplanes of agreed-upon specifications at a contract price. The department, having received many proposals from putative inventors of perpetual motion as well as flying machines, probably put the letter into the crank file. Having assumed that the Wright machine was not a developed device, the navy replied with stock paragraphs stating that funds would not be appropriated for experimentation.[9] Deeply disappointed by their own government's lack of interest in their "practical machine for use in war," and "seeing no way to remedy it," they made a formal proposition to the British government.[10] The Wrights' friend and adviser Octave Chanute, initially mortified by the failure of the U.S. War Department to seize the opportunity to inaugurate the era of the airplane, on reflection decided that the British would offer a better price, because they were less hampered in appropriating secret-service funds. Chanute, seemingly a man of peace, also reasoned that "your invention will make more for peace in the hands of the British than in our own, for its existence will soon become known in a general way and the knowledge will deter embroilments."[11] Chanute should be numbered among the countless persons who have argued throughout the endless history of armaments and war that new weapons, if frightening enough or in the hands of a particular country, will keep the peace. Predictions similar to his have been made about weapons ranging from machine guns to atom bombs.

The Wrights negotiated with the British War Office in 1905–6, but they finally decided that the British were more interested in keeping abreast of the progress that they were making than in contracting for a delivery of the airplane. The brothers tried the U.S. War Department again, stating explicitly that they were prepared to furnish a machine suitable for scouting purposes able to carry an operator, fuel, and supplies for a flight of a hundred miles. Again, a stock-phrased letter ignoring the Wrights' declaration insisted that the department would not support experimentation. Chanute concluded that "Those fellows are a bunch of asses."[12]

As in the case of Sims and the introduction of continuous-aim gunfire, a higher authority securely established outside the bureaucracy, challenged, even threatened, by innovation, intervened in behalf of the Wrights. Godfrey Lowell Cabot, of the prominent Boston family and a private investor, had written the brothers immediately after their success at Kitty Hawk asking if their machine could carry freight and expressing interest in a commercial venture. He wrote them again after he learned of the negative reaction of the War Department, asking for a prospectus, and whether they were interested in exploiting the machine for commercial possibilities. Cabot also informed Senator Henry Cabot Lodge, a relative, of the Wrights' difficulties with the War Department. The senator contacted the secretary of war and the department. Perhaps embarrassed by its prior performance, but reluctant to admit its mistakes, the War Department then stiffly approached the Wrights, who, still smarting from earlier treatment they considered insulting, responded coolly.[13] In December 1907, however, after the Wrights had made a number of demonstration flights in Europe, the War Department finally advertised for bids for an airplane, the specifications of which were those that the Wrights had informed the department that their plane could meet. The plane needed to carry the pilot and a passenger (their combined weight was specified), and to fly at forty miles per hour for at least ten miles. The Wrights had declared their price to be $25,000. When the advertisement for bids appeared, several newspapers and journals ridiculed the War Department for assuming that such an airplane could then be built. On 3 September 1908, when the first test flight was scheduled, fewer

MILITARY-INVENTION COMPLEX: WRIGHT PLANE BEING TESTED OVER FORT MYER, VIRGINIA, 1908

than a thousand spectators were on hand at Fort Myer, outside Washington, D.C. As Orville first rose to circle the field, the spectators gasped audibly, and when he landed the crowd "went crazy." The three or four "hard-boiled" newspaper reporters who rushed up to interview Orville had tears streaming down their cheeks.[14] In subsequent flights the airplane more than filled the specifications, but tragedy accompanied the triumph. A young army officer, Tom Selfridge, who accompanied Orville on a flight, died when the plane crashed after a propeller cracked. This did not deter acceptance of the airplane by the War Department, however. In Germany, France, and the United States, companies were organized

CRASH OF WRIGHT AIRPLANE AND DEATH OF LIEUTENANT T. E. SELFRIDGE, 17 SEPTEMBER 1908, AT FORT MYER

to manufacture the Wright brothers' invention. The Wright Company, incorporated in November 1909, with rights to Wright patents in the United States, opened handsome offices in New York City. Cornelius Vanderbilt and August Belmont were among the investors.

HIRAM MAXIM'S GUN

Several decades before the Wrights' association with the military, Hiram Maxim had also become involved. In Europe in 1881–82 Maxim reacted to the chance remark of an American he met in Vienna: "Hang your chemistry and electricity! If you want to make a pile of money, invent something that will enable these Europeans to cut each other's throats with greater facility."[15] Having previously invented an incandescent lamp that improved on Edison's filament, and having engaged in the installation of lighting systems not long after Edison had also begun, Maxim

MUCH-DECORATED SIR HIRAM STEVENS MAXIM, INVENTOR OF A MACHINE GUN

now dedicated his inventive talent to exploiting Europe's growing fascination with military inventions that facilitated throat-cutting or other violence. By 1885 Maxim had patented, set up for manufacture, and demonstrated the world's most destructive machine gun. Using the recoil from one cartridge to load and fire the next, the Maxim gun far surpassed in firepower and reliability rival weapons like the Gatling gun (1862). After several armies, including the U.S., had initially turned it down, arguing dully that the Maxim could not be supplied rapidly enough with ammunition in the field, the gun was widely adopted. The way in which it scythed down Zulus, Dervishes, and other colonial peoples soon won over disbelievers. As Hilaire Belloc expressed it:

> Thank God that we have got
> The Maxim gun, and they have not[16]

During World War I the German Maxim inflicted sixty thousand casualties at the Somme on 1 July 1916. Later Edison referred to Maxim, with whom he feuded, as a "death-dealer."[17]

Maxim, made wealthy by his invention, had turned in 1891 to the construction of a steam-driven airplane, which was completed in 1894. He never had the opportunity to sell this invention to the military, for it crashed after rising a few feet from guide rails to which it was tethered during a six-hundred-foot run. Because Maxim concentrated on the engine rather than on the more critical problem of control, some historians have doubted the importance of his contributions to manned flight. Maxim, however, should be remembered as an inventor with remarkable instincts for locating the most challenging inventive problems of his day—electric lighting, armaments, and flight.

ELMER SPERRY'S GYROS

Elmer Sperry's entry into armaments, or the military-inventor complex, illustrates the old adage that great oaks from little acorns grow. Fascinated by toy gyroscopes that he bought for his sons, Sperry alertly sensed in 1907 that the gyroscope had commercial possibilities. He relied as usual on articles in technical journals and patterns in patenting to locate the sites of inventive activity. Sperry found the inventor Otto Schlick in Germany using the gyro to stabilize a ship, and Louis Brennan in England gyrostabilizing a monorail car. Confident that he could improve on their designs by drawing on his long experience with automatic controls, Sperry first invented a gyrostabilized wheelbarrow for use by a circus clown on a tightrope. He then patented a gyrostabilizer for automobiles, which, without stabilization, had often overturned on the prevailing rough roads. Frustrated in his efforts to find users for these devices, Sperry also conceived of installing improved stabilizers on passenger ships to remove the ancient curse of seasickness, a distress he vividly recalled from his voyage to Europe in 1898. The shipowners, probably acclimatized to rolling and pitching ships, were not persuaded; the market did not respond.

The U.S. Navy, on the other hand, was interested, not because of a tender concern for seasick sailors, but because of a growing awareness that invention provided a competitive edge in the naval armaments race. Firing guns from a stabilized ship promised greater accuracy than from a ship tossed by rough seas. Gunnery crews practiced with varying success the art of compensating for roll and pitch, but the rule of the day, whether in industry or in the military, was to displace art and skill with the predictable routine of the precise machine. Furthermore, steam-driven iron ships rolled far more than the sailing ships they were displacing, sails having provided a stabilizing effect. The increasing range of the big guns also dictated greater accuracy in aiming them. The instability of ships in a world increasingly determined to achieve control and order presented inventors like Sperry with engaging critical problems.

Sperry knew from his close reading of the technical journals that both the German and the British navies were experimenting with stabilizers. In April 1908 he arranged a meeting in New York with Sir William H. White,[18] former director of construction at the admiralty and a designer of more than two hundred warships. Sperry wanted to describe to him his "active" gyrostabilizer, which improved upon earlier, passive stabilizers. He had drawn on his long experience with automatic feedback controls to transform the sluggish German device of Otto Schlick, which he called "a brute of a machine," into a nimble American version. (Sperry, too, was slight and spry.) White's reaction encouraged him but, when no contract was forthcoming, Sperry turned to his own navy. President Theodore Roosevelt had recently authorized six ships of the dreadnought class, and there were a few officers more open to invention and change, outstanding among them Captain (later Admiral) David W. Taylor. Having attained the highest academic record in the history of the Naval Academy, and having won similar distinction later at the Royal Naval College in Greenwich, Taylor concentrated on scientific studies of hull design and propulsion units. In 1914 he became chief of the Bureau of Construction.

The inventor and the naval officer found each other congenial. Sperry provided Taylor with models of gyrostabilizers to test on model ships in the waters of a model-ship basin. Together Sperry and Taylor exhibited a remarkable combination of ingenuity, experimental finesse, and scientific acumen. In a forty-page mathematical analysis of the behavior of

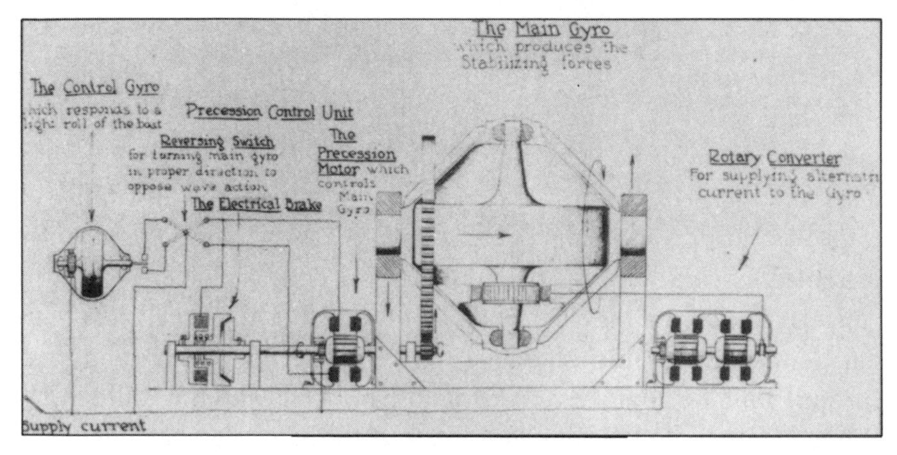

THE STABILIZER PLANT

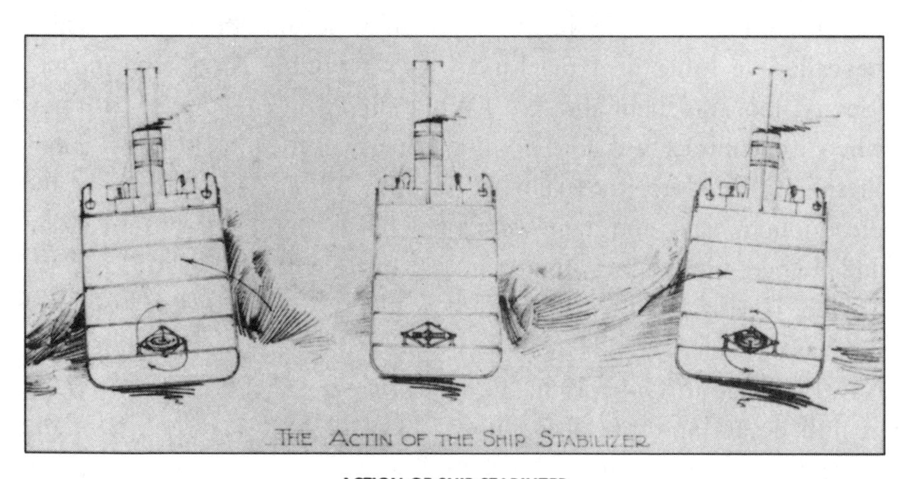

ACTION OF SHIP STABILIZER

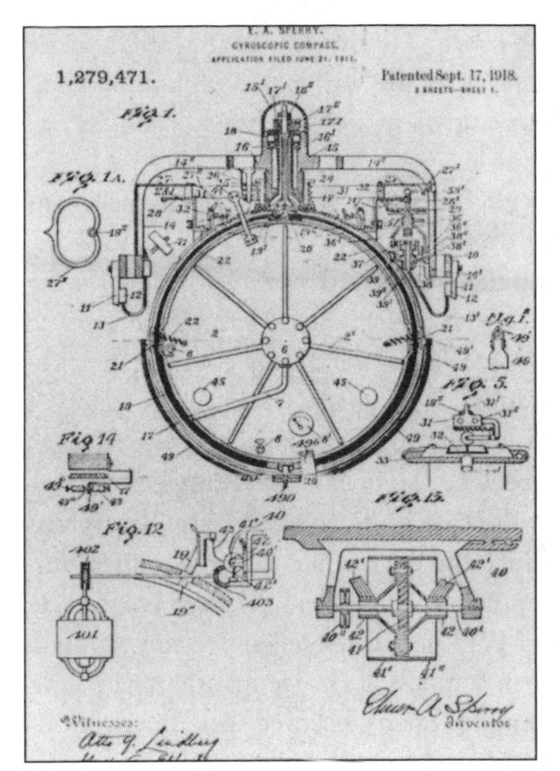

ELMER A. SPERRY APPLIED FOR A PATENT ON THE GYROCOMPASS IN 1911.

gyrostabilizers, Taylor derived a formula for determining the "roll quenching power" of various stabilizers when the characteristics of the ships and stabilizers were known.[19] This analysis provides an excellent example of what we mean when we refer to the marriage of technology and science in the modern era. Technology without science—the empirical approach—would have had to depend on endless tests of various stabilizers on various ships.

In 1912 the U.S. Navy installed a full-sized Sperry gyrostabilizer on the U.S.S. *Worden*, a 433-ton torpedo-boat destroyer. (The massive rotating wheel of the gyro weighed four thousand pounds and measured fifty inches in diameter.) The tests were successful, but not conclusive. A naval officer unofficially commented, "The stabilizer did pretty good work, cutting a total roll of thirty degrees down to about six degrees."[20] The creaks and groans of the heavy machine, however, disturbed the crew and strained the structure of the ship. During World War I, the

U.S. Navy installed, with mixed results, a twenty-five-ton Sperry stabilizer on a ten-thousand-ton transport. After the war, however, marine engineers installed simpler stabilizers employing gyro-controlled stabilizer fins extending from the bow of the ship.

Having proved himself particularly adept in finding inventive solutions to technical problems for the expanding navy, Sperry began to enjoy increasingly close relations with the navy immediately before and during World War I. The navy appreciated Sperry's style, because in essence he was an inventor of controls, feedback controls for concentrations of power. And naval ships and aircraft fell squarely into the category of concentrations of power needful of control. Not long after embarking on the ship-stabilizer project, he became the inventor and supplier of gyrocompasses for the navy. As in the case of the gyrostabilizer, Sperry was not the first inventor to introduce the device, but he made critical improvements in it at a time when its practicality had not been proved. He insisted that his gyrocompass of 1911 provided a reliable reference much more efficiently than the German compass of Dr. Hermann Franz Joseph Hubertus Maria Anschütz-Kaempfe. Sperry believed that American inventors were models of practicality, whereas the Germans were overdesigners, probably to prove themselves to their engineering and scientific peers.

After placing compasses aboard a number of American warships, Sperry followed his system-building bent to develop a network of repeater compasses that made compass reading from the master compass available at a number of stations on the ship. During World War I he incorporated the gyrocompass into a system for controlling the fire of the warships' guns. After the war his gyrocompass became the heart of an automatic ship pilot, the "Metal Mike," adopted by merchant ships throughout the world. About 1912 he also introduced the gyrostabilized stable platform for aircraft control that remains today, in greatly improved form, critical for the automatic control of airplanes and spacecraft.[21] During World War I he also came close to providing the navy with a gyroscopically controlled flying bomb. Later, Secretary of the Navy Charles Francis Adams observed, "It is safe to say that no one American has contributed

SPERRY SUPERVISING THE TEST INSTALLATION OF HIS GYROCOMPASS, 1911

so much to our naval technical progress."[22] Working closely with the navy, Sperry had established himself as the father of modern electromechanical, feedback guidance-and-control systems. He never doubted that his role as a provider of inventions for the military was socially and morally commendable.

FESSENDEN AND DE FOREST AND THE BUREAUCRACY

Other professional inventors found the adjustment to the bureaucratic, military organization more difficult. Reginald Fessenden, though attracted to the navy on account of its early use of wireless telegraphy and telephony, bridled under the regulations and practices the navy imposed. Like so many other inventors, Fessenden thought himself uniquely creative and deserving of extraordinary cultivation. He instinctively viewed rules and regulations as powerful and usually effective reinforcement of the status quo, and as a nemesis of the inventive spirit. From experience, the navy knew that the individual who wanted to work outside the rules and regulations was disruptive. The navy had to judge carefully whether or not the benefits to be derived from change outweighed the disorder and trials of transition. It also depended on organization, not individuals, for survival. It is ironical that the free-spirited Fessenden, with his wireless communications, was threatening to close one of the last breaks in the chain of command. Prior to wireless communication, a ship at sea could follow a course free of contact with land-based higher authority. The advent of wireless made possible coordination and control of naval forces at sea.

Despite the clash of values between the navy and the professional inventors, the navy persisted, if erratically, in its development of wireless communications. Concerned that the British might use the British-based Marconi Wireless Telegraph Company to establish a world monopoly in naval and civilian communications, and also aware that the country had become a world power dependent on military, especially naval, force, as demonstrated in the Spanish-American War of 1898, the U.S. Navy tested the Marconi system in 1899 and became interested in acquiring the apparatus. Marconi demanded that the navy purchase at least twenty sets at a total cost of $10,000 and pay the same amount annually as royalty. Not sympathetic to the idea that inventors should recover development costs, the navy found the price extravagant, and no purchase was made. For the next decade or so, navy negotiations with the British company made little progress.[23]

The American technical journal *Electrical World* took umbrage at the navy's dilatory ways:

> If the lack of such apparatus [wireless] in our army and navy is due to the neglect of the mossbacked bureaucrats who sent our artillery into

> action at Santiago with black powder, the public ought to know it, that the authors of the negligence may be properly pilloried.[24]

In 1902 the navy purchased for trials a few French and German wireless sets, and two of de Forest's. Generally, the de Forest system performed well, but the navy believed that it demanded more skill of the operators than the navy could expect. A year later, the navy angered de Forest and Fessenden and their nationalistic supporters by purchasing more wireless sets from Germany. These printed the incoming messages and demanded less skill of the operators. From the earliest days of telegraphy, almost a half-century earlier, the European style had been to print, rather than to depend on the ear of the operator. Soon, however, de Forest and the other Americans persuaded the navy to drop this requirement.[25] They persuaded the navy that relatively inexperienced and unskilled operators could use a non-printing receiver.

By selecting components, such as transmitters and receivers, from the systems of various inventors and composing them into a system, the navy avoided becoming hostage to a single inventor or manufacturer. Fessenden wanted his carefully integrated system of components to be taken *in toto*, for he, like Edison and other inventors, had carefully invented and designed components to interact harmoniously. Fessenden also found objectionable the navy practice of testing his apparatus using its own personnel, rather than his carefully trained assistants. He doubted the wisdom of the navy's insistence on equipment that could be used by large numbers of relatively unskilled and unspecialized men. In addition, the navy's tendency to purchase on the basis of price and ease of manufacture and installation, rather than incremental quality of performance, exasperated Fessenden. Like many other inventors, he highly valued craftsmanship, especially as expended on his inventions. Furthermore, before World War I the armed forces expected the inventor to bear the risks of development costs and to supply a device ready for testing aboard ship. If the device was rejected or there were mixed results, the inventor had no pecuniary compensation for his investment.[26] Perhaps most galling to Fessenden, who took professional pride as well as financial protection from his patents, was the navy's cavalier attitude toward the independents as it sought to avoid dependence on devices that could be monopolized.

The squabble between Fessenden and the navy came to a head when he found it encouraging de Forest to sell a copy of the Fessenden receiver, an electrolytic detector, to the service at a lower price. Furious, Fessenden notified the navy that it was purchasing pirated apparatus. He then filed a successful suit against de Forest for infringement, but the secretary of the navy airily dismissed the significance of patent suits, telling Fessenden that the navy was relieved of moral obligation to him because his price was too high. When the navy continued to buy from de Forest, Fessenden demanded that the secretary be impeached for buying stolen property. Fessenden and his company decided not to inform the government again of its inventions so that, as Fessenden wrote, "the government cannot steal them."[27]

De Forest went aground on the shoals of bureaucracy, too. In 1904 the navy decided that wireless stations were needed to help consolidate its growing domination of the Caribbean area, so it asked for bids on the installation of four wireless stations, in Key West, Puerto Rico, Cuba, and the Canal Zone. The specifications called for unprecedented performance: communication under all atmospheric conditions between stations one thousand miles distant in a tropical region notorious for lightning, violent storms, and relentless static. The de Forest company bid $65,000 and the Fessenden, which had not yet abandoned all hope with the navy, $324,000. Beginning work in January 1905, de Forest found the working conditions hellish. Instead of cooperation, he found hostility on the part of officials. Exasperated, he wrote, "If the Navy, through their cheap outfits and red tape delay our success, we will not let their still cheaper officers with more gold tape than brains throw the hooks into us." There was a bond to forfeit if work were not completed on schedule.[28] By 1906 the de Forest company had completed the stations, but de Forest had difficulty in maintaining communications farther than two hundred miles at night, and daytime communication was usually impossible. Discouraged, the navy nevertheless persisted in its efforts to establish itself as the provider of U.S. wireless transmission.

INNOVATIONS AND THE WAR OF ATTRITION

After World War I broke out in August 1914, Germany and then the other belligerents mobilized inventors, engineers, and industrial scientists to find substitutes for material resources in short supply, to develop weapons to break the terrible deadlock on land, and to sustain or destroy the supply lines extending across the seas. World War I is remembered as a war of technological surprises.[29] Those surprises most often recalled appeared amid the horrors of trench warfare on the Western Front. The machine gun frustrated the tactic of attack, helping to bring a stalemate between attacker and defender. This led to the war of attrition fought in the trenches. With the world's leading chemists, chemical manufacturers, and chemical-research laboratories, Germany drew on a long tradition of substituting science and technology for natural resources in short supply. The rapid development by chemists Fritz Haber and Carl Bosch of a process to substitute artificial, or synthetic, nitrogen compounds for natural ones relieved the dire shortage of nitrogen for explosives and fertilizers that followed the British blockade of natural nitrates from Chile to Germany.[30]

The Germans also relied on the productive combination of chemical science, engineering, and manufacture to transform synthetic dye into poison-gas technology. Again Fritz Haber played a leading role. Ironically for Haber, National Socialist racism later drove him out of the country. Other German scientists drawn into the poison-gas project included future Nobel Prize–winners James Franck, Gustav Hertz, and Otto Hahn.[31] The Germans first released large quantities of chlorine gas before Ypres on 22 April 1915, threw the enemy into disarray, but were unprepared to follow up on the breakthrough. Subsequently, chlorine and phosgene gases made from materials used for indigo and scarlet dyes poisoned countless soldiers on the Western Front. Gas warfare became an area of perverse innovation, with one side introducing new gases as the other found modes of protecting against older ones. Earlier, the synthetic dye producers had competed similarly to respond to fads for new colors. After spreading quantities of propaganda associating poison-gas development with the barbaric Huns' indifference to the rules of "civilized" warfare, the allies proceeded to develop their own gases. Haber and other Germans said that, in the face of the superior natural resources of the Allies, they had no choice but to resort to advanced technology: "In war," Haber observed, "men think otherwise than they do in peace."[32]

The British also actively innovated during wartime. Winston Churchill, as first lord of the admiralty, played an unorthodox entrepreneurial role by pushing the development and use of armored, armed, and tracked automobiles, or tanks, on the Western front. He transparently justified his involvement in the land war by calling the tank a land ship. Churchill said, "It was no use endeavoring to remedy . . . [the stalemate on the Western Front] by squandering the lives and valour of endless masses of men. The mechanical danger must be overcome by mechanical remedy."[33] Instead of launching a mass attack with tanks, the British used only twenty in their first appearance, in September 1916, during the Battle of the Somme. This inadequate and unimaginative innovation was compared by Churchill to the failure of the Germans to launch a mass attack of troops in April 1915 to exploit the confusion and casualties they caused by their initial use of poison gas. The first effective use of British tanks came on 20 November 1917, when five hundred tanks concentrated at Cambrai, broke the German lines, and made possible the taking of ten thousand prisoners. By war's end the British were mass-producing tanks, but it was the Germans who employed the weapon with tactical ingenuity at the beginning of World War II.

American inventors had introduced the airplane and taken a leading role in the development of wireless for their military before World War I, but the United States had quickly lost the lead in military aviation to France, Germany, and Britain. The Germans pushed not only heavier-than-air craft, but also the dirigibles of Count Ferdinand Zeppelin. In 1913 the U.S. ranked behind France, Germany, Russia, Britain, Italy, and Mexico in appropriations for military aviation. In 1914 the U.S. Census Bureau listed only sixteen aircraft manufacturers who produced only forty-nine airplanes of all types.[34] Early in the conflict the belligerents began to exploit the capacity of the airplane and of wireless to provide observation and communications on and above the battlefield, where highway, rails, and telephone and telegraph wires could not reach. During the last ten months of the war, Britain produced nearly twenty-seven thousand airplanes, and Churchill called for a massive air offensive involving tactical and strategic bombing.[35]

For the Allies, the greatest threat to survival from recently developed technology was the oceangoing, diesel, electric-powered submarine. American inventors Simon Lake, who designed the *Argonaut* (1897), a submarine able to operate in the open sea, and John Holland, who introduced the *Holland* (1898), a submarine equipped with internal-combustion engines for surface navigation and electric motors for submerged cruising, offered the United States an opportunity to seize the lead in submarine warfare. The Germans, however, finally, during the war, adopted the strategy of using large numbers of submarines to attack merchant ships and to threaten Britain's supply links with the United States and the British Empire. In 1916 the Germans began submarine warfare against armed merchant ships, and in 1917 increased their effort with an innovative naval technology to counter the Allies' superior industrial output. The heavy losses of shipping threatened to cripple the Allied effort until they countered with convoys, submarine chasers, and depth bombs, and with the mobilization of inventive, engineering, and scientific talent for antisubmarine warfare.

Across the Atlantic, the American military and government watched apprehensively as the war became one of attrition, mass production, and scientific and technological innovation. The outbreak of the war in Europe had closed the easy American access to European naval technology. Secretary of the Navy Josephus Daniels and many of his naval officers became concerned about the future state of American preparedness. The British navy had introduced the dreadnought-class battleship, the highest technological system of the day; the Germans were taking the lead in developing submarine warfare; the French showed the way in military aviation; and the British pioneered in highly complex naval gunfire-control devices. Informed critics claimed that the U.S. Navy had depended on the others, following a "Chinese plan of copying."[36] As Secretary Daniels and others responsible for preparedness began to assess the strengths from which the nation could draw, it is not surprising that they looked to American inventors. Daniels, given to easy enthusiasms, expected a ground swell of military inventions from concerned Americans. Because the American system of production was the world's most productive, he and the public also expected Henry Ford and other industrial entrepreneurs to bolster the nation's defenses with the mass production of machines of war.

NAVAL CONSULTING BOARD

Daniels, secretary of the navy from 1913 to 1921, earlier the editor of the Raleigh (North Carolina) *State Chronicle*, believed that American resourcefulness and ingenuity found their most intensive expression in Edison, whom he, like so many other Americans, idealized. After all, Edison, from a simple background, was plainspoken, practical-minded, self-educated, and fantastically successful. A politician who steered a course by the beacon of Edison and sang his praises, Daniels immediately heard a resounding chorus of public support. In May 1915 Daniels read with surpassing interest an Edison interview spread over two pages of the Sunday *New York Times* about his "Plan for Preparedness."[37]

Here Edison said that he reluctantly faced the eventuality that, if America were drawn into the conflict, he would have to come to the aid of the armed forces and invent terrible weapons of destruction to defeat despotism. He added that he would do this despite his pacifistic leanings: "Laying a kindly hand" upon a reporter's arm, he disclaimed, "you see, my boy, the dove is my emblem."[38] Edison drew his plan for preparedness out of an elaborate analogy. He envisaged the soldier "perspiring in the factory of death at the battle line."[39] He called, therefore, for labor-saving devices in this factory. Superior American machines (weapons) tended by natively gifted American mechanics (soldiers) would be the American line of defense. Expressing a characteristic American attitude, he said, "modern warfare is more a matter of machines than of men."[40] Appealing to the American distaste for taxes, he called for a small standing army and navy. He wanted skilled and experienced industrial workers, foremen, and engineers to be called up in peacetime for short terms as reservists and for emergency service in wartime. Regular officers should alternate between naval duty and work in industry, so that they would be familiar with the latest engineering and managerial advances. He wanted a naval laboratory where the emphasis would be on the construction of prototypes and experimentation with these. A master plan for mobilization would identify the manufacturers that could quickly transform laboratory-prepared blueprints into mass-produced weapons. Edison insisted that his plan would not involve vast expenditures.[41]

Warm in his adulation of Edison, possessed of great faith in American inventive genius, sensitive to the politics of expenditures and taxes, aware of the need to strengthen the navy, and cheered by the prospect of winning wars with machines rather than men, Daniels within two weeks of the interview asked Edison to be the navy's civilian adviser on problems of invention and development. "You are recognized," Daniels wrote, "as the one man who can turn dreams into realities," to originate "proper machinery and facilities for utilizing the natural inventive genius of Americans to meet the new conditions of warfare as shown abroad." He knew that if Edison accepted, "the country would feel great relief in these trying times."[42] Edison replied with a cheerful "aye aye, sir." This alliance of the uncritically enthusiastic politician and the inventor found an echo years later among politicians enthusiastic about the wondrous powers of atomic physicists.

The ingenious scheme that the secretary and the inventor concocted established an Edison-headed Naval Consulting Board. With a membership of the nation's "very greatest civilian experts in machines," the board was to originate inventions and critically examine and selectively develop those sent in from the "crank of today" who would be the "genius of tomorrow." Daniels anticipated receiving inventions of the caliber that a Fulton, a Morse, a Bell, or an Edison might propose. He was gratified by the enthusiastic letters that poured in from "every sort of people" when they learned of his plan.[43] Advised by Edison, Daniels asked each of eleven professional societies, representing thirty-six thousand engineers, inventors, and scientists, to recommend two members for the board.[44] The press and the people were disappointed, however, by the members selected. They expected Orville Wright, Glenn Curtiss, Charles Proteus

THOMAS A. EDISON (SECOND FROM RIGHT) AND THE NAVAL CONSULTING BOARD PROUDLY PARADE.

Steinmetz, Alexander Graham Bell, Reginald Fessenden, Nikola Tesla, and Henry Ford.[45] Instead they got, among others, Leo Baekeland, chemist and inventor of Bakelite; Willis R. Whitney, head of the General Electric Research Laboratory; Frank Julian Sprague, inventor of electric motors and vehicles; Sperry; and Edison's chief engineer, M. R. Hutchinson. *The New York Times* consoled itself by reflecting that "the twenty-three members were chosen for fitness rather than notoriety.[46]

Edison and Daniels deliberately omitted representatives of the American Physical Society (physicists) and the National Academy of Sciences from membership on the board. The academy, founded during the Civil War to provide scientific advice to the government, consisted of members elected largely from among academic scientists. Pressed about the absence of representatives from the American Physical Society, engineer M. R. Hutchinson replied that Edison desired "to have this Board composed of *practical* men who are accustomed to *doing* things, and not *talking* about it." Asked why the National Academy was unrepresented, a member of the board ventured, "because they have not been sufficiently active to impress their existence upon Mr. Edison's mind."[47] A gauntlet had been thrown down that several leading physicists were not loath to take up. A simmering struggle between independent inventors, on the one hand, and scientists, especially physicists, on the other—each group believing itself to be the source of innovation—came to the fore during the war.

The physicists were not to be denied. Their determination to be involved in national defense presaged their portentous achievements in World War II. Depressed by the omission of National Academy representation and of physicists on the well-publicized Naval Consulting Board, George Ellery Hale, foreign secretary of the academy, editor of the *Astrophysical Journal*, and director of the Mount Wilson Observatory, rose to the challenge.[48] Son of a wealthy Chicago family, graduate of MIT, connoisseur of English literature, collector of rare editions, and having access to private philanthropists such as Andrew Carnegie, Hale represented the new men of science with their growing power and influence, whereas Edison epitomized the waning influence of the independent inventor. Hale, eager to breathe new life into the academy and hoping to model its activities on those of the Royal Society in Britain and on the French Academy, regarded "himself as doing his best work as an initiator and promoter of scientific enterprises."[49] Daniels and the American public, however, were not yet ready to see scientists as a prime source of American inventiveness—or, as Hale put it, the public needed to be educated in order to appreciate the intellectual adventure of science.

In June 1915, following Edison's offer of his inventive genius to the nation, and after the loss of American lives following the German submarine sinking of the British *Lusitania* raised national anxiety, Hale wanted the National Academy to offer its services to President Woodrow Wilson. But associates saw the move as premature, probably because, in their pride, they wished not to appear political supplicants. Ten months later, however, the day after Wilson issued an ultimatum to Germany because of the sinking of the *Sussex*, academy members at the annual meeting unanimously accepted Hale's resolution that the academy place itself at the disposal of the government.[50] Persuaded by an academy delegation that it could organize an arsenal of science, Wilson approved the academy's establishing the National Research Council to promote cooperation among the nation's research institutions and among leading scientists and engineers. Hale righteously announced, "We must not prepare poisonous gases or debase science through similar misuse; but we should give our soldiers and sailors every legitimate aid and every means of protection."[51] Among those who agreed to serve on the National Research Council—the membership of which was wisely not limited to academy members, most of whom were academics—were Willis R. Whitney, Leo Baekeland, John Carty of AT&T, and Michael Pupin of Columbia University. Seeing the need for representatives from government services to match the influence of the Daniels-supported Naval Consulting Board, Hale persuaded the government to appoint representatives to the council, among them Admiral David W. Taylor (with whom Sperry worked so closely), Colonel George O. Squier of the Army Signal Corps, and Samuel Wesley Stratton, head of the Bureau of Standards. Squier, a regular-army officer, headed the aviation section of the U.S. Signal Corps, had a doctorate in physics from The Johns Hopkins University, and held numerous patents on electrical-communication devices, at least one jointly with Sperry. He took especial interest in the fledgling air arm of the army. Robert Millikan, the physicist who worked with him, described Squier as "a strange character who . . . was in no sense an organizer nor a man of balanced judgment, but . . . [who] had one great quality much needed at that time, namely, a willingness to assume responsibility and go ahead."[52]

In February 1917 after Germany resumed unrestricted submarine warfare, the navy asked the National Research Council to take on the critical task of developing submarine-detection devices. The threat posed by submarines loomed especially large both to the nation's leaders and the public. This assignment brought to the fore another energetic and effectual promoter of science. Robert Millikan, a professor of physics at the University of Chicago, a gifted experimentalist who determined precisely the charges on electrons, and later became a Nobel laureate, chaired the Research Council Committee for submarine-detection devices. Determined like Hale to thrust science into the mainstream of the preparedness effort, Millikan wrote to his wife, "This much is clear. If the science men of the country are going to be of any use to her it is now or never."[53]

Antisubmarine research and development precipitated not only cooperation but competition among inventors, engineers, and academic and industrial scientists. Each group believed that it had unique qualities to bring to the task, and each made claims on national resources to increase its numbers and activities. The Naval Consulting Board established a Special Problems Committee to respond to the submarine threat. Among the projects funded was one to develop a system involving antisubmarine nets, wireless-transmitter buoys, patrol boats, and depth charges. Developed by Elmer Sperry and his Sperry Gyroscope Company, and ready for testing in June 1917, shortly after the United States entered the war, the system was sensationally publicized by the Naval Consulting Board as a successful counter to the German submarine threat. The net snared a test submarine, but the wireless signaling buoys became snarled in the net and unable to call the patrol boats, which then could not accurately drop their mock depth charges. A review of the project found that most of the components of the net system—except for the wireless-transmission buoy, which did not work, anyway—had previously been tested by the allies.[54] Whitney, who headed a submarine-detection subcommittee of the Naval Consulting Board, took another tack by establishing a station at Nahant, Massachusetts, for development of sound-detection devices. There he assembled industrial scientists, among them Irving Langmuir of the General Electric Research Laboratory, who won the Nobel Prize in 1932. Others came from AT&T and the Submarine Signal Company of Boston. Academic physicists were excluded, for a navy representative decided that their presence would complicate the patent situation.[55]

The National Research Council's antisubmarine committee, headed by Millikan, took heart from the observation of the eminent British physicist Sir Ernest Rutherford that antisubmarine detection was "a problem of physics pure and simple."[56] Millikan set up a research-and-development facility for submarine detection at New London, Connecticut, and opened the doors to academic physicists by bringing in ten of the most competent in the country, drawn from Yale, Chicago, Rice, Cornell, Wisconsin, and Harvard. Whitney played the role of peacemaker by urging Secretary Daniels to fund the academics and the New London endeavor of the National Research Council, despite the prior establishment of the Nahant experimental station. He counseled that, "If a little of the energy which, I fear, may otherwise develop into hard feeling, could be utilized in the submarine work, time would be saved."[57]

The Nahant group developed and GE manufactured a stethoscopelike detector named the "C tube" after William Coolidge of the GE lab. The New London academic physicists introduced a detector of superior quality that was primarily the idea of Professor Max Mason of the University of Wisconsin. By July 1918 the U.S. had dispatched more than one hundred wooden sub-chasers with these detectors to the English Channel and the mouth of the Adriatic. No verified destruction of subs occurred in the channel, but submarines became bottled up in the Adriatic through a combination of the nets and sub-chaser-equipped detection devices. The convoy system, however, not detection, reversed the tide of submarine warfare in the Atlantic. But progress made by industrial and academic physicists in responding to the nearly intractable detection problem during the short time available before the war ended substantiated their argument that modern war required their services, not just those of inventors and engineers. Furthermore, they had experienced the seductive exhilaration of entering the corridors of political and military power.

Despite its inability to attract adequate government funds and its need to rely heavily upon private funding from the Carnegie Corporation and the Rockefeller Foundation, the National Research Council was eventually recognized as the government's scientific research arm, while the Naval Consulting Board became a source for inventions. The distinction, however, was deceptive: the scientists of the National Research Council proved themselves quite inventive, especially in antisubmarine work. The scientists were making progress in their struggle for public recognition and support, but the inventors of the Naval Consulting Board seemed to have reached a plateau. In the highly publicized battle against the submarine, the board had nothing to show that was as effective as the detection devices of the National Research Council. The board's program of evaluating inventions sent up from the grass roots, a project that Secretary Daniels had so enthusiastically proclaimed, proved a fiasco. Of the 110,000 "inventions" sent to the board, only 110 were deemed worthy of development by overworked board committees, and only one of these reached production before the war ended.[58] Most of the suggestions were old and previously adopted, were old and discarded, or could not be worked into the navy system. In short, the suggestions came from those who were not aware of the state of the art or of the outstanding technical problems in the expanding navy. As the inventor Hudson Maxim, brother of Hiram Stevens Maxim and member of the Naval Consulting Board, concluded, "this is an age of specialists, and it is necessary for any inventor, scientist, or engineer to devote a large amount of time and attention to the special requirements of naval and military matters before he can qualify himself to be of much use. . . ."[59] He singled out Sperry as one who qualified.

Like the Naval Consulting Board, Edison emerged from the conflict with his own luster somewhat dimmed. None of the "terrible weapons of destruction" he had promised were forthcoming. He, too, working on his own, had fashioned numerous antisubmarine devices using an old yacht to test them. Because he had been made ill by a laboratory explosion injuring his eyes, his wife, Mina, accompanied him. She was prone to seasickness, and the naval officers in charge must have found the duty on board trying. Edison complained that the navy ignored the some forty-five devices he fashioned, which were "all perfectly good ones, and they pigeon-holed every one of them. The Naval officer resents any interference by civilians. Those fellows are a close corporation."[60] He appeared stubborn and antiquated in a struggle over the design of the naval research laboratory that he proposed. He was not open to the ideas of others, but complacent in the defense of his own. Having been proved right so often in the past when others ridiculed or dismissed his inventive ideas, he seemed to have acquired the habit of dismissing criticism and of countering proposals.

In his initial plan for national defense in 1915, he called for a laboratory that would not only do research and experimentation, but build full-scale models of naval equipment for testing and redesign. He had in mind airplanes, range finders, submarine engines, small guns, "and everything relating to war machinery."[61] Once the design was approved, blueprints would be prepared and machine tools specified in peacetime to guide mass production in America's factories if the country should be drawn into war. Impressive in its simplicity, the plan revealed Edison's discomfort with the theory and mathematical analysis that industrial scientists had proved were effective in refining designs without the necessity of repeated construction of physical models. Edison's approach must have appeared inefficient and costly to scientists at AT&T, who, by relying heavily on theory and analysis, had designed loading coils. When Congress appropriated only $2 million instead of the $5 million that Edison requested, Willis Whitney and other members of the consulting board proposed a laboratory that would concentrate on research and experimentation instead of on the building of full-scale prototypes. Edison, adamant, remarked that "research work in every branch of science and industry, costing countless millions of dollars . . . has been on for many years. . . . Only a ridiculously small percentage has yet been applied. . . . It is therefore useless to go on piling up more data. . . ." In some curious lapse, he forgot how use-specific data needed to be for design and development. When the others would not support his plan for the laboratory, Edison appeared petulant and arrogant. He wrote to Daniels, "It is fixed in my mind, whether right or wrong, that the public would look to me to make the Laboratory a success, and that I would have to do 90% of the work. Therefore, if I cannot obtain proper conditions to make it a success, I would not undertake it nor be connected with it in the remotest degree. . . ."[62] Unwilling to override Edison, Daniels allowed a stalemate to persist and the naval research laboratory to remain unbuilt until ground was finally broken in December 1920.

AERIAL TORPEDO

Historians have found the Naval Consulting Board wanting and left the impression that the grass-roots invention fiasco, Edison's antiquated views, and the laboratory stalemate mirrored the ineffectiveness, even decline, of the independent inventor, and the rise of industrial research and academic research of interest to industry and the military.[63] This interpretation overlooks the innovative ideas of board members—the inventors Sperry, Hudson Maxim, and Frank Sprague among them—and the contracts for research and development that the board placed in industry to implement their ideas.[64] The board brought the General Electric laboratory into the submarine-detection project. It authorized funds for the Sperry Gyroscope Company to develop a long list of devices: the first depth charges made in the United States; a long-range, high-intensity searchlight; controls for naval torpedoes; airplane instruments; aerial bombsights; helicopters; mechanized gunsights for airplanes; and an aerial torpedo or flying bomb.[65] Obviously the board, Whitney, and Sperry did not see conflict of interest as a problem.

The board's most audacious and forward-looking project was the aerial torpedo. Rarely do we recall that a flying bomb, or cruise-missile device, was tested during World War I. Developed by the Sperry Gyroscope Company, the aerial torpedo anticipated by two decades the World War II German V-1 flying bomb. Both were winged missiles designed for about one thousand pounds of explosives and were guided and controlled by preset instruments. In both cases, the gyroscope was the heart of the instrumentation. The aerial torpedo had a piston engine, the V-1 a pulse jet. Sperry foresaw the potential of the aerial torpedo in 1918 when he wrote, "We have gone a long way toward completing the development of an extremely significant engine of war, it being nothing short of the coming gun." When launched from a distance of a hundred miles, Sperry predicted, it could destroy much of a city such as New York, heavily damage a fortification, and "turn a small town or a large munition plant practically upside down."[66] Sperry did not draw back from the killing and the massive destruction. Like so many other inventors and advocates of terribly destructive weapons, he went on to prophesy that the aerial torpedo would make "war so extremely hazardous and expensive that no nation will dare go into it."[67] Octave Chanute gave the Wrights similar assurances as they sought military contracts. Logical and clearheaded in his inventions, Sperry did not see the inconsistency in his sentiment: nations that could justify millions of deaths in the trenches were not likely to hold back in the use of a device as ingenious and destructive as a missile. Because of its secret nature, Sperry was not able to tell the world of the Naval Consulting Board's wartime aerial-torpedo project

LAWRENCE SPERRY AND EMILE CACHIN DEMONSTRATING STABILIZED FLIGHT, NEAR PARIS, JUNE 1914

until 1926. Then it caused a flurry of headlines: "Deadly Air Torpedo Ready at War's End: Elmer Sperry's Invention Is Told" (*The New York Times*); "Aerial Torpedo Explained: Steered 100 Miles Accurately by Gyroscope Compass, 10,000 Would Win a War" (Brooklyn *Daily Eagle*).

The aerial torpedo had a history. In June 1914, near Paris, a brave French mechanic named Emile Cachin walked out on the wing of a Curtiss airplane (flying boat) while the pilot, Lawrence Sperry, Elmer's son, raised his hands in the open cockpit to show that he had turned over control of the airplane to the Sperry automatic gyrostabilizer. For this, the Sperrys won a 50,000-franc prize in a French Competition for Safety in Airplanes. After the war began, the gyrostabilizer was directed toward destruction instead of safety. Lawrence, working in England with the British air force, tested a Sperry gyrostabilizer and a gyro steering device in conjunction with a bombsight. The gyros held the plane on course and in level flight while the bombardier kept the target in the sights up to the point of its release. These gyro devices anticipated the widely used Sperry bombsight of World War II.[68]

When Lawrence, an early engineer–test pilot, returned to the United States in 1916, he, his father, and other designers at the Sperry company combined a stabilizing gyro and a steering gyro to make an automatic pilot. (The wing-walking demonstration over the Seine had used only a stabilizing gyro.) In early testing, the automatic pilot flew a plane, in which Lawrence flew as a passenger, for more than thirty miles, probably the world's first automatic steering of an airplane. On another occasion, Lawrence, who cut a dashing figure as a handsome young aviator, took as passenger a glamorous young New York City society matron, known in the newspapers as the "blue streak" because of her spirited drives through Manhattan in a brightly colored automobile. Flying over Long Island Sound, he switched on the automatic pilot to show how he could fly with no hands—on the controls. The machine failed, however, and the plane plunged into the bay, seriously injuring his companion. Lawrence had other crack-ups on more routine flights, suffering, among other

LAWRENCE AND ELMER SPERRY, JUNE 1914

injuries, a badly broken nose. Grover Loening, another young aviation pioneer, called his friend Lawrence "a real genius, a terribly hard worker, and equally strenuous in his leisure."[69]

Lawrence and Elmer, preoccupied by the war in Europe and by preparedness at home, then saw the possibility of an aerial torpedo, or pilotless flying bomb. Their concept was simple but full of potential for further uses in both military and commercial aircraft. A Lawrence Sperry patent filed in 1916 described the aerial torpedo. The gyrostabilizer would maintain the plane in level flight, as Lawrence had demonstrated; the automatic steering gyro would hold the airplane on preset course; an altitude barometer would activate controls to level the airplane after its initial climb and to maintain elevation; and a simple engine revolution counter would cut off power and dive the aerial torpedo at its target after a predetermined distance. Servomotors activated by the various controls, and powered by small wind-driven propellers, moved the airplane's ailerons, elevator, and rudder. A windmill also drove the generators supplying electricity to the gyro motors. The aerial-torpedo concept was a milestone in the history of automatic feedback controls, long before MIT mathematician Norbert Wiener called the attention of the world to feedback controls with his book, *Cybernetics, or Control and Communication in the Animal and the Machine* (1947).

The Naval Consulting Board approved the aerial-torpedo project in April 1917, and the navy awarded a $200,000 development contract to the Sperry Gyroscope Company. The company agreed to test first the automatic controls on a Curtiss airplane, a flying boat, and several N-9s, and then to install the automatic controls in a Curtiss airplane especially designed for use as an aerial torpedo. Mass production was to follow. Tests of the automatically controlled N-9, with a test pilot along to control the aircraft during takeoff and to monitor the flight, were notably successful. In November 1917 the specially designed, simple, inexpensive Curtiss planes, equipped with two-cylinder engines, arrived at the Sperry testing grounds on Long Island, New York. Sperry engineers and mechanics installed automatic controls. In order to discover the numerous modifications needed, Lawrence flew and monitored—and crashed at least four times—a specially adapted aerial torpedo. March 6,

SPERRY AERIAL TORPEDO, 1917–1918

1918, brought "the first entirely successful flight of an automatic missile in the . . . [United States] if not in the world."[70] The plane climbed automatically from its launch and made a smooth, stabilized flight until the distance control automatically terminated the flight in the water at the preset distance of one thousand yards.[71] Despite this milestone event, further tests with the aerial torpedo in the summer and fall of 1918 proved discouraging, because of repeated structural failures, the malfunctioning launch device, and the precession of the gyros during the rapid acceleration of the launch. Wanting to try a modified launching catapult and controls on the less troublesome Curtiss N-9 plane again, the Sperry engineers launched it on 17 October 1918. With the distance gear set at eight miles, the N-9 rose smoothly, climbed steadily, and flew on a predetermined course eastward over the Atlantic—never to be seen again. This test suggested that, if a more suitably designed airplane had been used, rather than the simple one, a practical but expensive aerial torpedo might have been produced before the war ended.

The idea of a flying bomb captured the imagination not only of the Sperrys but also of Charles F. Kettering, another leading American inventor. Kettering, whose name is now associated with the cancer-research center in New York, had introduced, among other major inventions, the electric self-starter for automobiles, and an improved battery ignition. After World War I he became the first head of the General Motors research laboratory. In 1916, anticipating the mass production of airplanes if the United States should enter the war, he joined Orville Wright and others to establish the Dayton Wright Airplane Company in Ohio. Kettering also designed and manufactured the ignition for the Liberty airplane engine, one of the country's most successful wartime efforts at mass production. After witnessing Sperry aerial-torpedo tests in December 1917 as a member of an army committee, he was so enthusiastic that he agreed to develop a flying bomb for the army. Like so many other inventors, he believed he had spotted the weak points in another's design, and he set out to correct them. He criticized the cost and complexity of the Sperry system. "It's got to be simple," Kettering insisted. "We haven't got time to make it complicated."[72]

Kettering asked a team including Orville Wright to design the fuselage for the bomb-carrying airplane. To assist in designing the engine, he called on C. H. Wills, the former chief designer at the Ford Motor Company, whom Ford described as "the man the public think I am."[73] The final engine cost only $40. After failing to come up with a viable gyro control system, Kettering copied the Sperry design, with Elmer Sperry supplying detailed drawings and visiting Dayton, Ohio, to help with the installation. The other controls included an aneroid barometer to regulate altitude and a wing-mounted small propeller-driven revolution counter to measure distance. Kettering's staff designed a cheap, portable launching device.

Trials of the Kettering flying bomb began in September 1918 and continued through October. A full-scale flight on 2 October won the name of "The Bug" for the Kettering device. C. H. Wills recounted the event for Sperry. After an erratic flight pattern during the first few minutes

KETTERING FLYING BOMBS AT WRIGHT FIELD, DAYTON, OHIO, 1918

CRASHED KETTERING FLYING BOMB AT CARLSTROM FIELD, ARCADIA, FLORIDA, 1919

disrupted all the controls, the airplane began circles about a mile in diameter. "Mr. Kettering watched it until it was up about 12,000 feet and out of sight and with disgust said, 'Let the thing stay up there' and went home to bed."[74] Several weeks later another Bug performed well during a flight of five hundred yards to target, so the army ordered a hundred prototypes. But when the war ended in November, the government decided to combine the Sperry and Kettering projects. After competitive flight testing, the government scrapped the Kettering and continued the development of the Sperry-Curtiss aerial torpedo.

The U.S. military soon demonstrated its commitment to automatically controlled missiles. About a week before the armistice, the navy ordered five aerial torpedoes with controls designed by two former Sperry employees, Hannibal Ford and Carl Norden, to be installed on an airplane built by the Witteman-Lewis Company. The navy, as it had in the case of prewar wireless systems, seemed still reluctant to depend on one designer and supplier. Tests in 1920–21 proved the new design no more effective than the Sperry-Curtiss aerial torpedo, and not as promising as the Curtiss N-9 plane with Sperry controls. In 1920 the army brought the Sperry to the fore again by contracting with the newly founded Lawrence Sperry Aircraft Company to install three Sperry Gyroscope Company gyro control units in a Lawrence Sperry Company messenger airplane, and to outfit three army training planes as aerial torpedoes. In 1922 Lawrence filled the contract specifications and won a bonus for hitting targets at thirty-, sixty-, and ninety-mile ranges. During the ninety-mile flight to the target, however, remote wireless controls corrected automatic-control errors. (The contract had specified only that the torpedoes function without manual controls.) Many officers of the Army Air Service, including the forward-looking colonel, later general, William ("Billy") Mitchell, enthusiastically supported development of aerial torpedoes. Mitchell later described Lawrence Sperry as "one of the most brilliant minds and greatest developers in the world of aviation."[75] Lawrence died in 1924 while piloting an airplane of his own design across the English Channel. Elmer Sperry felt the loss not only personally but professionally, for Lawrence had been the engineer and test pilot for the early airplane controls invented by his father. Each anniversary of his son's death, Sperry made the sad notation in his diary, "the day Lawrence was lost."[76]

CONTROLS

A successful aerial torpedo depended on an adequate response to an obsessive twentieth-century concern—controls. Modern technology, especially military, has unleashed massive amounts of energy, which needs to be controlled for the achievement of goals, whether constructive or destructive. Sperry concentrated on control. Naval officers realized that not just airplanes but the great dreadnoughts could fulfill their potential only if means were invented to control and direct their enormous power. The navy wanted a control revolution to follow upon the improvements in steam turbines, steel armor, guns, and explosives. In 1900 the maximum range of battleship guns had been under four thousand yards; in 1910 it was extended to ten thousand. The U.S. Navy turned to Sperry for controls in 1911. He responded before World War I, as we have noted, with a gyrocompass that could take the bearings of the target far more accurately than the magnetic compass, and with a gyrostabilizer intended to quench the rolls of a ship at sea, so that the guns, in effect, were fired from a stable platform. In their search for effective gunfire, the navies of the West helped move society more rapidly into the era of automation and control.

Borrowing ideas from earlier work done for the British navy, Sperry and his engineers had by 1916 developed a patented system of gunfire control for the U.S. Navy. By 1920 nineteen U.S. dreadnoughts, eleven second-line battleships, and nine armored cruisers had "Sperry Fire-Control Systems." Far more complex and technically advanced than the submarine-detection gear that the physicists developed during the war, these control systems led the inventor and Annapolis graduate Frank Julian Sprague to declare Sperry "a man whose work has in the opinion of Naval Officers revolutionized navigation and gun fire."[77] Sperry's fire-control system provides a pertinent example of system building. First, repeater compasses driven by the master gyrocompass below deck were placed aloft, where bearings, or sightings, on the target were taken, and

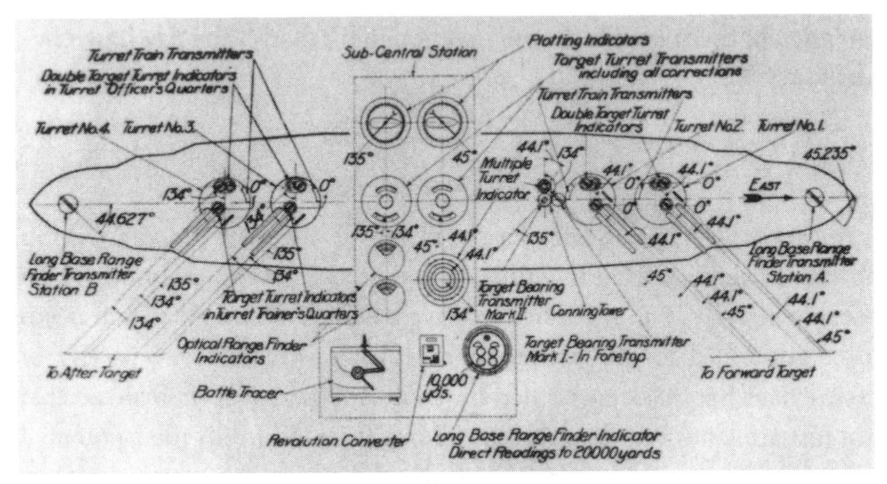

SPERRY FIRE-CONTROL SYSTEM FOR BATTLESHIPS

below, in the gunnery officer's control center, where calculations could be made. Sperry, finding that transmission of information from one station to another often brought distortions, then developed electrical means of sending the information. Commands from the gunnery officer to the turrets were displayed on an indicator for the gun crew. Another communication-and-display device indicated to the gun crew and to the gunnery officer the position of the rotating turret and guns in comparison to the position that had been ordered by the gunnery officer.

The Sperry gunfire-control system also placed an analogue computer, the "battle tracer," in the gunnery officer's control room. The use of this device during the war provides one more reminder of the early development of computer information and control long before the so-called computer revolution after World War II. The "battle tracer" automatically received information about the ship's course from the compass, the ship's speed from revolution counters on the propeller shafts, the target bearing and range from sighting devices aloft, and then combined these with other information about ocean currents. The output from the analogue computer consisted of a small ship model that moved along a chart continuously showing the ship's position, and an arm extending from the ship model that continuously marked on the chart the enemy, or target, ship position.[78]

MILITARY-INNOVATION COMPLEX

The Sperry Gyroscope Company and the U.S. Navy together formed an early, well-articulated instance of the military-innovation complex. Besides gyrocompasses, gyrostabilizers, fire-control systems, and automatic airplane controls, the company invented, designed, and produced aircraft instruments and high-powered searchlights that located targets in a way analogous to radar later. Most of the characteristics discernible today of interaction between the military and the prime contractors can be found in the Sperry-navy relationship during World War I. Sperry maintained a continuing dialogue with naval officers about the development of critical problems in the expanding naval ship system, especially in the dreadnoughts. He and the company responded knowingly with inventive solutions. Naval officers participated with Sperry and his engineers in the testing and development of prototypes, such as the first aerial torpedoes. The ultimate objective was the production contract. In 1918 a Sperry-company memorandum rightly concluded that "this company, practically from its inception, has been used by the Navy Department and later also by the War Department as nothing short of a 'brain mill' and experimental laboratory . . . for developing . . . intricate and even abstruse instruments of high precision."[79] The Sperry "brain mill" was much more a harbinger of the future military-innovation-industrial complex than the widely publicized Naval Consulting Board and the National Research Council.

4

NO PHILANTHROPIC ASYLUM FOR INDIGENT SCIENTISTS

The Naval Consulting Board dominated by Thomas Edison and other independent inventors did not fulfill the inflated expectations of Josephus Daniels, the public, or, probably, the board members themselves. We can see, in hindsight, that during the emergency years of the war the navy needed inventions to improve its existing systems more than radical breakthrough inventions, the stock-in-trade of independents. Design of antisubmarine weaponry required persons more highly trained in physics than was usual among independents. Industrial corporations wanted, as did the navy, inventions to improve their expanding systems. The organizational mentality favored incremental rather than abrupt change. This became especially true of the electrical industry, which the independents had helped found. Independent inventors began to fade from public view. When peace returned, the independents never again regained their status as the pre-eminent source of invention and development. No one else captured public attention as had Edison, Orville and Wilbur Wright, and Nikola Tesla. Instead, attention shifted to the industrial research laboratories as the presumed wellsprings of invention and discovery. Industrial scientists, well publicized by the corporations that hired them, steadily displaced, in practice and in the public mind, the figure of the heroic inventor as the source of change in the material world. Between the world wars, the industrial laboratories came to be seen as the source of "better things for better living." When, for purposes of publicity, Elmer Sperry, who had never before worn one, was asked to don a lab coat for a photograph, and when he, who had never used one, was then told to peer through a microscope, these attempts to change image clearly signaled that the heyday of the professional inventor was passing. The best address for inventors had become General Electric Research Laboratory, Schenectady, New York, not Edison Laboratory, Menlo Park, New Jersey. American invention was shifting from a revolutionary to an evolutionary mode.

Willis R. Whitney, who headed the General Electric laboratory, and Charles Proteus Steinmetz, who performed dramatic high-voltage experiments for the General Electric Company (GE), soon came to symbolize the creative power of industry and science. "Research and development" began to replace "invention" in everyday language. Perhaps Edison built his large laboratory at West Orange, New Jersey, and turned to large-scale industrial schemes because he had the prescience to see that industrial corporations, with their industrial laboratories, were displacing the independent inventor, who had to work on a small scale. He may also have realized that the era of the inventor-entrepreneur generalist must give way to teams of specializing scientists, engineers, and inventors working within an organizational framework. Independent inventors had manipulated machines and dynamos; industrial scientists would manipulate electrons and molecules. Independent inventors had boasted of their craft and art; scientists would take pride in the objectivity, universality, and transfer ability of their knowledge.

EDWIN ARMSTRONG: THE INDEPENDENT BELEAGUERED

Besides industry's need for the conservative inventions that improved large technological systems and often brought immense profits, industry's ability to draw on a battery of patent lawyers to challenge the patents of the independents and to reinforce its own helps explain the independents' demise. The history of Edwin Armstrong's engagements with AT&T and the Radio Corporation of America (RCA) provides a memorable example of difficulties faced by independents in a world populated by industrial corporations determined to control invention and development. Inventor of the frequency modulation and other major radio advances, Armstrong, like Edison, Sperry, and the other independents, was determined to survive as an inventor-entrepreneur. With a small laboratory and a few loyal assistants, he performed the various functions of inventing, patenting, developing, fund raising, and marketing. Earlier, in 1912, when he was only twenty-two years old and an undergraduate studying electrical engineering at Columbia University in New York, Armstrong had investigated the behavior of the poorly understood three-element tube. Not even its inventor, Lee de Forest, had yet fully appreciated its various implications. De Forest had first conceived of it as a detector of radio waves, but by 1912 he was trying, with limited success, to adapt it for use as a repeater or relay on long-distance telephone lines. Using an oscillograph in a laboratory at Columbia, and advised by one of his professors, Armstrong carefully analyzed the performance characteristics of the tube. An earlier eureka insight experienced while he was mountain climbing in Vermont now led him to feed the incoming signal on the tube's plate back to its grid, thereby establishing an amplifying feedback loop. He found that the tube with feedback and rightly tuned circuits acted as a remarkably powerful amplifier of the incoming signal. He could pick up stations loud and clear at distances then believed impossible. Armstrong had invented a feedback, or regenerative, circuit, a major invention in the history of radio. Continuing to investigate the feedback tube in a regenerative circuit, he also noted a hissing sound at the highest level of amplification. For detection of incoming waves he kept the amplification below this level, but he continued to explore, and decided, nearly six months later, that at the high, hissing level the tube acted as a generator of high-frequency waves. He had invented, then, both a feedback detector and a feedback generator, or oscillator, that became the principal mode of transmitting radio waves.[1]

Armstrong had his regenerative circuit notarized on 31 January 1913 but, preoccupied by the events of his graduation in the spring of 1913 and his duties as an assistant in electrical engineering at Columbia afterward, he did not file a patent application until 29 October 1913. Then he filed only for the detector, or receiver, circuit, an omission that caused him endless problems later, during litigation. Only in December 1913 did he belatedly file for a patent on the circuit as an oscillator, or a transmitter. Before his patents were successfully challenged in 1924, however, the feedback, or regenerative, circuit as a receiver and transmitter of radio waves had been recognized in the rapidly expanding world of radio as one of the major inventions, and Armstrong had become wealthy from licensing his patents. He maintained his independence and continued to invent long after the golden decades of the independent, professional inventor had passed. Among other improvements he invented the superregenerative circuit, and the superheterodyne receiver, which became basic for radio after 1930.

He became, however, embroiled in patent litigation, which his biographer, Lawrence Lessing, notes, "continued from 1920 to 1934 through more than a dozen courts and tribunals, piling up thousands of pages of testimony, wearing out nearly three sets of lawyers, costing well over a million dollars and rising in the end to a strange and terrible climax."[2] Armstrong, however, initially had little difficulty in establishing, with his January 1913 notarized sketch, the priority of his invention of a feedback circuit for radio-wave detection. In 1922 the U.S. Circuit Court of Appeals in New York City found that de Forest, who was backed by AT&T, had infringed on the Armstrong patent. In this instance, Armstrong was supported by the Westinghouse Company, which in 1920 had for $350,000 bought the rights to the feedback-detector patent, and also to patents on a superheterodyne circuit. Westinghouse agreed to pay $200,000 for the Armstrong feedback-transmitter (oscillator) circuit, if the patent held up in an interference case in which de Forest and Armstrong both claimed priority. De Forest did not yet hold a patent on a feedback circuit but, if the interference were decided in his favor, the Patent Office would issue him one, which, if sustained, would invalidate Armstrong's. The de Forest lawyers argued that he had made a more generic, or general, invention than Armstrong and that Armstrong's device was simply a special instance of application of the more general patent.

ARMSTRONG (FAR RIGHT) SERVED IN FRANCE AS AN OFFICER AND EXPERT ON WIRELESS IN THE U.S. SIGNAL CORPS DURING WORLD WAR I.

Patent lawyers, many of whom served the large corporations in the electrical field, and some inventors had mastered the subtleties of exploiting and defending an empire of patents. Patents, they knew, established the boundaries of intellectual property. Inventors and their patent lawyers, like gold prospectors, spoke of staking out claims on the metes and bounds of their property. Patent lawyers defended against incursions, or infringements, within the field laid out. A well-designed patent covered as much territory as possible without infringing on another's property. Well-advised inventors made claims as generic or extensive as the Patent Office and courts would allow. For instance, patent lawyers advised inventors to claim in their patents the principles that governed construction of a device, not simply the construction. Patents were seen as either basic or dependent. Dependent patents were valueless unless the owner also owned or had access to the basic patent. On the other hand, a virtually valueless basic patent could be enhanced greatly by a dependent patent that improved or extended the basic invention. So companies like AT&T and Westinghouse, and patent lawyers, negotiated endlessly to erect a monopolistic structure of interrelated basic and dependent patents that governed, say, long-distance telephony or radio broadcasting. Investment had to be made with great care, for Sperry's experienced patent lawyer estimated that "probably less than one per cent of the patents granted are very valuable, ten per cent to twenty-five percent of moderate value, and the remaining percentage of little value."[3] Armstrong dared to venture into this no-man's land, first with corporate backing and then on his own.

De Forest and his lawyers based his case on a circuit diagram and notebook entry dated August 1912, months before Armstrong's circuit diagram of January 1913. They claimed that de Forest's was a preliminary concept that implied the basic feedback circuit and that led directly to experiments in which it was demonstrated. Not until 1914 and 1915, however, did de Forest seek patents on the feedback circuit and on its use as an oscillator, or a generator of radio waves. This activity Armstrong and his lawyers took as evidence that de Forest did not see the implications of the feedback circuit until after Armstrong's was patented and well known. If, after the New York Circuit Court found de Forest infringing in 1922, Armstrong had not pressed for damages against the nearly bankrupt de Forest, the episode might have come to an end with Armstrong's priority established. Against the advice of the Westinghouse Company, Armstrong, proceeding alone, insisted on damages, even though he had little hope of economic gain. For Armstrong, however, his pride as an inventor was involved. The feud between the two men became even more bitter.[4] In 1920, when de Forest read a paper before the Franklin Institute in Philadelphia on the evolution of this three-element tube, or Audion, he recalled that "it was well received, except by one E. H. Armstrong who sought to show that it was he who had invented the feedback circuit.

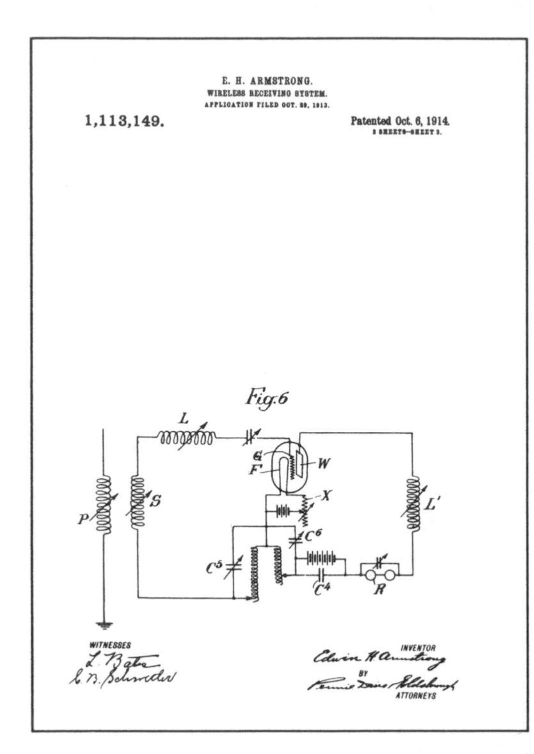

ARMSTRONG PATENT ON REGENERATIVE (FEEDBACK) CIRCUIT. APPLICATION 29 OCTOBER 1913

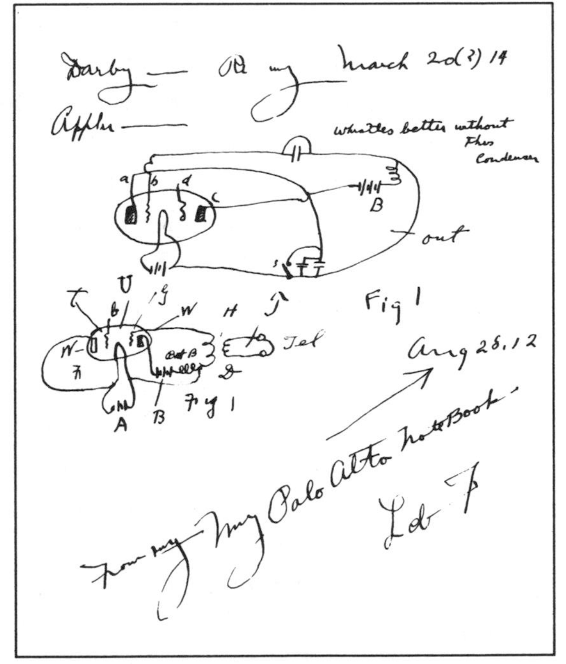

NOTEBOOK PAGE FROM 1912 THAT WAS BASIS FOR LEE DE FOREST'S CLAIM FOR PRIORITY IN INVENTION OF REGENERATIVE (FEEDBACK) CIRCUIT

'All de Forest invented was the Audion! We'll concede that,' he growled. Whereupon the chairman ordered him to 'sit down.' "[5] Earlier, in 1914, when de Forest demonstrated his Audion, Professor Michael Idvorsky Pupin, who had been Armstrong's mentor and friend at Columbia University, irately and loudly demanded of de Forest, "What right have you to have that here? That thing is not yours. That belongs to Armstrong."[6]

De Forest and his lawyers returned to the attack, insisting that Armstrong had filed two separate, relatively narrowly defined applications because he had not grasped the general character of the feedback circuit, a concept that covered both receiver and transmitter. They persuaded the U.S. Patent Office to issue in 1924 to de Forest two basic patents on the feedback circuit and on the circuit as oscillator, or generator of radio waves, for which he had applied in 1914 and 1915.[7] A crucial part of the de Forest case was a successful motion to have the wording of one application changed from "high frequency" to "electrical" oscillations. In so doing, the application and the subsequent patent became basic, and Armstrong's patents on high frequency, or radio, oscillations, could be subsumed under it as special cases. Armstrong argued, with substantial support, that de Forest, who had been concentrating on a search for a telephone amplifier and, therefore, on lower-frequency waves, had not seen the possibility of radio-frequency feedback circuits and had not developed a radio-frequency receiver or amplifier. He also attacked the motion that had allowed the change in wording in the de Forest application.

The U.S. Supreme Court twice found in favor of de Forest as the inventor of the feedback circuit, first in 1928 and again in 1934, after a circuit court had in effect reversed the 1928 decision. (De Forest in an aside noted that the circuit-court judge later went to the penitentiary.)[8] Armstrong and his lawyers agreed that he had infringed the de Forest patents of 1924 as issued, but insisted that these were void because Armstrong had priority. Not only Armstrong, but also Irving Langmuir of the General Electric Research Laboratory and Alexander Meissner, an Austrian radio engineer, claimed priority in the invention of the feedback. In 1934 the Supreme Court stated that Armstrong's 1913 development of the feedback circuit as an oscillator was "a brilliant concept, but another creative mind [de Forest's], working independently, had developed it before."[9]

Armstrong, a man of great integrity, decided in 1934 to return the medal awarded to him earlier by the American Institute of Radio Engineers for the discovery of the regenerative, or feedback, circuit. Appearing at the annual convention in Philadelphia, Armstrong had a speech in his pocket to announce his decision. Before he could speak, the president addressed him directly from the chair and reported that the members of the institute's board had unanimously reaffirmed the citation and the bestowal of the Medal of Honor. The convention gave Armstrong a standing ovation.

In 1933, Armstrong had patented another major invention. Ever since his graduation from Columbia, he had investigated from time to time the problem of static interference in radio transmission. Numerous inventors saw it as a major reverse salient on the expanding radio front and tried to find a solution to critical problems associated with it. Armstrong had experimented with ways of reducing static when he was assistant to his distinguished mentor at Columbia, Professor Michael Pupin, who had earlier sold his loading-coil patent to AT&T. Their efforts fruitless, Armstrong in 1922 said, "The biggest problem that I can see is the elimination of static. . . . It is the only one I ever encountered that, approached from any direction, always seems to be a stone wall."[10] By 1925 he caught a glimpse of light at the end of the tunnel as he began to suspect that the existing system of radio transmission by wave-amplitude modulation could not be corrected, but that a new system needed to be introduced. He investigated wave-frequency modulation (FM) as an alternative to amplitude modulation (AM). Frequency modulation for transmission had been tried before, but rejected because unacceptable distortion resulted when FM was kept to a narrow band of frequencies, as was customary. In 1932 he had the radical idea of transmitting with frequency modulation over a wide band of frequencies. On finding that transmission of unprecedented clarity, he proceeded to patent a new system with transmitters and receivers.

In so doing, he was, with his radical invention, flying in the face of a moving front of great momentum. Large manufacturers of radio equipment such as RCA, existing broadcasting stations, and countless engineers and workers skilled in the existing art all had financial and personal stakes in further developing the amplitude-modification system. Armstrong was particularly piqued by the rejection of his proposed system by the mathematician John Carson, who was employed by AT&T. As was the case with Edison, Sperry, the Wright brothers, and other major independent inventors, mathematics was not Armstrong's forte. He had a lifelong quarrel with mathematicians. Armstrong believed that invention followed from experiment and from reasoning with physical metaphors, not, as he saw it, from the sterile manipulation of mathematical formulae.[11] He referred to a cult that believed that invention involved dealing with a bewildering maze of symbols and curves. Armstrong opined that only the open-minded and greatest mathematicians understood that mathematics could only propose a theory, not prove it. Physical experimentation was necessary for this. After he had proved FM, Armstrong later took pains to point out that he had proceeded in exactly the opposite direction from that suggested by mathematical theory.[12]

He demonstrated the effectiveness of FM by a dramatic public display before the Institute of Radio Engineers in November 1935. Armstrong gave a lecture on the new system and then had receivers, hidden during the talk, turned on to bring in with startling fidelity a broadcast from a transmitter he and a friend had set up in Yonkers, New York. Despite these and other demonstrations, RCA, the major radio manufacturer and major owner of the National Broadcasting Company (NBC), the principal chain of broadcasting stations, did not exercise an option to purchase the rights to Armstrong's patents. Armstrong was particularly discouraged because David Sarnoff, president of RCA, with whom he had been on friendly terms and who had been personally involved in 1922 in the purchase of rights to a super-regeneration circuit invented by Armstrong, did not respond positively to FM. Armstrong invited him to see and hear the "little black box" to eliminate static that Sarnoff had been demanding. Although Sarnoff was impressed, he noted that the little black box was a room full of equipment that would substitute a new system for the one in which RCA was so heavily invested. Armstrong could not sympathize with the perfectly rational position that Sarnoff was taking as head of a

ARMSTRONG (RIGHT) WITH GUGLIELMO MARCONI, 1932. BEHIND THEM IS THE SHACK IN WHICH MARCONI BUILT HIS FIRST WIRELESS STATION IN THE UNITED STATES.

company with a large investment in an existing and more than adequate radio system. Here was a classic case of the independent inventor's radical attitude toward invention and development clashing with the conservative approach of the large corporation.

After two years of desultory encouragement from RCA, Armstrong in 1935 drew on his own funds to develop the new system. He established laboratory facilities at Columbia University with two young engineering graduates as assistants. His demonstrations persuaded the General Electric Company, then embarked on a campaign to take some of the radio business from RCA, to take out a license to build FM receivers. He spent heavily on a transmitting station overlooking the Hudson at Alpine, New Jersey, which finally went into full operation in 1939. As small entrepreneurs established FM stations independently of the major AM broadcasting networks, the demand of FM sets increased and other manufacturers took out licenses. RCA, however, proceeded on a modest scale to support research and development on its own FM system and went ahead energetically to push television.

The onset of World War II retarded further spread of both FM and TV, other than for military purposes, but after the war Armstrong was ready to push the expansion of FM through the acquisition of additional broadcasting frequencies that would allow more stations. Then he found himself colliding head-on once again with RCA and others supporting AM and TV. In June 1945 the Federal Communications Commission, headed by a former legal counsel for RCA, ordered that FM broadcasting be moved out of the 50-megacycle band it had been using and into a new frequency band between 88 and 108. This action made obsolete about fifty prewar transmitters and five hundred thousand FM sets. Television, championed by RCA, was moved into the vacated frequencies.[13] Armstrong found the reasons advanced for the decision without merit. He proceeded to rally the FM forces and to reorganize for the new transmission band. Despite opposition, FM spread, as a new generation of listeners, captured by high-fidelity recordings and players in the late 1940s, increasingly requested FM radio. By 1949 over six hundred FM stations were on the air. Armstrong liked to compare the ongoing "battle of the frequencies" with the "battle of the systems" between direct and alternating currents a half-century earlier.

Scarred but undaunted by the patent struggle over the feedback circuit, Armstrong decided in 1948 to go to court, claiming that RCA and NBC had infringed and induced others to infringe five of his basic FM patents. Thus he arrayed himself against a battery of lawyers and one of the country's largest corporations. The struggle became relentless and demanding. The preliminary fact-finding by the court lasted for five years. Armstrong, sixty-three years old, was persuaded that RCA intended to draw the case out until he either went broke or died. He testified endlessly, and each evening he took home the records to study and prepare for the next day. His wife and his friends noted the strain and urged him to let up and to consider a compromise, for which RCA seemed inclined. He refused. A particularly dramatic moment occurred in 1953 when Sarnoff testified that he and Armstrong had been friends and he hoped that they still were, but then declared that RCA and NBC had done more than Armstrong to develop FM. According to a lawyer present, Armstrong's eyes flashed a flame of pure hatred.[14]

By January 1954, after five years of litigation, Armstrong was exhausted physically, and his financial resources were also nearly depleted. He was depressed by his conviction that he had expended his declining energies on noninventive activities and that he would never again invent a device of importance. His relations with his wife were strained, because he refused to heed her advice and withdraw from the legal battle. In the late fall she retired to Connecticut with her sister to wait out the struggle. Sometime on 31 January 1954 Armstrong wrote to her, lamenting how he had hurt "the dearest thing in the world to me."[15] (They had no children.) The next morning he was found face down ten stories below their New York apartment, completely and neatly dressed in overcoat, hat, scarf, and gloves.

His widow, Marion, having accepted a compromised offer of $1 million from RCA, then decided to press the case against twenty-one manufacturers that had taken licenses from RCA. Friends advised against this trying and expensive process, but in 1967, after the last of the contesting manufacturers, Motorola, was found infringing, she vindicated her husband posthumously. All twenty-one suits were decided for him for a grand total in settlements or damages of over $10 million.[16]

INDUSTRIAL RESEARCH LABORATORIES: BELL TELEPHONE

Armstrong's struggle with industrial corporations, their industrial research laboratories, and the batteries of lawyers ended with the independent, or at least his survivors, triumphant. In general, however, as we have noted, the independents were losing place to the industrial laboratories as the center of inventive activity in the United States. The history of the rise of the laboratories began about the turn of the century with the increasing emphasis on invention, research, and development by AT&T and with the founding of General Electric Research Laboratory. Their successes, measured in profits to the corporations, spurred the founding of countless other industrial laboratories. A major reason for the gradual displacement of the independents who had modeled their behavior on that of young Edison was the need of corporations like Bell and General Electric to preside over the expansion of existing technological systems. To this end, they wished to choose the problems that inventors would solve, problems pertaining to the patents, machines, processes, and products in which the corporations had heavily invested.

The American Bell Telephone Company with its long-distance arm, AT&T, was one of the earliest of the large innovation-oriented corporations to make the gradual transition from reliance on the professional inventor to reliance on the industrial scientist and the industrial research laboratory. From its beginning, the company's principal assets were patents, initially those of Alexander Graham Bell. Company strategy turned about patent acquisition as a means of obtaining a monopolistic position as supplier of telephone service in the United States. By supplying unique telephone services protected by patents, the company sought to avoid competition. Until 1893–94, when the basic Alexander Graham Bell patents expired, the Bell company repeatedly brought successful infringement suits—at least six hundred—against would-be competitors. On one occasion, when the Supreme Court decided in favor of the Bell company by only a four-to-three vote, the small group of Bostonians who controlled the company breathed sighs of relief, but realized as never before how dependent they were on secure patents and skilled patent lawyers, not to mention the inventors of the patented devices. The strategy of infringement suits was so successful, however, that a Bell patent attorney observed in 1891, "The Bell company has had a monopoly more profitable and more controlling—and more generally hated—than any ever given by any patent."[17]

Once the basic patents expired, the Bell company had to decide in 1894 whether to depend on the purchase of patents from independent inventors or to cultivate invention and patenting by employees in the engineering department of the company. As yet there was no research laboratory staffed by industrial scientists and engineers. Until after the turn of the century, the company depended mostly on independent inventors instead of its own employees. In 1884 the Bell company had established a "mechanical department" by absorbing Charles Williams's model shop in Boston, the same shop in which Bell and Edison experimented. The company greatly increased the size of its engineering department after 1894, though it was charged mostly with the responsibility of designing apparatus and installations, improving equipment and services, testing supplies and materials, and adapting inventions for commercial use—not creating them. Some of the problems taken up at Bell before 1900 included transmitting current with maximum efficiency, switching, diminishing interference and cross-talk, replacing single-wire grounded circuits by two-wire all-metallic circuits, and placing wires underground in response to legislative pressure.[18] These improvements in the system required knowledge of physics, but Hammond V. Hayes, AT&T's chief engineer and a Ph.D. physicist, could still write to the president of the company in 1906:

> Every effort in the Department is being executed toward perfecting the engineering methods. No one is employed who, as an inventor, is capable of originating new apparatus of novel design. In consequence of this it will be necessary in many cases to depend on the acquisition of inventions of outside men. . . . The very fact that any great invention at the present time must in all probability come from some man of unusual scientific attainments would render a laboratory under the guidance of such men a most expensive and probably unproductive undertaking.[19]

A few years earlier, Hayes had fired John Stone Stone, a physicist-engineer from The Johns Hopkins University, because he was too much an inventor and too little an engineer. Bell managers and owners at that time believed that the universities should do research and that inspired, independent, and virtually unmanageable men of creative temperament —the independents—should invent, but outside the industrial labs.[20] The company's directors insisted that AT&T, the company subsidiary which presided over major technical developments in long-distance transmis-

MICHAEL PUPIN
(1858–1935)

sion, could maintain its dominant position in telephony by acquiring, often for only a few hundred dollars, "a thousand and one little patents and inventions," most of them from inventors outside the company.[21] The achievements of Edison, Elihu Thomson, Tesla, and the other independents had made their mark. The managers believed that those within the company could make improvements within the existing systems, but not accomplish the radical, breakthrough inventions that led to new systems.

Professor Michael Pupin, university professor and independent inventor, helped confirm the assumption that major inventions could come from outside the company. In 1900 he sold a patent to Bell that saved the company an estimated $100 million over a twenty-five-year period. Pupin, a Columbia University electrical-engineering professor, fulfilled the heroic image projected by the myths surrounding the earlier independents. Born into a peasant family, Pupin immigrated to the United

States in 1874 and worked his way through Columbia before he became, in 1901, a professor of electromechanics there. He recounted his American success story in a widely read book, *Immigrant to Inventor* (1923). In it he recounted a romanticized and simplified story of his invention of the loading coil, the idea for which, he emphasized, came to him about 1894, while he was mountain climbing in Switzerland. Armstrong, we should recall, also experienced a eureka moment while he was mountain climbing. In 1900 Pupin sold the patent on his alpine inspiration to AT&T for $185,000 plus $15,000 every year the patent was in effect, which amounted to $255,000 during the seventeen-year life of the patent. The company agreed to pay Pupin even before an interference case involving his patent was decided. This event was especially remarkable since the interference was with a similar loading-coil patent of George Campbell, an AT&T scientist. Bell patent lawyers feared that the flamboyant and gregarious Pupin would win the interference in competition with their own scientist, Campbell, because of Pupin's appeal as an independent inventor, an immigrant who had made good, contending with a large, monopolistic, "generally hated" corporation. Pupin did win the interference in 1903, in part because the company's patent lawyer, W. W. Swan, handled the Campbell patent application ineffectively. Assuming that patent examiners and judges preferred, as he did, the mechanical to the mathematical description and explanation, Swan had Campbell omit his thorough mathematical description of the invention. The opinion in favor of Pupin, who had included mathematical description in his application, read:

> Campbell's policy of carefully shielding the intelligence of those skilled in the art from the knowledge he had acquired by his experiments and of eliminating mathematics from a specification, where every sentence implies an equation cannot be too strongly condemned.[22]

The eagerness of AT&T not to lose the Pupin patent to a competitor stemmed from its prediction that loading coils would save $1 million on New York City circuits alone, and would also give Bell a substantial, patent-protected lead in long-distance transmission.[23] In use, the loading coil did solve a critical problem and made possible the extension of AT&T's long-distance line beyond a twelve-hundred-mile circuit, like the one from Boston to Chicago. Installation of loading coils on the telephone lines doubled this practical distance and lowered the cost of the lines. Loading coils proved to be the single most important invention during the forty years between Bell's original invention and the appearance of electronic amplifiers. Although the success of Pupin, an outside inventor, seemed to confirm Bell's policy of relying on them, Bell was to learn some more important lessons from the episode. Both Pupin and Campbell could not have invented and developed loading coils without a fundamental knowledge of physics and a highly developed competence in mathematics. Both depended heavily on the fundamental scientific investigations of electrodynamics, and on the wave theory of transmission lines done by Scottish physicist James Clerk Maxwell and the British physicist-engineer Oliver Heaviside. By 1893 Heaviside had suggested self-inductance to reduce distortion in signal transmission, the essence of the loading-coil inventions of Campbell and Pupin. Not only was knowledge of fundamentals needed, such as those expounded in Maxwell's *A Treatise on Electricity and Magnetism*, but Campbell, in determining the number and spacing of loading coils along telephone lines, did not proceed empirically, or by hunt-and-try, but on the basis of a thorough understanding and mathematical description of the characteristics of loading coils and telephone transmission lines. His theoretical understanding resulted in the most efficient practice and in great savings in costly copper for coils and lines.[24] A combination of imaginative experimental skills guided by enlightened theory became a hallmark of leading industrial laboratories.

The relationship between fundamental science and money was thoroughly understood by the new management of AT&T, installed after a banking syndicate headed by investment banker J. P. Morgan took over during the financial panic of 1907.[25] In 1900 AT&T became the central holding company of the Bell system. Morgan, who had played an instrumental role in the formation of both General Electric and U.S. Steel, intended for AT&T to control the entire U.S. telecommunications system, both telephone and telegraph. As president, Morgan chose Theodore

N. Vail, whose stated goal for AT&T became "One Policy, One System, Universal Service," which, in other words, meant domination if not monopoly of long-distance telephone service in the United States. Previously Vail had established a reputation for standardizing and systematizing business services. From 1878 to 1887, he had been general manager of the Bell Telephone Company and had established AT&T, where he became president, until he chose retirement in 1887. After becoming president in 1907, Vail named John J. Carty chief engineer. Lacking formal training in engineering, Carty had nevertheless worked his way up in the company from switchboard operator. He had received more than twenty-four patents while doing so. As chief engineer, he served more as a spokesman and an effective advocate of scientific research in the company than as a researcher himself. He saw particularly clearly the necessary role of science and scientists in order to fulfill the company policy of dominating long-distance telephony by means of patent-protected unique and improved service.[26] He further realized that such utilization of science saved the corporation money. Corporation financiers and executives no longer saw scientists as simply otherworldly and long-haired, as Americans had tended to do during the heyday of Edison.

In 1907 the company's telephones totaled about half those in service in the United States, but the competitors mostly offered only local, not long-distance, service. Engineer Carty, instinctively realizing that system building required precisely, even dramatically, defined goals, announced that AT&T would have a coast-to-coast telephone line in operation for the 1914 Panama-Pacific Exposition. (The opening was delayed until 1915.) Although the Pupin-Campbell loading coil made possible extension of service to Denver, Colorado, without a device to amplify telephone signals, coast-to-coast transmission was impossible. Carty declared that the problem required "especially exhaustive and complete laboratory investigation."[27] After unsuccessfully trying amplification by means of arc, magnetic, and electromechanical devices, Carty, advised by his assistant Frank Jewett, a former instructor in physics and electrical engineering at MIT with a doctorate in physics from Chicago, decided that the development of an amplifier using electron discharge was the critical problem. Since electron discharge was a recent discovery and still poorly under-

JOHN J. CARTY OF AT&T, WHO LED THE TRANSCONTINENTAL TELEPHONE PROJECT

stood, Jewett recommended the employment of skilled physicists familiar with recent advances in molecular physics. Jewett, who had studied under physicist Robert Millikan at the University of Chicago, and who knew that Millikan had been working on electronic discharge, asked him to let AT&T have "one or two, or even three, of the best of the young men who are taking their doctorates with you and are intimately familiar with your field."[28] Carty and Jewett organized a new arm of the engineering department in 1911, the "research branch," to succor the young physicists. A subtle change had occurred: instead of talking of *inventing* an amplifier, as would have been common earlier, now they anticipated applicable discoveries flowing from fundamental research.

The completion of the coast-to-coast project and the development of an electronic amplifier making that possible provide a remarkable example

FRANK JEWETT BECAME PRESIDENT OF BELL TELEPHONE LABORATORIES IN 1925.

of historical continuity. The inventors did not suddenly step offstage to give way to the industrial scientists, nor did their artful probing of invention abdicate to the systematic assault of science on the unknown. An independent inventor's creation, refined by industrial scientists, solved the problem. In October 1912 Jewett, who learned that de Forest's invention of the three-element electronic vacuum tube had amplifying possibilities, asked scientist H. H. Arnold of the research branch to analyze its performance. De Forest had invented the tube as a receiver, not as an amplifier. Persuaded that the crude device was potentially a superior amplifier, AT&T purchased the rights to de Forest's patent. During the next three years, twenty-five researchers and assistants worked on the Audion, or triode-amplifier project. Understanding the principles of electronic amplification, which de Forest did not, they transformed "the weak, erratic, and little-understood audion into the powerful and reliable triode amplifier that the Bell system needed."[29] Carty unsparingly drove himself, the transmission engineers, and the research branch. On 25 January 1915 Bell spoke from New York to his erstwhile assistant Thomas Watson in San Francisco on the occasion of the Panama-Pacific Exposition. The physicist Millikan observed that "the electron—up to that time largely the plaything of the scientist—had clearly entered the field as a patent agent in the supplying of man's commercial and industrial needs."[30]

WHITNEY AND GENERAL ELECTRIC RESEARCH LABORATORY

The heads of American industrial research laboratories were becoming the organizers of American invention, and the industrial scientists who served under them the sources of creativity. Among the laboratory heads, Whitney of the GE laboratory was the most celebrated and "undoubtedly the best beloved of all the remarkable men who are the crowned heads of research."[31] He, like Edison, had a vision of the sequestered place where the inquiring mind could explore the frontiers of knowledge to seek solutions to practical problems. Edison's vision, however, had taken shape as a variation of the mechanical-model maker's workshop and the chemical experimenter's cabinet; Whitney's concept found expression as a variation of the university research laboratory. The ideals of academic research had fired Whitney's imagination in 1894, when he was working for his Ph.D. in Germany at the University of Leipzig. Having taken leave from his position as an instructor in chemistry at the Massachusetts Institute of Technology, he became a doctoral student of Wilhelm Ostwald. The renowned Leipzig professor of physical chemistry represented well those characteristics that the academic world, especially the American, celebrated in the late nineteenth century as the epitome of research and scientific learning. Ostwald "overflowed the boundaries of chemistry and physics into philosophy, literature, psychology, linguistics, and landscape painting." Another American student of Ostwald's described him as "a bear, I tell you, in knowledge. He has got not only chemistry, but physics, down cold."[32] In the new German research tradition, Ostwald surrounded himself with students and assistants who did highly specialized dissertations that were expected to become the building blocks of an imposing edifice of knowledge. German universities championed the freedom of their professors to pursue their research wherever their interests dictated, which in many cases was a search for fundamental and pure knowledge without commercial or practical import. At the same time, an increasing number of science professors also chose to do fundamental research of commercial interest to industry. In so doing, they provided the world with a model that deeply influenced Whitney and other pioneers of American industrial research.

In the 1860s heads of German dyestuff companies, many of whom had studied chemistry in the universities, increasingly asked university professors to apply their growing theoretical understanding of chemical compounds to invent and analyze new organic dyes that could be used in the competitive struggle for the market in Germany as well as abroad. The academic-industrial alliance succeeded so well with synthetic aniline and alizarin dyes that the chemical firms, led by Friedrich Bayer & Co., established company laboratories and employed research scientists to place the research and development of dyes on a systematic and continuous basis. The German patent-law reform of 1877 also encouraged industrial research, for it assured the companies protection for their inventions and the opportunity to monopolize markets with them. The German industrial laboratories and the inventive activity of academically trained research scientists, both in industry and in the universities, alerted the world to the commercial benefits of the alliance of science and industry.[33] Not only the example set by Edison and Menlo Park, but also the German example helped stimulate Whitney and others to establish American industrial laboratories after the turn of the century.

After his return to MIT, Whitney chose research problems of interest to industry. He asked pertinent questions, but he did not relentlessly push to publish his results in scientific journals, a practice that would have hastened his move up the academic ladder.[34] In 1899 he joined Arthur A. Noyes, a senior MIT professor of chemistry, to develop a process for recovering valuable industrial solvents, and to establish a small but highly profitable chemical plant to exploit the process. By early 1901 Whitney had received more than $20,000 from the plant, a sum about ten times

YOUNG WILLIS R. WHITNEY
(1868–1958)

greater than his MIT salary. At a time when bridges were being erected between academia and industry, the commercial success of his initial venture must have inclined young Whitney to cross over. The opportunity came as a result of initiatives taken by Steinmetz, GE's chief consultant, research engineer, and mathematician. Born in Breslau, Germany, the son of a minor railroad official, Steinmetz had studied mathematics at the university in Breslau and taken courses in mechanical engineering at the famed Polytechnical Institute in Zurich before emigrating to the U.S. in 1889. Within a decade he had become chief consulting engineer at the GE works in Schenectady. Steinmetz, whose intellect, physiognomy, eccentric habits, and socialistic politics have made him a memorable figure in American history, introduced American engineers to advanced mathematical modes of analyzing alternating-current light-and-power systems. These modes greatly enhanced the problem-solving abilities of engineering colleagues at General Electric. On the lecture platform, waving his cigar and filling a blackboard with mathematical symbols somehow related to the solution of current industrial problems, Steinmetz left an indelible impression on a generation of electrical engineers. An inveterate cigar smoker, an abrasive practical joker, eventually a socialist candidate for public office in Schenectady, Steinmetz, severely hunchbacked, was a brightly plumed maverick among engineers and industrialists and in comparison with the relatively staid Whitney. The managers of General Electric provided Steinmetz with a semiautonomous consulting-engineering department and the privilege of smoking his cigars where others were denied the opportunity to indulge the habit.

Informed about the notable successes of industrial researchers in Germany, he saw a pressing need for General Electric to emulate the German model. By 1900 Steinmetz knew that European scientists, especially physical chemists, some of them working in industrial laboratories, others in universities, were patenting incandescent-lamp filaments that would soon threaten the company's profitable near-monopoly in the American lamp business. He wrote to GE management proposing a chemical-research laboratory that would concentrate on the lighting problem but would also do general chemical research "in spare time." Appreciative of the ultimate practicality of fundamental research and wishing to free the laboratory from the fetters of short-term problems, he also insisted on an experimental laboratory "entirely separate from the factory."[35] He wanted a lab where "anyone who mentioned orders or selling would be thrown out."[36] Steinmetz took care to send a copy of his proposal to Albert Davis, the company patent attorney, who was well aware of the vulnerable patent position of GE since the expiration of the Edison basic lamp patents in 1894. Following a policy not unlike Bell Telephone's before 1900, GE had tried to hold its advantage by purchasing the patents and consulting services of independent inventors, but in so doing the company sometimes found itself paying more for patent rights than for the salaries of a small laboratory staff. The case of Pupin and Bell must have come to mind. Thomson, the one-time independent inventor and later founder of the Thomson-Houston and the General Electric companies, also endorsed the proposal, and defined the laboratory as one dedicated to the "commercial application of new principles, and even for the discovery of those principles."[37] Thomson, unlike the dogmatic

Edison, readily acknowledged the importance of theoretical understanding.

Steinmetz, wanting a chemist with a practical turn of mind to head the lab, turned to physics professor Charles R. Cross of MIT for recommendations. Cross recommended Whitney, whom he evaluated as one of MIT's best instructors in chemistry. When General Electric approached him, Whitney had doubts, fearing that a scientist in industry would be heavily burdened by assigned routine. But because of his taste for, and success with, the solution of practical problems, and because of slow promotions and low salaries at MIT, he accepted a trial, two-day-a-week position to establish the laboratory in Schenectady while continuing at MIT. His salary for two days a week was the equivalent of a full professor's annual salary, a disparity that Whitney later used to lure other academics to Schenectady. (The MIT president had humiliated Whitney by turning down his plea for a $75 annual raise shortly before the GE offer came.) Within a year Whitney perceived his new position as an outlet for his characteristic enthusiasm and unfulfilled ambition. Eight months of work there also convinced him that GE would not burden him with routine assignments: "There is no evidence," he wrote, "on the part of the officers of the Company, of impatience or a wish to interfere at all in my work." "The only thing I want now," he wrote to a friend in 1901, "is to accomplish some great thing for the 'General Electric'. They are giving me free hand here to spend and experiment, as well as I am able, and I shall die with a ten-ton shadow on my opinion of Whitney if I don't do some good work here."[38] Although he did do some good work, his best work was not done as a research scientist but as a laboratory head, cultivating the talents of others, who had greater aptitude for research than he.

Whitney discovered his limitations as a researcher after he chose as a research problem the development of a more efficient incandescent lamp, the problem that GE most wanted solved. In Europe, scientists highly trained in physical chemistry, the same field as Whitney's, had set the pace. The Austrian Carl Auer von Welsbach, who had been trained at the University of Heidelberg by the renowned chemist Robert Bunsen,

CHARLES STEINMETZ AND THOMAS EDISON

had introduced in 1891 the gas mantle, an intense, steady, and cleanly burning gas lamp, modifications of which are still used today. Besides the gas mantle, he invented the osmium-tungsten filament lamp, one of the major improvements in incandescent lighting. Welsbach's knowledge of the relatively new science of physical chemistry gave him a fundamental

STEINMETZ'S CAMP ON THE MOHAWK RIVER TO WHICH HE WITHDREW TO THINK THROUGH MANY TECHNICAL AND SCIENTIFIC PROBLEMS

understanding of metallic-filament materials that other inventors, ignorant of physical chemistry, could not match. Walther Nernst and Werner von Bolton, German physical chemists, had also invented practical metallic-filament lamps. Nernst, a professor at the University of Göttingen, turned over his patent rights on a filament of refractory metallic oxides to Allgemeine Elektrizitäts-Gesellschaft, which produced seven and a half million Nernst lamps before 1907. Bolton, who like Whitney had studied under Ostwald at Leipzig, assigned his patents to Siemens & Halske, the German electrical manufacturer by whom he was employed as a research scientist.

Dozens of scientists and inventors were developing metallic filaments about the same time, because of information available to them from the advancing field of physical chemistry, the availability of the electric furnace, and the plentiful supply of rare earths, and because increasingly precise cost accounting by the electrical utilities had identified the inefficient carbon filament as a costly component in electric-lighting systems.[39] The price of electric current in Europe was higher than in the United States, so the European inventors and the companies rightly anticipated that consumers would pay more for incandescent lamps that operated more efficiently. The tantalum lamp of von Bolton, for instance, could be produced at commercially attractive prices and offered an efficiency fifteen percent better than the best General Electric carbon filament.[40]

When the Westinghouse Company obtained the rights for the Nernst patent in the United States, the threat to GE's position in the lamp industry was unmistakable. With nothing more to offer than its carbon filament, the company faced the probability of losing its predominant and profitable place in the lamp market. General Electric had to respond with comparable filaments invented by Whitney and his associates in the new lab, or to purchase, despite the price, European patents, a policy that Whitney's lab had been set up to avoid. In 1904 Siemens & Halske offered the General Electric Company the von Bolton lamp, but at a price so high that Whitney responded with an improved carbon filament, the GEM, and then with an intense effort to invent a tungsten filament. When progress proved slow with tungsten, Edwin Rice, GE vice-president and former Thomson lab assistant, concerned about the European lead, sent Whitney to Europe in 1906 to learn firsthand about the situation. After finding their hotel room in Germany lit by a Welsbach electric lamp, and after visiting the Auer company, which manufactured it, Whitney recommended that GE purchase the rights to the process. He also recommended acquiring the rights to buy the same tantalum that Siemens & Halske used in its von Bolton filament. The only consolation for Whitney in these transactions was that GE could use the evidence of some progress with tungsten at his lab in the negotiations over price. Nevertheless, the General Electric Company had to pay Siemens & Halske $250,000 for the right to purchase tantalum wire, and the Auer

FIRST HOME OF GENERAL ELECTRIC RESEARCH LABORATORY

(Welsbach) company $100,000 for the right to its process. "These were the expenses that the Laboratory had been founded with the purpose of preventing."[41]

Whitney assumed personal responsibility for adapting the Auer process for manufacture and use under American conditions, where, for instance, lamps operated with lower voltages and on alternating current instead of direct. But by 1907 he had experienced a series of frustrations. A colleague commiserated: "I am almost sorry that Whitney ever took hold of the German method . . . because he's having an awful time with it."[42] Then the company management told Whitney to drop the project in which the company—and he—had invested so much, in favor of another European one that Whitney had rejected in 1906. To add to his sense of failure, the company, because of the economic panic of 1907, decided to reduce by one-third the laboratory staff, which had swollen to 150 persons. A few days after receiving this bad news, Whitney entered the

GE LABORATORY STAFF, C. 1904: NUMBERS 5, 6, AND 7 IDENTIFY EDNA MAY BEST (GE'S FIRST WOMAN CHEMIST), CHARLES STEINMETZ, AND WILLIS R. WHITNEY.

hospital close to death; his collapse was attributed to appendicitis left too long untreated, but was surely compounded by physical and mental strain. A long rest followed, including a three-month convalescence in Florida. He seriously considered resigning his post as research director and beginning anew as a surgeon.

Then colleagues whom he had brought to the lab and encouraged and supported so well achieved several major lamp-filament breakthroughs. Whitney had persuaded William D. Coolidge, a physicist, who like Whitney had a doctorate from Leipzig and had been an instructor in physical chemistry at MIT, to become assistant director of the laboratory. In 1910 he successfully enticed Irving Langmuir, a twenty-nine-year-old chemist who had taken a Ph.D. at the University of Göttingen under Walther Nernst, the inventor of the lamp. Langmuir, carrying a heavy teaching load at a low salary at the Stevens Institute of Technology in New Jersey, longed for time to do research. In 1909 he asked for a summer of research support at the GE laboratory, anticipating that Whitney would then offer him a long-term position. The offer was forthcoming with an unexpectedly high salary, but Langmuir intended to remain at GE only while "looking around for a really good position in a university." He stayed on and transferred to the study of manufactured objects the techniques he had learned in the university for studying objects found in nature.[43]

Both Coolidge and Langmuir responded to Whitney's encouragement that they do research on tungsten. In 1909, within two years of Whitney's great despair, Coolidge found the combination of hot swaging and gradual cooling that produced a ductile tungsten,[44] and several years later Langmuir, who had acquired a taste for the well-funded research at GE, perfected the gas-filled Mazda incandescent lamp, named for the Persian god of light. This tungsten lamp virtually assured GE of a monopoly on the lamp industry once again. By the 1920s it was making profits of more than $30 million a year, a thirty-percent return on the investment in the lamp business.[45] In 1928 the General Electric group had ninety-six percent of the U.S. incandescent-lamp sales.[46] Langmuir, who continued to investigate the fundamental mechanisms of surface chemical reactions that had led to the Mazda lamp, won the Nobel Prize for this work in 1932. With his laboratory generating other profitable inventions, developments, and patents in X-ray, electronic tubes, and other GE-supported fields, Whitney with reason stood out as the doyen of laboratory directors. As early as 1910 he claimed that annual sales of products originating in the lab amounted to $2.4 million and that, if only seven percent of these sales were credited to the research laboratory, it had more than paid for itself.[47]

Whitney had sacrificed his career as a research scientist in order to lead others.[48] He had not fulfilled the academic ideals of learning and teaching that he had idealized in Germany. After his illness in 1907, he saw his role clearly and symbolized his awareness by posting outside his office a sign, "Come in, rain or shine," above an always-open door. Such a policy virtually precluded the time for reflection and experimentation needed for science. Saying, "I would rather be a little Moses than a big Jeremiah," he gave himself without apparent resentment to furthering

WILLIAM COOLIDGE PATENT ON METHOD OF FORMING TUNGSTEN INTO A FILAMENT

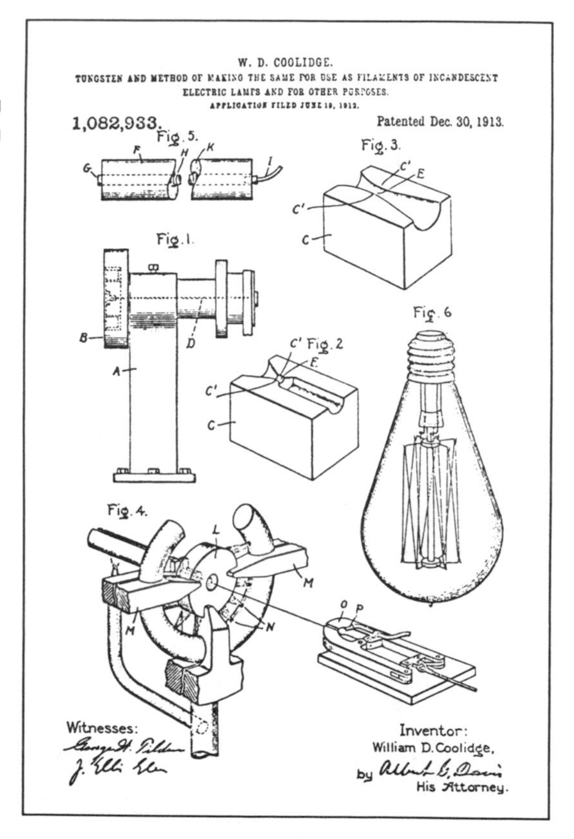

talents for research greater than his. Langmuir wrote to his mother shortly after coming to GE that he was seeing a great deal of "Dr. Whitney." Whitney confided to his diary that he tailored his leadership according to the personality, encouraging to excess those who tended to be doubtful and giving detailed criticism to the more tough-minded whose "probably greater interest and faith will sooner yield results."[49]

As laboratory director, Whitney also sensitively responded to his scientists' ambivalence about leaving the groves of academe for the corridors of General Electric. Scientists still respected the strong tradition of pure research that extended back at least to the towering nineteenth-century figure of the British scientist Michael Faraday, who had refused to profit from the commercial application of his discoveries. Because of the frank materialism and drive for profit in American business, American scientists were especially sensitive to the difference, no matter how fundamental, between pure and industrial science. When Jewett, future head of Bell Telephone Laboratories, first joined the Bell company, his university mentor Albert A. Michelson "thought that . . . I was prostituting my training and my ideals." Earlier, in 1883, physicist Henry Rowland of The Johns Hopkins University, said, "We are tired of seeing our professors degrading their chairs by the pursuit of applied science instead of pure science."[50] Possibly to counter this accusation, Jewett stressed that, although industrial science was not free of economic motives and therefore not pure, it could and often was fundamental in that it sought for basic explanations.

Acutely aware that academic scientists enjoyed higher status than their industrial counterparts, Whitney tried to establish a universitylike atmosphere at the GE laboratory. In this way he hoped to lure scientists away from the campus. At the same time, he realized that the value system in an industrial laboratory had to be different. On several occasions he observed "that the Laboratory is not primarily a philanthropic asylum for indigent chemists" and that tangible and negotiable results were in order.[51] "I often think," he wrote, "though probably wrongly, that men in academic position ought not to be paid as much as men in industrial position. On the average, the men in the latter case work harder, overcome more obstacles, worry more, and compromise with their natural desires much oftener."[52] Maintaining a laboratory that partook of some of the qualities of both a university in exile and a component in the General Electric system, he urged his scientists to have "fun" in their researches. For Langmuir and Coolidge, after their remarkable successes, he drew only broad guidelines for research. He tried to lead rather than direct others down paths paralleling the company's interests; he instituted a weekly colloquium; and he commended scholarly publication. Though he was unable to promote pure research, Whitney said that the best way to run a laboratory was "to turn scientists loose, to encourage their curiosity."[53] This was patently an exaggeration, for Whitney worked comfortably with a management committee that informed him about the production and product improvements needed by the company and, in responding to pressing practical problems, he often formulated short-term

GENERAL ELECTRIC LABORATORY STAFF MEMBERS ATTENDING SEMINAR

projects for his scientists. The lesser-known scientists in the lab were far more likely to feel this gentle pressure than stars like Langmuir and Coolidge. He insisted that careful laboratory notebooks and journals be kept to facilitate patenting; patents obviously scored more with him than publications. He attracted to the lab scientists who were inventive and practical-minded but had no taste for the entrepreneurial activity and the financial risk taking in which the independent inventors had had to involve themselves.

Whitney's early emphasis on the improvement of the incandescent-lamp filament provides a prime example of an industrial laboratory's customary policy of gravitating toward improvements of existing systems rather than toward the inauguration of new ones. Improvements often brought cost reductions in the manufacturing processes and the products in which the companies had invested so heavily. On occasion, however, industrial labs took off in a direction related only remotely to the main interests of the company. The GE Research Laboratory under Whitney, for instance, branched out from electric light and power to develop elec-

IRVING LANGMUIR (THIRD FROM LEFT), WHITNEY (SEATED), AND COOLIDGE (FAR LEFT) IN GENERAL ELECTRIC LABORATORY, C. 1912

tronic tubes for wireless telegraphy and telephony. This can best be explained by the tendency of research projects to develop a momentum as if they were their own sweet beasts. The twisting trail to electronic-tube research at GE began when Langmuir asked whether the Edison effect blackened incandescent lamps. In 1880 Edison had by chance observed that the inside of his lamps became blackened by what were apparently particles of carbon. After some experimentation, he decided that these were being discharged from the negative side of the hot carbon filament of the bulb. At the time electrons had not been identified, but Edison in fact was observing an electron discharge. He did not, however, pursue his discovery to the point of designing a useful discharge tube—a task left, for more than two decades, to others, including British engineer Sir William Preece and de Forest.

In contrast to many other investigators, Langmuir decided that the Edison effect was not blackening the GE tubes, but he did not drop the matter, for his curiosity about the nature of discharge led him to investigate whether the discharge was in fact electrons. The chemist Frederick Soddy had raised doubts about the Edison effect, claiming it was an electrochemical discharge. After experiments and reflection, Langmuir decided that the emission was electronic and was increased or decreased by a space-charge effect discovered by a Colgate University physicist. Langmuir also found that by placing a third electrode in the bulb he could control the electron current, and he was able to predict the rate of flow. He now had a thorough understanding of the de Forest three-element tube. After an engineer at the laboratory heard of Langmuir's work, he saw the commercial implications for wireless and introduced Langmuir to Ernst Alexanderson, GE's expert on wireless systems. GE then put a research team to work on developing a vacuum-tube wireless system in order to compete for leadership in the wireless field.[54]

Whitney's General Electric Research Laboratory was one of several of the company's laboratories. Because the research laboratory was associated with science and with a Nobel Prize–winning scientist at a time when conventional wisdom increasingly and mistakenly saw invention and technology as simply applied science, it was by far the best known. The relatively unknown work done at GE's high-voltage laboratory, on the other hand, provides more insight than do the investigations at Whitney's laboratory into the way in which modern technological systems, such as electric light and power, evolve by means of research and development.

During the 1920s urban electric-power systems throughout the country were expanding and becoming regional. The development of the steam turbine and high voltages for transmission helped make this possible. As the systems spread, however, the breakdown of insulation that protected high-voltage transformers and other components became increasingly common. Electricity would discharge in a lightninglike way across the insulation and burn out the equipment. Such effects frustrated the move to higher voltages and larger regional systems with their attendant economy. Not only this, but natural lightning increasingly often struck the far-flung transmission lines. Charles Peek, Jr., a 1905 engineering graduate of Stanford University who became a member of Charles Steinmetz's consulting-engineering department at GE, took the lead for the company in studying the effects of high voltages, natural and otherwise, on insulation and equipment. To do this, Peek used electrical-impulse generators that provided crackling, booming lightning effects. Publicity often gave the credit for this spectacular research to Steinmetz, because he had become an integral part of the company's public image. But Peek's investigations resulted in more than twenty scholarly papers between 1911 and 1931. Through these he described with mathematical equations the behavior of natural lightning, and similar sparkovers resulting from the breakdown of insulation. With the data, design engineers at the General Electric Company and elsewhere could protect against insulation failure in economical ways and further extend the scope of regional transmission systems.[55] Peek was an inventive engineer and scientist. The problems on which he worked related specifically to the improvement of the technological system over which the company employing him presided. In this, he differed from the equally creative independent inventors.

DU PONT

The AT&T and GE laboratories were most comfortable making improvements along existing lines and drawing on accumulated experience. The history of research at the Du Pont Company, a name become synonymous with the rise of the American chemical industry to world prominence during the twentieth century, provides a notable example of a major laboratory's inability to invent and develop processes and products in which the company and the laboratory had had no substantial experience. Du Pont was unable through its own research and development to make a place for itself in the dynamic and profitable field of organic chemicals.

By 1917 the management of the Du Pont Company, with its long history of gunpowder-and-explosives manufacture, decided to turn the firm into a diversified manufacturer of various chemicals, including organic dyestuffs.[56] To branch out in such fashion would have helped the company counter charges that it was an explosives manufacturer, a monopolistic trust, and an exploiter of armaments races and war. Management was also concerned that after World War I the company's human, physical, and financial resources could not find adequate scope in supplying only the explosives market. The decision to develop and manufacture dyestuffs was a bold venture, because the Germans had long dominated the field and had carefully guarded as secrets the nuances of manufacture. Care was taken by the Germans to see that thousands of dyes on the market could not be identified with the numerous patents that described their manufacture. Some German companies even took out evasion patents to lead would-be imitators down dead ends.[57] To be competitive, organic-dyestuff manufacturers needed regularly to introduce dyes with new colors and characteristics. With the half-century history of the immensely innovative and profitably involved German chemical companies in this field as a precedent, Du Pont launched an intensive effort to hire knowledgeable organic chemists, built the Jackson Laboratory for research in organic chemistry, and by 1917 assigned about half of the company's more than a thousand research scientists, engineers, and technicians to the Jackson Lab and to another special laboratory for dye applications.[58] By 1921, as much as $20 million had been invested, and management forecast that another $20 million might be needed in dyestuffs before profits would emerge. The venture into dyestuffs became an endless maze of unexpected and intractable problems.[59] The company realized that its reach had exceeded its grasp. Chagrined, Du Pont found that within the company it did not have the knowledge—theoretical, technical, or tacit—to succeed in manufacturing dyestuffs. The company had to swallow its pride and turn to German dye chemists for help.[60] Two Du Pont chemists observed, "No matter how much we may dislike to be followers and not pioneers, we must, in the first few years, confine our efforts in this field largely to the manufacture of colors that have already been produced by foreign manufacturers."[61] Not until 1928 did Du Pont begin to earn a modest return on an investment reputed to have been $40 million. The money could have been more profitably invested elsewhere, but perhaps the accumulation of useful scientific and technical knowledge would not have been as great.

From the dyestuffs venture, Du Pont management learned that it was easier to acquire research-and-manufacturing capability by purchasing it from other firms and individuals experienced in a field than to generate it internally. The AT&T and GE companies had long followed a policy of acquiring patents, inventors, and small innovative companies. So Du Pont also acquired capabilities in diverse fields of chemistry by purchasing other firms that then became industrial departments of the company. During the 1920s the company moved into the production of rayon fibers, cellophane films, and synthetic ammonia through the purchase of French technology. It also moved into the manufacture of tetraethyl lead (anti-knock compound for gasoline) and Freon refrigerants, compounds invented outside the company.[62] An agreement between Du Pont and ICI, the large British chemical manufacturer, allowed each company information about the other's research-and-development projects.

Du Pont achieved a notable success in the invention and development of immensely profitable nylon. The dyestuff episode revealed the doubtful validity of the often heard argument that, if enough money is spent on research and development, problems will be solved and goals reached. The nylon case demonstrated how one gifted young scientist could sometimes make an enormous difference in a highly organized research-and-development enterprise.

The history of nylon is a cautionary tale for those dedicated to the proposition that money and organization, not the individual, are the keys to success. The story of nylon begins in 1926, when the history of research at Du Pont took a new turn. Charles M. A. Stine, head of the company's central research organization, known as the chemical department, succeeded in defining a new role for the central research laboratory. This new role contrasted with that of other Du Pont laboratories, which were closely affiliated with manufacturing departments. Stine proposed a "radical departure" for the central establishment through concentration on fundamental research. He argued that fundamental research would allow the company to analyze its various development and manufacturing processes and generate general knowledge about these. His definition of "fundamental"—as contrasted with "pure"—research agreed with that of other laboratory heads, such as Frank Jewett. Stine believed that company technology could be made more technically and economically efficient once the researchers had achieved fundamental understanding of the chemical processes involved. AT&T had demonstrated the value of fundamental research in its analysis of loading coils, and GE in its analysis of vacuum tubes. Stine pointed out that, since many Du Pont products, such as paints, plastics, rayons, and cellophane film, were polymers, or high-molecular-weight compounds, the company needed to do fundamental research directed toward a better understanding of polymer chemistry. He also argued that, if a commitment to fundamental research were publicized, Du Pont could attract higher-caliber research scientists and engineers from the universities. He, like Whitney, had to lure scientists out of the universities.

Allocated substantial funds, Stine built an additional laboratory for the researchers whom he hoped to recruit. Old hands labeled the new venture Purity Hall. Frustrated in his attempts to recruit older, established academics away from the universities, he succeeded in attracting younger ones. Among them was Wallace H. Carothers, a chemistry instructor at Harvard University, who was promised the opportunity to do "pure" research in the field of polymers, a chemical field in which the methods of investigation were based mostly on experience, not theoretical understanding. Stine assured Carothers that he could continue to work on problems of his choice, but that the growth of his group would depend on his "capacity for initiating and directing work that we consider worthwhile undertaking."[63] Before accepting the position, Carothers demurred, partly because he questioned how he might adjust to Du Pont, particularly since he was given to "neurotic spells of diminished capacity."[64] After taking up his position, he found that "a week of industrial slavery" elapsed without breaking his "proud spirit," for he was occupied by "thinking, smoking, reading, and talking" from 8:00 AM to 5:00 PM.[65] He found that for funds the sky was the limit. Carothers soon showed in a number of now classic papers that polymers were not mysterious aggregates but ordinary molecules, only longer. After three years, in 1930, chemists in the group Carothers directed unexpectedly discovered neoprene, a synthetic rubber, the first wholly synthetic fiber.[66] Only after ten years of development did neoprene become a profitable product.[67] In a manner characteristic of the scientist with academic values, Carothers published

WALLACE CAROTHERS IN DU PONT LABORATORY

the findings, established his reputation as a leader in polymer chemistry, and then moved on in search of additional understanding and more discoveries in polymer chemistry. When Stine moved upward in the management of the company, Carothers faced frustrations, for Stine's replacement in central research embarked on a policy of less fundamental research and more research directed toward the interests of the company. Repeatedly urged to concentrate on applications with commercial possibility, Carothers in 1933 agreed to renew work on the development of synthetic fibers; within the year he found the way to nylon. A commercial success at a time when American industry was only beginning to pull out of the Great Depression, the nylon project became a model, or paradigm, for Du Pont management. It wanted more "nylons." Nevertheless, the company, probably because of a short memory, went all out to develop the fiber for market and virtually abandoned fundamental research. Carothers, who had come to Du Pont to do pure research, began to feel that the company severely limited his choice of problems and that he "now had to regard scientific contributions as an occasional and accidental by-product. . . ."[68] He considered but rejected an offer to become chairman of the chemistry department at the University of Chicago. Over the next two years, he fell into repeated bouts of depression, forcing him to have recourse to psychiatric care. He never recovered from a severe breakdown in 1936. The death of his sister and an obsession that he had failed as a scientist compounded his problems. On 29 April 1937, at the age of forty-one, he committed suicide by taking cyanide in a Philadelphia hotel room.[69] His death was lamented by the science community as the loss of a potential Nobel laureate.

SURVIVAL OF THE INDEPENDENT

The rise of the large, innovation-prone electrical and chemical manufacturers in the late nineteenth and early twentieth centuries changed the character of scientists and science and threatened the survival of the independent inventors. An increasing number of physicists and chemists entered industry. Besides General Electric, Bell, and Du Pont, Kodak, Standard Oil of New Jersey, General Motors, and other manufacturers established major laboratories. Before World War I there were at least one hundred industrial laboratories in the United States; by 1929 there were more than a thousand.[70] By 1920 physicists employed in industrial research laboratories made up a quarter of the membership of the American Physical Society, the leading professional organization. Between the two world wars, GE and AT&T alone employed about forty percent of the membership of the organization. These well-paid physicists, experimenting with excellent apparatus and aided by assistants, were choosing practical and fundamental research problems. The solutions to these problems were incorporated into the great body of scientific knowledge, giving it a cast and a direction unlike those that would have been taken if science had simply been pure.

To promote the industrial laboratories and further enhance the prestige of the industrial scientists, proponents trivialized the image of Edison, the symbolic figure among the independent inventors—the sons felt compelled to destroy the fathers. Writing or speaking to company management, investors, and the public, heads of the rapidly growing number of industrial research laboratories often caricatured the Edison method as hunt-and-try Jewett, as head of the Bell Telephone Laboratories after it was formed in 1925 to consolidate the various research facilities of the AT&T and Bell system, stressed that Edison flourished before 1900, when his untrained, intuitive brilliance and persistence suited the state of the development of technology.[71] Ph.D. Jewett observed that Edison was bypassed and outclassed by formally trained scientists and engineers. The heads of the industrial research labs portrayed their laboratories as using science methodically and economically. Promoters of the laboratories realized that, until Edison and his method could be considered antiquated, the possibility remained that pragmatic corporation executives might ask why young Ph.D.s from the universities were being hired at substantial salaries to do what Edison and other professional inventors were known to have done so well—introduce fantastically remunerative inventions.[72]

Without question the so-called Edison method, which had been shared by most of the independent inventors of the late nineteenth century, was more empirical, or hunt-and-try, than the approach of scientists working in industrial laboratories. As in the development of the loading coil and vacuum-tube repeater at Bell, and the tungsten filament at the General Electric Research Laboratory, the scientists preferred to work out, and work with, theoretical explanations for the devices and processes they had invented or were developing. At the same time, however, that the laboratory scientists used science more effectively, they introduced a conservative cast to invention, as compared with the more radical approach of the independent inventors. A few of those presiding over the newly organized laboratories and research scientists realized that the more radical approach of the independent inventors was still needed. Vannevar Bush, who, as the Massachusetts Institute of Technology vice-president and dean of engineering, was familiar with institutionalized research and development, predicted that the day of the professional inventor was not by any means past. The independent inventor, he continued:

> . . . has a much wider scope of ideas and . . . often does produce out of thin air a striking new device or combination which is useful and which might be lost were it not for his keenness. . . . New ideas are coming forward with as great frequency today as they ever have, and while a great research laboratory is a very important factor in this country in advancing science and producing new industrial combinations, it cannot by any means fulfill the entire need. The independent, the small group, the individual who grasps a situation, by reason of his detachment is oftentimes an exceedingly important factor in bringing to a head things that might otherwise not appear for a long time.[73]

Jewett, despite his attitude toward the Edison method, also believed that:

> It is inevitable that the great bulk of what you might call the run-of-the-mine patents in an industry like ours will inevitably come from your own people. . . . I think it is equally the case that those few fundamental patents, the things which really mark big changes in the art, are more likely to come from outside than from the inside. . . . There are certain sectors where the independent inventor cannot operate. . . . There are certain sectors where . . . the chances are ten to one that the fundamental ideas are going to come from outside the big laboratories simply because of the nature of things.[74]

In Germany, Carl Duisberg, a German research chemist and director of Bayer, a leading German chemical manufacturer, decided that the laboratories routinized invention. He characterized the inventions of industrial research laboratories as establishment or institutional inventions that had "*von Gedankenblitz keine Spur*" (no trace of a flash of genius).[75]

As Bush and Jewett anticipated, professional inventors continue in the twentieth century, with less publicity, and working at the peripheries as well as on the frontiers, to produce, in contrast to their small number, a disproportionate number of the radical, breakthrough inventions. Among inventions attributed to independent inventors are air conditioning, automatic transmission for automobiles, power steering, the helicopter, catalytic cracking of petroleum, Cellophane, the jet engine, Kodachrome film, magnetic recording, Polaroid Land camera, quick-freezing, and Xerography.[76] Recently, the general, all-purpose digital electronic computer and the laser have been added to the list.

THE SYSTEM MUST BE FIRST

Since 1870 inventors, scientists, and system builders have been engaged in creating the technological systems of the modern world. Today most of the industrial world lives in a made environment structured by these systems, not in the natural environment of past centuries. Charles Darwin helped explain the influences of nature; Sigmund Freud tried to comprehend the psychological forces crackling like electrical charges within and all around us; but as yet we reflect too little about the influences and patterns of a world organized into great technological systems. Usually we mistakenly associate modern technology not with systems but with such objects as the electric light, radio and television, the airplane, the automobile, the computer, and nuclear missiles. To associate modern technology solely with individual machines and devices is to overlook deeper currents of modern technology that gathered strength and direction during the half-century after Thomas Edison established his invention factory at Menlo Park. Today machines such as the automobile and the airplane are omnipresent. Because they are mechanical and physical, they are not too difficult to comprehend. Machines like these, however, are usually merely components in highly organized and controlled technological systems. Such systems are difficult to comprehend, because they also include complex components, such as people and organizations, and because they often consist of physical components, such as the chemical and electrical, other than the mechanical. Large systems—energy, production, communication, and transportation—compose the essence of modern technology. As Alan Trachtenberg has pointed out, Americans take the "West" and the "machine" as symbols providing perspectives on their early and recent history.[1] After a century of system building, they might well see the "system" as their hallmark.

Many modern technological systems are extensions of the inventions of Edison, Sperry, Tesla, and other independent inventors. These inventors conceived of technical systems as consisting primarily of mechanical, electrical, and chemical components, such as cams, gears, springs, valves, dynamos, incandescent lamps, antennae, belts, pipes, and transmission lines. The more entrepreneurial among them also integrated organizations into the nascent technological systems. Industrial research scientists have been responsible for many of the extensions or improvement of these systems. About the turn of the century, persons possessing similar system-building drives rose to prominence, but their goals were more complex than those of the inventors and industrial scientists. These system builders have left their mark on modern technological society by creating technological systems of immense size that embody not only technical components but mines, factories, and organizations such as business corporations, banks, and brokerage houses. In addition, system builders established large bureaucracies of labor and white-collar workers to tend the systems. Many of the system builders were trained and gained experience as engineers, managers, and financiers rather than as inventors or industrial scientists. As we shall see, they found that a nation committed to mass consumption, freedom of enterprise, and capitalism particularly suited their goal of technological-system building, whether it was socially benign or destructive. Some were motivated by desire for power and money, but they shared a drive to order, centralize, control, and expand the technological systems over which they presided. In seeking the creators of modern industrial America, we must consider the system builders as well as the independent inventors and the industrial scientists.

Henry Ford's production system remains the best known of the large technological systems maturing in the interwar years. Contemporaries then usually perceived it as a mechanical production system with machine tools and assembly lines. But Ford's system also included blast furnaces to make iron, railroads to convey raw materials, mines from which these came, highly organized factories functioning as if they were a single machine, and highly developed financial, managerial, labor, and sales organizations. Other systems contemporary with his were more advanced than Ford's. But Ford's attracted the attention, because the public was better able to comprehend the mechanical. It was fashionable to say that no one comprehended the ethereal force, electricity—but gears, one could feel and see.

Electrical light-and-power systems, such as those managed and financed by the system-building utility magnate Samuel Insull of Chicago, incorporated not only dynamos, incandescent lamps, and transmission lines, but hydroelectric dams, control or load-dispatching centers, utility companies, consulting-engineering firms, and brokerage houses, as well. When Ford placed a mechanical assembly line in motion, the public was greatly impressed, but electrical systems transmitted their production units too rapidly to perceive: 186,000 miles per second, the speed of light. The concepts motivating and guiding Insull and the other electrical-system builders were more subtle and abstract than those driving Ford and his mechanical engineers. Concepts of electrical circuitry rather than mechanical gadgetry shaped the ways in which builders of electrical systems thought and acted; they manipulated interactions, not the simpler linear relationships of cause and effect. The builders of electric-power and chemical-process plants also envisaged flow rather than the movement of batches of materials and mechanical parts. Instead of being the age of the machine, the interwar years emerged as the apogee of the age of electric power and chemical process. The machine as symbol of an age applies better to the British Industrial Revolution of more than a century earlier.

The wave of system building that crested in the United States during the first half of this century had been building up for decades. As early as the mid-nineteenth century, British engineers and industrialists began to refer to a system of manufacture in the United States characterized by the use of highly specialized machine tools and the arrangement of machines, tools, gauges, and other devices in factories to facilitate the flow of production.[2] The next generation, both British and American, spoke of a unique and fruitful "American System of Manufactures."[3] They realized that the American system involved more than interchangeable parts, special-purpose machine tools, and factories laid out for the smooth flow of work; they understood that the American commitment to an economic democracy as well as to a political one brought a new and unprecedentedly large market for mass-produced goods and services for masses of the population. American values and the market influenced by them were also part of the system.[4] Europeans were well aware of the differing character of their own and the American markets. In the late nineteenth and early twentieth centuries, European products were priced and designed as luxury goods. Europeans expected high unit profits on a small turnover. Insull often showed charts graphically demonstrating that in London the price for each kilowatt hour of electricity was high, the profit margin great, and the kilowatt hours produced low, whereas in Chicago the opposite was true. So he continually lowered prices to increase sales and gross profits. In Germany the electric light-and-power utilities catered far more to industry than to a residential market that could only afford mass-produced low-cost electricity.

Of the American system builders, none took on a more difficult and controversial task than Frederick Winslow Taylor. Ford directed his ordering-and-controlling drive primarily to production machines; Insull focused his on ensuring the large and steady flow of electrical power; Taylor tried to systematize workers as if they were components of machines. Ford's image was of a factory functioning as a machine; Insull envisaged a network or circuit of interacting electrical and organizational components; and Taylor imagined a machine in which the mechanical and human parts were virtually indistinguishable. Idealistic, even eccentric, in his commitment to the proposition that efficiency would benefit all Americans, Taylor proved naïve in his judgments about complex human values and motives. In the history of Taylorism we find an early and highly significant case of people reacting against the system builders and their production systems, a reaction widespread today among those who fear being co-opted by "the system."

TAYLORISM

Taylor was not the first to advocate a so-called scientific approach to management, but the enthusiasm and dedication, bordering on obsession, with which he gave himself to spreading his views on management, his forceful personality, and his highly unusual and erratic career filled with failure as well as success have left a strong, indelible impression on his contemporaries and succeeding generations. More than a half-century after his death, many persons in Europe, the Soviet Union, and the United States continue to label scientific management "Taylorism." Labor-union leaders and radicals then and now find Taylor convenient to attack as a symbol of a despised system of labor organization and control. In the early decades of this century, Europeans and Russians adopted "Taylorism" as the catchword for the much-admired and -imitated American system of industrial management and mass production. The publication in 1911 of Taylor's *Principles of Scientific Management* remains a landmark in the history of management-labor relations. Within two years of publication, it had been translated into French, German, Dutch, Swedish, Russian, Italian, Spanish, and Japanese. In his novel *The Big Money* (1936), John Dos Passos gave a sketch of Taylor, along with ones of Edison, Ford, Insull, and a few others, because he believed that they expressed the spirit of their era. Dos Passos noted that Taylor never smoked or drank tea, coffee, or liquor, but found comparable stimulation in solving problems of efficiency and production. For him, production was the end-all, whether it be armor plate for battleships, or needles, ball bearings, or lightning rods.[5]

Taylor's fundamental concept and guiding principle was to design a system of production involving both men and machines that would be as efficient as a well-designed, well-oiled machine. He said, "in the past, the man has been first; in the future the system must be first,"[6] a remark that did not sit well then with workers and their trade-union leaders and that today still rankles those who feel oppressed by technology. He asked managers to do for the production system as a whole what inventors and engineers had done in the nineteenth century for machines and processes.

Highly efficient machines required highly efficient functionally related labor. When several Taylor disciples, including later U.S. Supreme Court Justice Louis D. Brandeis, sought a name for Taylor's management system, they considered "Functional Management," before deciding on "Scientific Management."[7] Taylor and his followers unfeelingly compared an inefficient worker to a poorly designed machine member.

Taylor developed his principles of management during his work as a machinist and then as a foreman in the Midvale Steel Company of Philadelphia. Son of a well-to-do Philadelphia Quaker family, a graduate of Phillips Exeter Academy, and a U.S. champion doubles player in tennis, Taylor was not the typical machine-shop worker. Without doubt he was the only shop-floor worker at Midvale who was a member of the exclusive Philadelphia Cricket Club. Taylor's physician had recommended manual labor after the deterioration of his eyesight during his last year at the academy precluded his entry into Harvard University. Taylor's father expected his son to follow him into the law profession, but the son chose to remain a blue-collar worker. At Midvale he came under the protective wing of its president, William Sellers, one of the nineteenth century's most influential inventors of machine tools, a mechanical engineer and an industrialist who insisted that every member of a machine designed by him or his associates must be functional, or contribute efficiently to the end for which the machine was intended. Taylor afterward referred to Sellers as "undoubtedly the most noted engineer in this country in his time," "a truly scientific experimenter and a bold innovator," and "a man away beyond his generation in progress."[8] When Taylor moved up the ladder and became a foreman and later chief engineer at Midvale, he deeply depended on Sellers's backing as he experimented with fundamental changes that went against the grain of traditional work practices.

Worker soldiering, variously called "stalling," "quota restriction," "goldbricking" by Americans, *Bremsen* by Germans, and "hanging it out" or "Ca'canny" by the English and Scots, greatly offended Taylor's sense of efficiency. Having concluded that workers, especially the skilled machinists, were the major industrially inefficient enclave remaining after the great wave of nineteenth-century mechanization, Taylor proposed to eliminate "soldiering." He later wrote that "the greater part of systematic soldiering . . . is done by the men with the deliberate object of keeping their employers ignorant of how fast work can be done."[9] The machinists at Midvale, for example, were on a "piecework" schedule, so they were determined that the owners not learn that more pieces could be turned out per hour and therefore demand an increase in the number of pieces required. They did not trust the owners to maintain the piece rate and allow the workers, if they exerted themselves, to take home more pay. The workers believed that the increased effort would become the norm for the owners. We can only conjecture about the natural rhythm and reasonableness of the pace that the workers maintained over the long duration; Taylor believed that they were soldiering. Nevertheless, he also showed that he was determined that the diligent worker be rewarded with a share of the income from more efficient and increased production. To his consternation, he later found that management and the owners also soldiered when it came time to share the increased income. Taylor was no close student of human nature; his approach was, as he described it, scientific.

After being put in charge of the machinists working at the lathes, Taylor set out to end soldiering among them. His friends began to fear for his safety. As Taylor recalled, the men came to him and said, "Now, Fred, you are not going to be a damn piecework hog, are you?" To which he replied, "If you fellows mean you are afraid I am going to try to get a larger output from these lathes," then "Yes; I do propose to get more work out."[10] The piecework fight was on, lasting for three years at Midvale. Friends begged Taylor to stop walking home alone late at night through deserted streets, but he said that they could shoot and be damned and that, if attacked, he would not stick to the rules, but resort to biting, gouging, and brickbats. At congressional hearings in 1912—about thirty years later—he insisted:

> I want to call your attention, gentlemen, to the bitterness that was stirred up in this fight before the men finally gave in, to the meaness of it. . . . I did not have any bitterness against any particular man or men. Practically all of those men were my friends, and many of them are still my friends. . . . My sympathies were with workmen, and my duty lay to the people by whom I was employed.[11]

In his search for the one best way of working, of deciding how and how fast a lathe operator should work, he used a method that he considered scientific. He believed values and opinions of neither workers nor managers influenced his objective, scientific approach. Beginning in 1882, first he, then an assistant began using a stopwatch to do time studies of workers' motions. Timing was not a new practice, but Taylor did not simply time the way the men worked: he broke down complex sequences of motions into what he believed to be the elementary ones and then timed these as performed by workers whom he considered efficient in their movements. Having done this analysis, he synthesized the efficiently executed component motions into a new set of complex sequences that he insisted must become the norm. He added time for unavoidable delays, minor accidents, inexperience, and rest. The result was a detailed set of instructions for the worker and a determination of time required for the work to be efficiently performed. This determined the piecework rate; bonuses were to be paid for faster work, penalties for slower.[12] He thus denied the individual worker the freedom to use his body and his tools as he chose.

Taylor stressed that the time studies, with their accompanying analysis and synthesis, did not alone constitute scientific management. He realized and insisted that, for the work to be efficiently performed, the conditions of work had to be reorganized. He called for better-designed tools and became known for his near-fixation about the design of shovels. He ordered the planning and careful management of materials handling so that workers would have the materials at hand where and when needed. Often, he found, men and machines stood idle because of bottlenecks in complex manufacturing processes. Taylor even attended to lighting, heating, and toilet facilities. Seeing inanimate machines and men together as a single machine, he also looked for ways in which the inanimate ones failed. Believing that machine tools could also be driven faster, he invented a new chromium-tungsten steel for cutting tools that greatly increased their speed. As we would expect, he did not leave decisions about even the cutting speed of the machine tool or the depth of the cut to the subjective judgment of the machinist. In his book *On the Art of Cutting Metals* (New York, 1907), he described his thousands of experiments that extended over twenty-six years.

FREDERICK W. TAYLOR INSPECTING CONSTRUCTION

As a system builder seeking control and order, Taylor was not content to redesign machines, men, and their relationship; he was set upon the reorganization of the entire workplace or factory as a machine for production. Stimulated by his example, individuals with special education, training, and skill contributed to the establishment of "the new factory system."[13] To understand this achievement, we need to consider the way in which the work process was carried out in many machine shops, engineering works, and factories before Taylor's reforms. After the concern received an order, copies specifying the product and quantity to be made were sent to the foremen. They carried most of the responsibility for the production process. Once draftsmen had prepared detailed drawings, foremen in the machine shop, the foundry, pattern-making shop, and forge determined the various component parts needed, ordered the raw materials, and wrote out job cards for the machinists. The machinists then collected drawings, raw materials, and tools, and planned the way in which the job for a particular component part would be done. When the machinists had completed the particular job, they reported to the foreman for another. The foremen had overall supervision, but there was little scheduling and, therefore, little planned coordination of the various jobs. Components sometimes reached the assembly point, or erecting shop, haphazardly. Because of lack of planning, scheduling, and close monitoring of the progress of work, raw materials were often not on hand. How the workmen might then use their time is not clear, but proponents of Taylorism leave the impression that they were idle.

Taylor found the disorder and lack of control unbearably inefficient and declared war on traditional methods responsible for these. His reform specified that an engineering division take away from the foremen overall responsibility for the preparation of drawings, the specification of components, and the ordering of raw materials. Upwardly mobile young graduates from the rising engineering schools were soon displacing their "fathers," the foremen. The planning department in the engineering division coordinated deliveries of materials, and the sequence in which component parts would be made. The planning department prepared detailed instructions about which machines would be used, the way in which machinists, pattern makers, and other workmen would make each part, and how long the job should take. Careful records were kept of the progress being made in the manufacture of each part, including materials used and time consumed. Unskilled workers moved materials and parts around shops so that they would be on hand where and when needed. By an elaborate set of instruction cards and reports, the planning department had an overall picture of the flow of parts throughout the shops, a flow that prevented the congestion of the work at particular machines and the idleness of other machines and workmen. The reports of worker time and materials consumed greatly facilitated cost accounting.

The complexity and holism of Taylor's approach was often ignored because of the widespread publicity given to some of his simplest and most easily reported and understood successes. Taylor often referred to the "story of Schmidt," who worked with the pig-iron gang at the Bethlehem Steel Corporation in Pennsylvania. When Taylor and his associates came to Bethlehem in 1897 to introduce their management techniques and piecework, they found the pig-iron gang moving on the average about twelve and a half tons per day. Each man had repeatedly to lift ninety-two pounds of iron and carry it up an inclined plank onto a railroad car. After careful inquiry into the character, habits, and ambition of each of the gang of seventy-five men, Taylor singled out a "little Pennsylvanian Dutchman who had been observed to trot backhome for a mile or so after his work in the evening about as fresh as he was when he came trotting down to work in the morning."[14] After work he was building a little house for himself on a small plot of ground he had "succeeded" in buying. Taylor also found out that the "Dutchman" Henry Noll, whom Taylor identified as Schmidt, was exceedingly "close," or one who placed "very high value on the dollar." The Taylorites had found their man.

Taylor recalled the way he and Schmidt talked, a story that tells us more of Taylor's attitudes than of what actually transpired:

> "Schmidt, are you a high-priced man? . . . What I want to find out is whether you want to earn $1.85 a day or whether you are satisfied with $1.15, just the same as all those cheap fellows are getting?"
>
> "Did I vant $1.85 a day? Vas dot a high-priced man? Vell, yes, I vas a high-priced man."
>
> ". . . Well, if you are a high-priced man, you will load that pig iron on that car to-morrow for $1.85. You will do exactly as this man tells you to-morrow, from morning till night. When he tells you to pick up a pig, and walk, you pick it up and you walk, and when he tells you to sit down and rest, you sit down. . . . And what's more, no back talk."[15]

Taylor found it prudent to add:

> This seems to be rather rough talk. And indeed it would be if applied to an educated mechanic or even intelligent laborer. With a man of the mentally sluggish type of Schmidt it is appropriate and not unkind, since it is effective in fixing his attention on high wages which he wants. . . .[16]

Perhaps Taylor, the upper-middle-class Philadelphian, forgot that the Pennsylvania Dutchman was not so mentally sluggish that he could not save for land and build a house. Schmidt moved the forty-seven tons of pig that the Taylorites had decided should be the norm, instead of the former twelve and a half tons, and soon all the gang was moving the same and receiving sixty percent more pay than other workmen around them. We are not told whether Schmidt was still able to trot home and work on his house.

Numerous other examples of Taylor's methods increasing worker output and production abound, but there is also abundant evidence of failures. Ultimately his efforts at Bethlehem Steel exhausted him, and the head of the company summarily dismissed him. Taylor had come to the steel company with the full support of Joseph Wharton, a wealthy Philadelphian into whose hands the company had passed. Wharton wanted a piecework system installed in the six-thousand-man enterprise. Taylor warned that his system would be strongly opposed by all of the workmen, most of the foremen, and even a majority of the superintendents. Bold and determined, he forged relentlessly ahead, introducing a planning department and new administrative roles for the foremen. Instructions for routine were codified with time cards, work sheets, order slips, and so on. As worker resistance stiffened over several years, Taylor became rigid, even arbitrary, in dealing with labor and management. His achievements were impressive, but "as time went on, he exhibited a fighting spirit of an intensity almost pathological," an admirer wrote.[17] Taylor's communications to the Bethlehem president were tactless and peremptory (he believed Wharton would shelter him). He complained of poor health and nervous strain. He thought that some of the major stockholders opposed him because he was cutting the labor force, and they were losing rents on the workers' houses. The curt note dismissing him came in April 1901.

Many workers were unwilling, especially the skilled ones, to give control of their bodies and their tools to the scientific managers, or, in short, to become components in a well-planned system. An increase in pay often did not compensate for their feeling of loss of autonomy. Taylor's scientific analysis did not take into account worker independence and pride in artful craftsmanship—even artful soldiering. Perhaps this was because Taylor, despite his years of experience on the shop floor, did not come from a blue-collar worker culture.

HENRY NOLL, WHOM TAYLOR MADE FAMOUS AS "SCHMIDT"

Samuel Gompers, a labor leader, said of Taylorism and similar philosophies of management:

> So, there you are, wage-workers in general, mere machines—considered industrially, of course. Hence, why should you not be standardized and your motion-power brought up to the highest possible perfection in all respects, including speeds? Not only your length, breadth, and thickness as a machine, but your grade of hardness, malleability, tractability, and general serviceability, can be ascertained, registered, and then employed as desirable. Science would thus get the most out of you before you are sent to the junkpile.[18]

One of the most publicized setbacks for Taylorism took place at the Watertown Arsenal when Carl G. Barth, a prominent Taylor follower and a consultant on scientific management, tried to introduce the Taylor system. Serious trouble started in the foundry when one of Barth's associates began stopwatch-timing the men's work procedures. The skilled workers in the shop discovered that the man carrying out the study knew little about foundry practice. The foundrymen secretly made their own time study of the same work process and complained that the time specified by the "expert" was uninformed and represented an unrealistic speed-up. The Watertown project was also flawed because Taylor's practice was to reorganize and standardize a shop before doing time-and-motion studies, and this had not been carried out at Watertown Arsenal. On the evening after the initiation of the stopwatch studies, the workers met informally and in a petition to the commanding officer of the arsenal they stated:

> The very unsatisfactory conditions which have prevailed in the foundry among the molders for the past week or more reached an acute stage this afternoon when a man was seen to use a stop watch on one of the molders. This we believe to be the limit of our endurance. It is humiliating to us, who have always tried to give to the government the best that was in us. This method is un-American in principle, and we most respectfully request that you have it discontinued at once.[19]

A TAYLOR PLANNING DEPARTMENT RATIONALIZED THE WORK PLACE AND PROCESS.

When stopwatch timing continued, the molders walked out on 11 August 1911.

Promised an investigation of the "unsatisfactory conditions," the molders returned to work after a week, but the publicity given a strike against the U.S. government intensified, fermenting union opposition to scientific management, specifically Taylorism, at Watertown and at another U.S. arsenal, at Rock Island, Illinois. August brought the formation of a special congressional committee of three to investigate scientific management in government establishments. The committee took extensive testimony from Taylor, among others. He became so exercised by hostile questions that his remarks had to be removed from the record. The report of the committee did not immediately call for any legislation. In 1914, however, Congress attached to appropriations bills the proviso that no time studies or related incentive payments should be carried out in government establishments, a prohibition that survived for over thirty years. Yet Taylorism involved, as we have seen, more than time studies and incentive payments, so work processes in government establishments continued to be systematically studied, analyzed, and changed in ways believed by management experts to be scientific.[20]

Worn down by the never-ending opposition and conflict, Taylor moved in 1902 to a handsome house in the Chestnut Hill area of Philadelphia. He no longer accepted employment or consulting fees but announced that he was ready to advise freely those interested in Taylorism. When he had serious inquiries from influential persons, he often invited them to his home, Boxly, lectured to them, often for an hour or two, and then arranged for them to visit several plants in Philadelphia. Among the plants was the Link-Belt Company, where the Taylor system had been successfully implemented. He might also show a particularly welcome guest his success at Boxly in the use of systematically organized labor—including his own—to landscape the grounds, or the specially designed golf clubs that he used on the local course.

Free from the confrontations in the workplace, Taylor dedicated himself to showing that his philosophy of management would ultimately promote harmony between management and labor. He argued that increasing production would increase wages and raise the national standard of living. His principles of scientific management struck responsive chords in a nation intent on ensuring economic democracy, or mass consumption, through mass production and also on conserving its natural resources. Taylor wrote that maximum prosperity could exist only as a result of maximum productivity. He believed that the elimination of wasted time and energy among workers would do more than socialism to diminish poverty and alleviate suffering.

Because of his firm belief that his method was objective, or scientific, he never fully comprehended the hostile opposition of aggressive, collective-bargaining labor-union leaders. He found the unions mostly standing "for war, for enmity," in contrast to scientific management, which stood for "peace and friendship."[21] Nor could he countenance unenlightened and "hoggish" employers who either found his approach and his college-educated young followers unrealistic or were unwilling to share wholeheartedly with the workers the increased profits arising from scientific management. He considered the National Manufacturers Association a "fighting association," so he urged his friends in scientific management to cut all connections with it and its aggressive attitudes toward labor unions. Firmly persuaded that conflict and interest-group confrontations were unnatural, he awaited, not too patiently, the day when management and labor would realize, as he, that where the goal was increased productivity there were discoverable and applicable scientific laws governing work and workplace. Scientific managers were the experts who would apply the laws. He wrote:

> I cannot agree with you that there is a conflict in the interests of capital and labor. I firmly believe that their interests are strictly mutual, and that it is practicable to settle by careful scientific investigation the proper award that labor should receive for the work it renders.[22]

Their interest was not only a mutual but a national one—production and democracy. Production and Democracy. Taylor's times were not ones of affluence for workers, so his means to the end of mass production, thereby raising living standards of the masses, seemed in accord with democratic principle. Within a few years, Vladimir Lenin argued that Taylor principles accorded with socialism, as well.

Taylor became nationally known when Brandeis, the Boston "people's lawyer," argued in 1911 that scientific management, especially Taylorism, could save the nation's railroads so much money that the increased rates that the railroads were requesting from the Interstate Commerce Commission would not be needed. Since the rate hearings were well publicized, writers from newspapers and magazines descended on Taylor to find out about his system and then, at his suggestion, visited Philadelphia plants to see firsthand Taylorism in practice. The favorable publicity induced Taylor to write that "the interest now taken in scientific management is almost comparable to that which was aroused in the conservation of our natural resources by Roosevelt."[23]

Taylor rightly associated his scientific management with the broader conservation movement that had attracted national interest and support

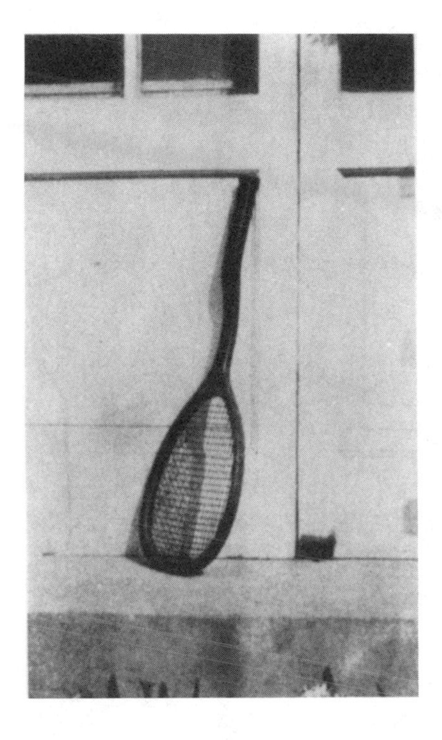

TAYLOR INVENTED A TENNIS RACQUET THAT WAS UNDOUBTEDLY LABOR-SAVING.

during Theodore Roosevelt's terms as president, 1901–08. This progressive program for conservation focused on the preservation and efficient utilization of lands and resources. Like scientific management, it advocated that decisions about conservation be made scientifically by experts. Like Taylor, the progressive conservationists did not countenance as inevitable conflict of interests among ranchers, farmers, lumbermen, utilities, manufacturers, and others. To the contrary, they believed that such conflict was regressive, that it must be displaced by a scientific approach expected to bring harmonious and rational compromises in the general interest. This approach expressed a technological spirit spread by engineers, professional managers, and appliers of science, a belief that there was one best way. College-educated foresters, hydraulic engineers, agronomists should be, the progressives argued, the decision makers about resources; professional managers about the workplace.

Taylor and the growing number of his followers wrote books, published articles, gave lectures, and acted as consultants. He authorized C. G. Barth, H. K. Hathaway, Morris L. Cooke, and Henry L. Gantt to teach

LILLIAN GILBRETH (1878–1972)

his system of management: "All others were operating on their own."[24] Frank Gilbreth, among those who operated "on their own," became well known for his *A Primer of Scientific Management* (1914) and for the use he and his wife, Lillian Gilbreth, made of the motion-picture camera to prepare time-and-motion studies. Her contribution to scientific management has yet to be generally acknowledged. She, not her husband, had a Ph.D. degree in psychology (Brown University, 1915). Perhaps because of her study of psychology, she sensitively took into account complex worker characteristics. The Gilbreths' articles on scientific management show the influence of her concern that the worker should not be seen simply as a component in a Taylor system. After her husband's death, she continued her consulting work and served as a professor of industrial management at Purdue University.[25]

FORDISM

Ford denied that Taylor and his disciples had inspired him when he presided over the creation of a massive system of production. Flow characterized his automobile system, too, but moving assembly lines, conveyor belts, gravity feeds, and railroads, not workers and foremen, constituted the materials-handling network. Ford, unlike Taylor, did not need detailed schedules and routing instructions to direct the movements of materials and work across the shop floor. Ford and a few like-visioned mechanics and self-educated engineers created at his Highland Park plant a system of mass production unlike any the world had ever before seen. They established a finely directed, controlled, and steady flow of energy and materials on a scale then unprecedented. At Highland Park, from about 1910 to 1913, Ford experienced spontaneous teamwork and brilliant *ad hoc* innovation. He displayed the unconscious and inspiring leadership that he longed later—in vain—to recapture. The few years when Ford and a band of enthusiasts, sure-handed, keen-eyed, and ingenious, created the assembly line recall the similar creative exhilaration of Edison and his men at Menlo Park.

Historians and biographers have offered countless explanations for the remarkable achievement of Ford and his men at Highland Park. Siegfried

Giedion, the historian of mechanization, attributes Ford's introduction of the moving assembly line—the best-known component of his mass-production system—to an analogy that Ford drew with the moving disassembly lines of Chicago meat packers. Others believe that he knew about the moving-line production used in the manufacture of tin cans. Some conclude that Ford must have been aware of the various types of moving conveyors, such as the gravity slides used for centuries in flour mills, when he introduced conveyor systems to feed the moving assembly lines. The critical idea of the layout of machine tools to facilitate the flow of production through the factory may have come to Ford first through persons familiar with the best practices in New England machine shops. His insistence on interchangeable parts continued in a long tradition reaching back early in the nineteenth century in the United States to manufacturing at army arsenals.

The list of likely precedents and stimuli can be extended but, unaccountably, the Ford historians have overlooked another likely explanation for Ford's fixation on the flow of production. From 1891 until 1899 he worked for the Edison Illuminating Company of Detroit, becoming the chief engineer of the company's Washington Boulevard electric-power station. Even though Ford's job gave him responsibility for technical, not organization and economic, problems, as a notably alert and curious man he probably absorbed the fundamentals governing the production and consumption of electricity. From Alex Dow, who headed the utility company after 1896, Ford could have learned much, for Dow was destined to be recognized as one of the most innovative U.S. utility managers. Ford learned, perhaps by osmosis, that electricity continuously flows and cannot be stored. For this reason it was essential that demand and supply move hand in hand. (Later he insisted that dealers take his cars as they moved off the assembly lines.) He also saw that electricity supply involved a seamless network, or system, of interconnected machines, transmission, and communication facilities. Electrical engineers usually referred to their "systems." Progressive utility managers advocated the economies of large-scale production machines and power plants, low prices to encourage mass consumption, the cultivation of a widespread market, and continuous flow of production to reduce costs.[26] When Ford spokesmen reflected on the production system that evolved at Highland Park and River Rouge, a system that also depended on constant flow, mass demand, and mass supply, they said that Ford's guiding principles were power, accuracy, economy, system, continuity, and speed.[27] A newsman describing the new Ford plant at Highland Park about 1914 spoke for other reflective observers when he identified its salient feature as "System, system, system."[28] The Ford and electric-utility approaches are too much alike to ignore the strong possibility that he absorbed some of the electric-utility style of production when he was an engineer at the Edison company in Detroit. The search, however, for priority and key individuals becomes less important if we remember that mass-production and mass-consumption principles permeated the American industrial and social environment about the turn of the century.

From about 1909 through 1913, Ford and his mostly young assistants engaged intensively in designing the Model T and the system for producing it. Afterward the inventive process of improvement, especially in the system of production, continued. Because so many writers have studied and written Ford history, we can identify the contributions of some of his mechanics and engineers and need not fall into the error of portraying Ford as a heroic figure leading but not learning from others. Ford had an uncommon gift for, or was simply lucky in, attracting mechanics who considered creative work play. Charles "Cast-iron Charlie" Sorensen had been a foundryman and brought ingenious ideas from that experience; Walter E. Flanders, a machine-tool salesman whom Sorensen believed to be a "roistering genius," brought to Ford the lore and craft of the Yankee mechanic thoroughly familiar with machine tools, the critical components in the manufacturing process. Flanders, who as a machine-tool salesman had acted as something of a cross pollinator in moving from company to company, taught Ford that the essence of the motorcar business should be the fusion of the art of buying materials, the art of production, and the art of selling.[29] When Ford purchased the Keim company to acquire its labor-saving techniques involving stamping instead of casting parts, he also acquired the services of William Smith, its superintendent and part owner of the company. For a time William Knudsen, who later became head of General Motors, helped design the

SOME OF THE CREATORS OF THE FORD ASSEMBLY LINE, 1913. CHARLES SORENSEN IS SEATED AT FAR LEFT.

Ford system of production. The list is long, and the men on it changed as other companies wishing to learn Ford methods hired them away.

HIGHLAND PARK AND RIVER ROUGE

The design and layout of the Highland Park plant, which first produced the Model T, has attracted more general interest than the designing of the car itself. Albert Kahn, who became the most noted factory architect of the day, designed the plant that came to be known as the Crystal Palace because of its great expanse of windows. Ford, his engineers, and his mechanics laid out the machinery. Memoirs of participants tell us that they had no hard-and-fast responsibilities, no well-defined chain of com-

RIVER ROUGE PLANT

mand, no painstakingly worked-out set of instructions—they simply threw themselves wholeheartedly into solving the problems of production. Ford was the leader in the sense that he, much more than any other automobile manufacturer in Detroit or in the world, possessed the unswerving and overarching commitment to the mass production of an automobile for mass consumption. The Ford men became known for designing the best special-purpose machines in the world, laying them out along with their materials-handling network for a smooth flow of parts through the plant. In 1913, *annus mirabilis*, the dramatic step forward in production technique came when Ford and his associates introduced moving assembly lines for magnetos, engines, and transmissions. By early 1914 the chassis was also moving along complex assembly lines. With various conveyor systems carrying subcomponents to the assembly lines, with railroad lines constantly moving materials into the plant, and with dealers throughout the country supplying eagerly waiting Americans, the Ford system could be portrayed metaphorically as a great flowing tide of production.

In planning the great River Rouge plant, which displaced Highland Park in the 1920s as the heart of the Ford system, Ford again fulfilled his near-fixation on flow. He worked once more with Kahn the architect, and with Sorensen, Knudsen, and others of his lieutenants. Ford, no verbal or blueprint man, insisted on having scale models of machine tools, conveyors, windows, pillars, and floor space, so that these could be moved around to test ideas about production.[30] Today this can be done with complex computer models that reveal where there will be obstructions to flow. Between 1922 and 1926 Kahn designed and had constructed at the Rouge site a coke-oven plant, a foundry, a cement plant, an open-hearth steel plant, a motor-assembly building, and several other plants. Not only for its engineering aspects, but also for its aesthetic inspiration, River Rouge was the most important industrial complex of its day.[31]

The massive array of facilities at Highland Park and River Rouge existed because of Henry Ford's determination to sell his Model T to average Americans, especially to millions of farmers. Without warning, one morning in 1909 Ford announced that in the future his Ford Motor Company would build only the Model T. In the five succeeding years of increasing production efficiency and savings, he cut the price of the basic car from $900 to $440, well below the price of the nearest comparable automobile. The average monthly total of unfilled orders swelled to almost sixty thousand.[32] In 1921 the Ford company had a fifty-five-percent share of the automobile market. Production of the Model T climbed to its peak in 1923 with production of two million cars and trucks. Before changing over to the Model A in 1927, the Ford company had produced more than fifteen million Model T's. At the start, production took twelve and a half hours for one car; by 1925 cars rolled off the assembly line at half-minute intervals. Allan Nevins and Frank Hill, authors of the seminal work on Ford and the company, wrote:

RIVER ROUGE BLAST FURNACE

. . . by 1926 the entire productive activity of the company had been impressively developed. Raw materials were now flowing from the iron mines and lumber mills of the Upper Peninsula, from Ford coal mines in Kentucky and West Virginia, and from Ford glass plants in Pennsylvania and Minnesota, much of the product traveling on Ford ships or over Ford-owned rails. Ford manufacture of parts had been ex-

> panded—starter and generators, batteries, tires, artificial leather, cloth, and wire had been manufactured by the company in increasing quantities. The Rouge was producing coke, iron, steel, bodies, castings, engines, and other elements for Highland Park and the assembly plants, and also manufacturing the full quota of Fordsons [tractors].[33]

When he and his band of bright young mechanics and engineers were designing and redesigning the legendary Model T and creating the famed production system, Ford flourished as a visionary and a problem solver. In the 1920s, after the mammoth River Rouge plant had been placed in operation and while the Model T continued to be produced year after year, Ford became part of the management problem rather than of the solution. Even though he had an instinctive grasp of the fundamentals of mass production and mass consumption in a capitalistic society, he did not understand or appreciate managerial organization and the essential managerial practice of cost accounting. He even reduced the administrative staff needed for information and control of a large organization. As a result, production and sales were poorly coordinated. He made annual decisions about the price of the Model T on crude estimates of profits. The myth persists that the decimated accounting staff estimated costs by sorting bills into broad cost categories, estimating the average amount of each bill in the category, or pile, and then measuring the height of the pile to get a total. An aging Ford was trying to lead an aging bureaucracy manufacturing an aging automobile as if it were a team of enthusiastic mechanics and engineers solving the problems of a rapidly evolving system of mass production to bring a new automobile into the world.

In recent years historians and biographers have found Ford and his company in decline as interesting as Ford on the rise. His reluctance to abandon the Model T from 1908 to 1927 has become part of the legend of Ford "the destroyer"[34] and despotic obsessive.[35] By the time he accepted the changeover to the Model A, the Ford share of the automobile market had dropped to thirty percent. Anecdotes abound about his ossifying and then deteriorating personality, and about the chaotic managerial policies

ASSEMBLY-LINE WORKERS, RIVER ROUGE

he tolerated, even encouraged, in the name of flexibility but exploited in the spirit of an authoritarian figure who intervened at will in the absence of a managerial structure and routine. William Knudsen, considered one of the ablest production men in the company, discussed with Ford's son

Edsel possible improvements in the Model T. Learning of this, the infuriated Henry Ford regularly countermanded Knudsen's orders and humiliated him. Knudsen resigned in 1921 and moved to General Motors, where he quickly took over the Chevrolet division that within a few years took the major market share from Ford. Later, when Edsel Ford and the company's chief engineer had designed and built a model of a six-cylinder engine as a needed replacement for the Ford four-cylinder, Henry Ford requested the chief engineer to accompany him to see a new scrap-conveyor. Riding on the conveyor, destined for destruction, was the new engine. Also cited as an example of Ford's irrational behavior was his dismissal of experts. He wrote in *My Life and Work* (1922), "We have most unfortunately found it necessary to get rid of a man as soon as he thinks himself an expert—because no one ever considers himself expert if he really knows his job."[36] Sorensen, who echoed Ford's state of mind, said, "When one man began to fancy himself an expert, we had to get rid of him. The minute a man thinks himself an expert, he gets an expert's state of mind, and too many things become impossible."[37] Another Ford company executive recalled that if Henry Ford wanted a job done right he would always choose the man who knew nothing about it.[38] One after another the able men left, until, in the 1930s, Henry Ford was reduced to relying on Sorensen, who was unwilling to counter Ford's obvious mistakes, and Harry Bennett, a one-time prizefighter who used bullies to control the plant and keep out the union.[39]

Despite Henry Ford's behavior tactics, Edsel, a mild-mannered and intelligent man experienced in the automobile industry, and many of those who had worked with Henry Ford in the exhilarating days when the Model T and Highland Park were created, remained loyal as long as they could to him and to the legend. In 1926 Ernest Kanzler, a talented Ford production chief and brother-in-law of Edsel, tried tact, flattery, adulation, and reason to persuade Henry Ford to consider the manufacture of the six-cylinder engine. In a seven-page memorandum to him cautiously proposing changes, Kanzler expressed concern that even to suggest change might affect "your feeling for me, and that you may think me unsympathetic." He added diplomatically, "Please, Mr. Ford, understand that I realize fully that you have built up this whole business

FORD EXECUTIVES, C. 1925: E. C. KANZLER (FRONT ROW, THIRD FROM LEFT), SORENSEN (SECOND ROW, FIFTH FROM LEFT), EDSEL FORD (THIRD ROW, SECOND FROM LEFT), HENRY FORD (THIRD ROW, SEVENTH FROM LEFT)

. . . that all the company's successes . . . will really be your personal accomplishment . . . even after your lifetime." Kanzler then risked mentioning that among most of "the bigger men in the organization there is a growing uneasiness. . . . They feel our position weakening and our grip slipping." "The buoyant spirit of confident expansion is lacking." Not long after, Henry Ford fired Kanzler.[40] In the same year, the Ford company share of the market had fallen to about one-third. The fabled Model T, despite changes, no longer fulfilled the dreams of a car-hungry public. Ford jokes about the Model T took a less tolerant form:

> The Ford is my auto / I shall not want another.
>
> It maketh me to lie down beneath it, / It soureth my soul.

> It leadeth me in the path of ridicule, / For its name sake.
>
> Yea, though I ride through the valleys, / I am towed up the hills, / For I fear much evil.
>
> Thy rods and thine engines discomfort me.
>
> I anoint my tires with patches; / My radiator runneth over.
>
> I repair blowouts in the presence of mine enemies.
>
> Surely, if this thing followeth me all the days of my life
>
> I shall dwell in the bug-house forever.[41]

A few of the widely told anecdotes about Henry Ford also took on derogatory twists that suggested lack of learning and eccentricity of intellect. Ford, for example, told reporters he believed in reincarnation, and as evidence noted that chickens had once run into the paths of oncoming automobiles but more recently had stayed by the side of the road. He explained that the roadwise chicken "had been hit in the ass in a previous life."[42]

His growing eccentricity and gradual transformation from inspired system builder to arbitrary, mean-minded, and ineffectual manager offer a major explanation for the decline of his company.[43] In fact, the change was not so much in Ford's personality as in the company problems he faced: not in the man, but in his environment. His solutions were for past problems. Ford's autocratic behavior and his dismissal of experts can be seen, on the one hand, as indicating the increasing rigidity and domineering nature of an aging man. Another explanation, however, reveals his longing for the exhilaration of creative activity and problem solving that he had known earlier as an inventor and a system builder. When he and his team were creating the Model T and the Ford system of production, there were no lines of authority, routine procedures, or experts. Theirs was a resourceful, ingenious, hunt-and-try probing into the unknown future. Edison, whom Ford revered and with whom he enjoyed a close friendship, also rejected experts, especially those with university

HENRY FORD AND THOMAS EDISON

degrees. Edison associated them with inappropriate theories drawn from past experiences. Ford and Edison understood that there were no experts about the unknown; no theories, only hypotheses or metaphorical insights, about the uninvented. Edison even withheld information about prior work on a problem from his assistants, fearing that sharing such information would move them into a particular track and close their minds. He subjected his expert assistants with advance degrees to crude ridicule, hoping to destroy what he considered their smug and unwarranted self-assurance. In common in Ford and Edison's attitudes, we find prejudice and ignorance but also a shrewd understanding of the freewheeling nature of invention and innovation. Unfortunately for Ford and his company, he continued to advocate a leadership style suited for times of invention and great change long after the Ford company had become an extremely large and a relatively stable managerial and technical system with high inertia. Ford would not, or could not, make the transition in leadership style from the inventive stage to the managerial. His attack on a bureaucracy was, in the context of a bureaucracy, irrational behavior. For the company, it would have been a blessing if he had resigned after the Highland Park system of production had stabilized around 1915. Elmer Sperry, on the other hand, regularly left the companies he had helped to found with his inventions. Perhaps Ford sensed his incapacity, even distaste, for management when he considered selling the Ford Motor Company three times between 1908 and 1916. In the style of the inventor, he spoke of entirely new ventures he would undertake if he were relieved of the burden of routine.[44] He found himself bored and constrained by the very system of production he had enthusiastically created. His lack of self-awareness is ironic in view of what he wrote in *My Life and Work*: "Business men go down with their businesses because they like the old way so well they cannot bring themselves to change."[45] He could not understand that he, too, was resisting change from his own old way, invention and innovation, to an appropriate style of management dependent not on radical invention and innovation, but on incremental, slow-paced improvement, growth, and systematization.

Ford's unwillingness to make substantial changes in the Model T can be better understood if we remember that for years he held the large share of the market by regularly reducing the cost of production and the price of the automobile. To reduce costs, his policy was to introduce changes in the mode of production rather than in the product. He wrote:

> Our big changes have been in methods of manufacturing. They never stand still. I believe that there is hardly a single operation in the making of our car that is the same as when we made our first car of the present model. That is why we make them so cheaply. The few changes that have been made in the car have been in the direction of convenience . . . [or] added strength.[46]

He believed that a change such as a six-cylinder engine would force price increases. Ford also confessed that "one idea at a time is about as much as any one can handle." The Model T with a four-cylinder engine was his idea, one that he intended to perfect.[47] Despite his unwillingness to make Model T changes, his longing for the creative experience was demonstrated by other ventures on which he embarked after 1913. Ford the system builder flourished again when he, with Sorensen at his side, planned and constructed the River Rouge plant after World War I. He also revealed his imagination and foresight when he tried, in the early 1920s, to create an industrial complex in the Mississippi Valley and a decentralized system of production at waterpower sites outside of Detroit. The contradictions and complexities in Ford's behavior can be better understood only if we perceive the irony of the creative person engaged in system building. Thomas Mann, in his novel *Doctor Faustus*, captured the ironical essence when he had his protagonist Adrian Leverkühn, creator of a twelve-tone musical system, express a longing for a method of composition. This was a request to which Mephistopheles eagerly assented, for the composer would create a system that would then become an iron cage preventing his further free expression.

Henry Ford's relations with labor as well as with his managers deteriorated in the later years. Like other system builders of his era, he insisted on control, order, and system for workers while fighting it off for himself. For machines he designed, this was no emotional, psychological burden. For men, especially the workers, it was different. Men tending the machine tools and on the assembly line had to conform to the rhythm and logic of machine production. Ford, like Taylor, saw them as components in the system of production, but, also like Taylor, he believed that a well-functioning worker should receive some part of the cost savings for

WORKERS AT FORD COMPANY LEARNING ENGLISH

which he was responsible. Ford wanted to spread the national income among the grass roots, and he also wanted workers to stimulate the market for automobiles and other mass-produced goods. When possible, however, he and his engineers replaced the workers with more easily directed machines. Machines, unlike men, could be designed especially for the function to be performed. They did not strike, and independent thinking did not lead them to vary their work methods from those prescribed by production engineers and planning departments. On the other hand, Ford had to consider that the workers could be laid off in a recession but investment in machines was fixed. Creative and skilled work was done by a relatively few engineers and tool-and-pattern makers; the mass of Ford labor was unskilled and included thousands of newly immigrated Hungarians, Poles, Serbians, Armenians, Bohemians, Russians, Rumanians, Bulgarians, and Italians. A touching photograph from the early years at Ford shows workers diligently studying elementary English during their lunch period. They learned their simplified, routine, and highly specialized tasks in several days, and they, like the parts of the Model T, were easily replaceable.

The workers, however, found the mechanized system of production in place at Highland Park by 1913 so wearing and depersonalizing that the turnover reached 380 percent. If the company wished to add a hundred workers, 963 had to be hired. Signs of unionism alarmed Ford and his staff. The flow of production was adversely affected. In 1914 Ford, disturbed not only by the labor turnover and unionism but also by the great discrepancy in spread between the salaries of his executives and the wages of the workers, decided on an unprecedentedly high five-dollar-a day wage. Job applicants lined up before the factory gates. An anonymous wife of a Ford worker wrote to him:

> The chain system you have is a slave driver. My God!, Mr. Ford. My husband has come home & thrown himself down & won't eat his supper—so done out! Can't it be remedied? . . . That $5 a day is a blessing—a bigger one than you know but oh they earn it.[48]

The decline of the Ford company cannot be attributed solely to Henry Ford. The company's decline was relative to that of other automobile manufacturers, especially the General Motors Corporation under the leadership of Alfred Sloan. By 1927 General Motors had forty-five percent, the leading share of the automobile market, which it did not relinquish. In 1931 the Ford market share had dropped to twenty-six percent, with losses amounting to more than $50 million.[49] Sloan, president of General Motors, introduced consumer credit in 1919, used-car trade-ins, a closed car, and an annual model.[50] (Ford dismissed the costly annual model change with thinly veiled contempt, recalling that earlier it had been common among bicycle manufacturers.[51]) General Motors also depended on general-purpose instead of special-purpose machine tools, thereby facilitating basic model changes such as the replacement of a four- by a six-cylinder engine. Additionally, Sloan, following managerial practice at the Du Pont Company, installed a multidivisional, decentralized management structure that was becoming a model for large industrial firms. Sloan carried Ford's policy of low inventories to a much higher stage of managerial expertise when he introduced control of factory flows. These flows were based on statistical feedback derived from dealer information sent every ten days about orders, deliveries, and new and used cars on hand. General Motors also developed the art and science of long-range forecasting of sales and of allocation of resources.

AUTOMOBILE PRODUCTION AND USE SYSTEM: PETROLEUM REFINING

At the peak of his powers, Ford controlled a highly complex system of automobile manufacture that spread throughout the United States and into other regions of the world. Despite the extent of its holdings, Ford's motor company was only a component in a greater production system, or network, that involved an even larger array of organizational, production, supply, and service activities.[52] Besides the automobiles, this automobile production-and-use system involved physical components like the automobile, roads, and service stations, as well as people and organizations like automobile manufacturers, the suppliers of raw materials and components to the manufacturers, the unions organizing the automobile workers, the dealers selling the automobiles, the suppliers of gasoline, the operators of a network of service stations, the public authorities constructing highways, the organizations financing car purchases, and numerous other organizations such as advertising agencies stimulating the market. No individual or organization could centrally organize and control the automobile production-and-use system, but in diverse and complex ways a level of coordination was achieved. Ford's visible hand coordinated the system he created; the invisible hand of the market, along with a variety of informal institutional and personal ties, coordinated automobile manufacture and petroleum refining.

Karl Marx in *Capital* showed how increased production by weavers during the British Industrial Revolution stimulated an increased output of spinners and how, in turn, technical improvements and output in spinning forced the development of weaving. The systematic interaction followed from the overarching goal of cloth production that inextricably linked spinning and weaving. The increased production of cloth also put pressure on the British chemical industry to find ways to improve the quality and to increase the quantity of bleaches and dyes. Similarly, in the twentieth century, automobile production has been inextricably interwoven with the refining of gasoline. In both cases, the concepts of mass production and flow dominated the producers.

During the nineteenth century kerosene for lamps was the chief product of the petroleum refiners. The gasoline fraction, or component, of petroleum was a waste product. The dramatic rise in automobile use after the turn of the century brought a dramatic change—the refiners needed to find ways to increase the yield of gasoline, the lighter fraction of petroleum. Between 1909 and 1913 Dr. William M. Burton, a Ph.D. in chemistry from The Johns Hopkins University and refinery superintendent with Standard Oil Company (Indiana), developed the process of thermal cracking. In contrast to an earlier distilling technique that involved heating petroleum in open vessels to drive off the various fractions, beginning with the lighter ones and proceeding to the heavier, Burton heated the raw petroleum in a closed vessel. The pressure that built up in the vessel broke down, or cracked, the heavier molecules of the heavier fraction—once used for kerosene—into lighter, or gasoline, molecules. When he received the Perkin Medal in 1921, Burton recalled that his success in almost doubling the yield of gasoline from a barrel of crude came from his foolishness in heating oil under pressure despite the obvious danger of an explosion. By 1920 other cracking processes stimulated by Burton's success rendered his obsolete, but the Burton process brought Standard Oil of Indiana profits of $150 million.[53] This success also brought other refiners to invest in chemists and chemical engineers. (When young Burton had first arrived at Standard Oil of Indiana with his Ph.D., his superintendent asked him where his tools were.)

Other refiners burdened by license fees for the Burton process turned to engineers and research scientists for alternative ways of responding to the steadily rising demand for gasoline. Because of dire predictions in the early 1920s of an early exhaustion of the world's supplies of crude oil, the need to increase the amount of gasoline from a barrel of crude appeared critical. In Germany scientists and engineers sought ways to derive gasoline from coal. Authorities estimated that petroleum supplies would be exhausted within fifteen years. Others saw a good possibility for improving the yield from the Burton process by increasing the rate of flow of materials through the refinery. Petroleum refiners, emulating Ford and Taylor, became obsessed with the need to systematize and increase the flow of production. Burton's was a batch process, in contrast to continuous flow. Petroleum was statically contained in tanks, or stills, as it was processed. Afterward refinery workers removed the products and recharged the still with another batch. As had been the case with the steam engine before James Watt introduced his separate condenser, the cylinder—or still, in the case of the Burton process—had to be alternately heated and cooled, with attendant wastes. The answer to the static nature of the batch process was one of continuous flow, in which petroleum passed through a stage where various transformations, or unit operations, occurred because of heating, cooling, and so on. Similarly, Ford had moved the evolving automobile past workers at fixed stations, where they carried out particular functions. No wonder that an American engineer visiting China about this time thought the principal differences between the two countries were that in the United States everything and everyone was in motion.

The Universal Oil Products Company introduced the continuous cracking process of a young inventor named Carbon Petroleum Dubbs. (Jesse Dubbs, a Massachusetts Institute of Technology graduate and also a pioneer in the petroleum industry, had given his son this name.) J. Ogden Armour, whose family fortune had been made in meat packing, invested in the Dubbs process because he saw a parallel between the earlier introduction of a cost-saving continuous disassembly line in the abattoir and continuous cracking in the petroleum industry.[54] The Dubbs process involved a number of stations in sequence. Heat and pressure remained constant at each station as the oil passed through and the products of the cracking process were sequentially removed. Having proved more economical than the Burton process, by 1924 the Dubbs process was displacing it throughout the industry. With an increasing allocation of money for research and development throughout the petroleum industry, the Burton and Dubbs processes proved to be only the beginning of a series of improvements resulting in greatly increased gasoline yield from the stock of crude oil.[55]

Before 1930 the oil refiners and their chemists and engineers, worried about an energy crisis, concentrated on increasing the yield rather than the quality of gasoline. But after the discovery in late 1930 of the rich East Texas oil fields, their emphasis shifted to quality. The East Texas fields had more than three thousand producing wells by December 1931. Greater production kept the ever-increasing number of automobiles on the road, but improved quality allowed higher-compression engines to move the autos down the highways faster and more efficiently. Even before the bountiful oil of East Texas solved the oil shortage, however, Charles Kettering and his gifted associate Thomas Midgley, Jr., cooperated with the automobile manufacturers to reduce engine knock and allow higher engine compression. During World War I Kettering, whom we last encountered watching his flying bomb soar out of control to dizzy heights, had also tried to improve the quality of gasoline for airplane engines and thereby to reduce knocking—a sharp ringing sound also familiar to early automobile owners when their engines were working hard.[56] Not only annoying and an indication of inefficient combustion, knocking could become physically destructive. Employed by Kettering, who headed an independent research lab, Midgley, a mechanical-engineering graduate of Cornell University, soon after the war took the lead in the search for a gasoline additive to reduce knocking. The resulting search, which continued after Kettering became head of the newly formed research-and-development division of General Motors in 1919, provides an insight into the combination of empirical method and systematic research common in the industrial laboratories of the day. Midgley attributed his eventual success "in part to luck and religion, as well as to the application of science."[57]

Probing the unknown, Midgley and Kettering assumed that the fuel's low volatility caused the knocking. Then Midgley invented by analogy. He remembered that one of the first-blooming spring flowers was trailing arbutus, which had red-backed leaves, and he assumed that the early blooming resulted from red's absorption of the early-spring heat. So he tried dyeing gasoline red with iodine, the only red coloring substance in his storeroom. He hoped that this would increase volatility and reduce knocking. Iodine reduced knocking, but not because of its red color. It also, as Midgley observed, had a "slight" drawback: iodine changed the cylinder into a salt factory, with the cylinder walls as the raw material. Over several years Midgley tried more than thirty-three thousand compounds in his search for an antiknock additive. He called this hunt-and-try an Edisonian approach, a misnomer common among research scientists who did not know that Edison's approach at Menlo Park often involved a theory-based systematic approach as well as hunt-and-try. Then, by chance, Kettering read a newspaper article reporting a universal solvent, which, he and Midgley noted with amusement, was delivered in a glass bottle. Interested in the inflated claims and open to the slightest leads, they tried selenium compound, the solvent, and found that it did reduce knock. Used as a gasoline additive, however, selenium had an extremely unpleasant odor. He found that, after a day in the lab experimenting with the selenium compound, he had to forgo family, friends, and the evening movie.

Next he resorted to what he called applied science. He made a pegboard of the chemist's periodic table of the elements and began testing various soluble compounds of other elements in the table in the vicinity of selenium. In the board he inserted wooden pegs of a length corresponding to their antiknock properties.[58] Midgley said that he had turned a wild-goose chase into a "scientific fox hunt."[59] The hunt ended with tetraethyl lead, which worked perfectly after a few additives were found to prevent harmful deposits. Developing and producing leaded gasoline in quantity required a complex system involving universities, chemical companies, automobile manufacturers, and petroleum refiners, among them the Dow Chemical Company, General Motors Corporation, Du Pont Chemical Company, Standard Oil Company of New Jersey, Brown University, and the Massachusetts Institute of Technology. A newly formed enterprise, Ethyl Gasoline Corporation, a company formed by General Motors and Standard Oil of New Jersey, marketed the leaded gasoline.[60]

With production, the specter of lead poisoning then arose, and a number of physicians warned of the risk. In 1924 the U.S. Bureau of Mines experimented for months by exposing animals several hours each day to exhaust from an engine running on leaded gasoline. They found "no indication of plumbism in any of the animals used."[61] Then, however, forty-five persons handling the concentrated tetraethyl lead at a pilot plant fell ill, and four died from lead poisoning. In 1925 sales were halted. The U.S. surgeon general appointed a committee to investigate the potential hazard. Chemical authorities decided that, when the distribution and use were controlled by proper safeguards, there was no hazard from gasoline containing tetraethyl lead.[62] "Ethyl" no-knock gasoline then became common at service stations. Decades later the additive was targeted as an environmental hazard.

In the 1930s a French independent inventor working outside the mainstream of petroleum technology developed another major improvement in the quality of gasoline. Eugène Jules Houdry, born near Paris in 1892, received a technical education at the Paris Conservatoire National des Arts et Métier. Auto racing fascinated this son of a wealthy steel manufacturer, and in 1922, when he attended the five-hundred-mile Memorial Day Race at Indianapolis and inspected Ford's plant in Detroit, he concluded that American automobiles were of excellent construction but that the gasoline used was of poor quality. He then realized that advances in automotive-engine design depended on simultaneous advances in petroleum refining.[63] In 1925, drawing on his own and his wife's inherited wealth, he began the search, not for an additive like lead, but for a new way of refining petroleum that would produce a gasoline of high quality. After thousands of experiments, he and his associates found in 1927 that using activated clay as a catalytic agent during the refining process produced such a gasoline. Technical problems remained, but Houdry approached the Standard Oil Company of New Jersey. Finding the Houdry process technically unrefined, the Standard Oil engineers turned down the outside inventor's process. They probably did not know that there was a long history of industrial labs' buying crude devices, like de Forest's three-element tube or Pupin's loading coil, and then greatly improving their technical and economic efficiency. Turning to other

U.S. refiners, Houdry ultimately secured support from the Vacuum Oil Company, and from Sun Oil Company, a relatively small firm known for its innovative spirit. By 1936 Houdry-process gasoline was in the service stations. Then, in 1938, Standard Oil of New Jersey, in alliance with Standard Oil of Indiana, the giant German chemical company I. G. Farben, and M. W. Kellogg Company, a U.S. engineering-construction firm, responded with the formation of Catalytic Research Associates, whose goal it was to improve on the Houdry process without infringing its patents and without paying license fees. A landmark event then, though it is almost forgotten now, the formation of Catalytic Research Associates probably represented the largest *ad hoc*, or single-purpose, concentration of scientific and engineering manpower in the world prior to the establishment of the Manhattan Project. Four hundred engineers and scientists were involved at Standard Oil of New Jersey, and six hundred in other companies.[64] In 1941 the crash program brought forth the "fluidized catalytic cracking process" that exploited the principle of continuous flow more fully than the Houdry process and produced a gasoline of high quality.

INSULL THE SYSTEM BUILDER

Petroleum refiners and automobile manufacturers managed smoothly flowing processes of production, but their products and rates of production did not compare in subtlety and velocity with that of an electric light-and-power system. The latter's product, the moving electron, traveled at the speed of light. An automobile-production system loosely linked countless machines and processes by clanking conveyors, flapping belts, and heavy, traveling cranes. An electrical-supply system was a seamless web of whirring machines and humming transmission lines. Insull, as a master system builder, presided over one of the world's largest and most complex power systems. Ford, his mechanics, and engineers had to allow for the irrational and unmanageable nature of thousands of workers; Insull and his associates felt omnipotent as they manipulated pliable machines and processes in great power plants attended by only a man or two. Insull and other heads of major urban and regional electric light-and-power utilities created systems of mass energy production that preceded the better-known Ford system and anticipated its essential characteristics. Throughout the world, Insull and Commonwealth Edison of Chicago, his principal utility, were respected as setters of standards of efficiency and growth, until Insull's utility holding companies began to collapse during the Great Depression. In the presidential campaign of 1932 Franklin Delano Roosevelt, aware that Insull had become a symbol of financial manipulation to the public, delivered a long attack on him and electric-utility holding companies. He spoke of the "lone wolf, the unethical competitor, the reckless promoter, the Ishmael or Insull whose hand is against every man's. . . ."[65] With reason, Insull's biographer, historian Forrest McDonald, characterizes him as "America's most powerful businessman of the twenties—and its most publicized business villain in the early thirties."[66] Today a respected historian refers to him, because of his financial manipulations, as a "notorious crook."[67] Yet Insull was found not guilty of the charges of financial chicanery for which he was indicted. Before his fall he was an impressively accomplished technological-system builder. In 1934, when he was on trial for using the mails to defraud in connection with his bankrupt holding company, Insull denied that he was a predatory holding-company tycoon, insisting that he, like Edison his hero, was a creative man enthusiastically committed to managing an expanding and productive technology.[68]

Insull learned system building in the Edison school. Edison, he said, "grounded me in the fundamentals. . . . No one could have had a more considerate and fascinating teacher."[69] From a middle-class, dissenting-Protestant background, Insull had emigrated from England, where he had been secretary to Colonel George E. Gouraud, Edison's business representative. On arriving in the United States in 1881, when he was twenty-one, Insull became Edison's personal secretary. From then until 1892, when he became manager of the Edison General Electric plant in Schenectady, New York, Insull witnessed and took part in the formative years of the electric utility and manufacturing industry. He closely observed Edison and helped him create his electric-light system, build the path-breaking electric central station on Pearl Street in New York, and

SAMUEL INSULL, 1885

establish the various plants to build incandescent lamps, electric generators, and distribution cables. Insull sat in on countless meetings where engineers, mechanics, business entrepreneurs, financiers, managers, and others pooled their knowledge and resources to solve the problems of expanding electric light-and-power systems. He absorbed the creative, problem-solving, systematizing, and expansionistic approach of the system builder. He learned from Edison how to solve problems by weaving a web of ideas, artifacts, and people. Insull pleased Edison, and would have Ford, because he was no expert, no specialist. If he had attended an engineering school and been trained as a specialist grounded in science, as was increasingly the case in the expanding engineering colleges of his day, he might never have absorbed the problem-solving approach of an Edison or a Ford. None of the three rigidly sought a mechanical, electrical, or chemical solution. Instead, all ranged widely for answers without respecting disciplinary boundaries. If a technical response did not work, they resourcefully turned to one commonly labeled political or economic.

In 1892, after the Edison company merged with the Thomson-Houston Electric Company, Insull left the newly formed General Electric Company to head the Chicago Edison Company, a small Chicago electric-supply utility. He accepted the challenge of building the system, but with certain conditions. He would not be involved in financing the company, and the directors and stockholders would supply sufficient capital at all times; in addition, the company would build a new power plant and pay for it by issuing $250,000 in stock, all of which would be sold to him. (He borrowed the entire amount from the wealthy Chicago merchant Marshall Field.[70])

Within three decades he had absorbed about twenty other Chicago electric utilities to form the Commonwealth Edison Company, which monopolized the Chicago market and became known as early as 1910 as the world's leading utility. By the 1920s he had interconnected the Chicago system with others in urban and rural areas to create a regional electric-supply network. Then he established the Middle West Utilities Company, a holding company with electric-supply properties throughout the nation. This holding company brought him to the attention of Roosevelt and others campaigning against private utility holding companies. Like other system builders, Insull strove to merge, couple, link, centralize, and control all of the institutions and artifacts that he needed to solve the problems of supplying electricity at low cost to a mass market. To create the electrical empire that he managed in the 1920s, he combined or coordinated physical artifacts, such as electric generators, transformers, and transmission lines, with organizations like utility companies, investment banks, and state regulatory agencies. He saw to it that all of these components interacted effectively, he insisted, to produce electricity efficiently. His critics saw him as creating a monopoly primarily to produce profits. He countered with statistics showing that his companies followed the best American practice and took small unit profits on massive sales of kilowatt hours, while others took large profits on small sales.[71]

Insull's policies show how he simultaneously manipulated a broad range of technical, economic, and political factors. With the technical advice of Sargent & Lundy, a forward-looking firm of consulting engineers, Insull and his staff pioneered in the introduction of steam turbines to replace reciprocating steam engines in central stations. Because turbines represented much larger concentrations of power than the reciprocating steam engines they replaced, Insull and his engineers had to extend the area served by a single massive central station. To do this, he used his considerable political power in Chicago to bring about the enlargement of the franchise held by his company. He also had the imagination to call for state, rather than city and county, regulation of electric supply. This allowed him to extend the area served by his growing electric-supply system to the borders of the state, not simply to the borders of the political jurisdiction of Chicago. Having drawn on his technical and political resources, he then turned to the Chicago stock-and-bond market. The Chicago investment-and-brokerage firm of Halsey, Stuart and Company became a part of the Insull empire. Because of Insull's reputation for management, and because of the profits and expansion of his company, its securities could be sold at lower interest rates. Lower interest rates, in turn, meant lower-cost electricity. He thrived before the onset of the Great Depression, for financing seemed always available, markets insatiable, and cost reductions through improved technology unending.

By the mid-1920s his electricity-and-gas-supplying system—his empire—consisted of Commonwealth Edison, a $400-million company serving electricity in Chicago; Peoples' Gas, a $175-million Chicago gas utility; Public Service of Northern Illinois, a $200-million company supplying gas and electricity in three hundred communities around Chicago; Middle West, a holding company with several hundred subsidiaries representing an investment of $1.2 billion supplying electricity and gas in five thousand communities spread over thirty-two states; and Midland, another holding company, representing an investment of $300 million, supplying gas and electricity in Indiana communities. These and several other enterprises that he controlled and managed amounted to nearly $3 billion in utility properties, with six hundred thousand stockholders and about five hundred thousand bondholders supplying about four million customers with about one-eighth the electricity and gas consumed in the

United States. Insull's personal stock and bond holdings in this empire, however, were not impressive by the standards of his day: his net worth in 1926 at age sixty-seven was about $5 million. "His friends and enemies would have been shocked that it was so little; he could easily have made twenty times that, had he been willing to work for the sake of acquiring it, but since 1912—by which time he had a million pounds sterling—he had simply not been interested in accumulating money."[72] Manipulating and controlling an immense system of things, institutions, and people may have filled his psychological needs more fully than money making.

The technical intricacies and organizational complexities of Insull's creation were too abstract for the public and press to comprehend or to visualize, as they could Henry Ford's automobile empire. Therefore Ford, not Insull, became the world-famous system builder, and he remains so today. The public could see the assembly lines moving, the blast furnaces pouring forth metal, the machine tools cutting, shaping, and turning at Ford's River Rouge plant, but the electricity flowing out of Insull's central stations over thousands of miles of transmission lines to power countless motors driving factories and railroads remained too ephemeral for the public to envisage. Not only did Insull and his lieutenants create a system for mass-producing energy, but they also articulated the concepts of mass production more subtly, more extensively. Today Ford's mechanistic concepts seem familiar, fairly simple, and bear the patina of the era; Insull's remain vital, complex, and applicable in an era that remains essentially electrical. Before the Ford system of mass production was analyzed and widely publicized by Horace Arnold, a technical writer, in *Engineering Magazine* (1914)[73] and more than a decade before the *Encyclopedia Britannica* (1926) published a widely quoted article attributed to Henry Ford on "Mass Production,"[74] Insull summarized his ideas of mass production in a series of public addresses (1897–1914).[75] Other thoughtful utility managers in the United States simultaneously grasped many essential principles of electricity supply and learned from one another, but Insull, with the help of his accounting and planning staff, articulated these principles, so that he became a spokesman for his peers in the United States and abroad. (In the 1920s the British government asked him to preside over the planning of their national electric supply grid.)

By the turn of the century Insull had absorbed the spirit driving the rapid industrialization—a second industrial revolution—in his adopted country. He believed in mass production and mass consumption, and accepted the conditions of capitalism. Translating these into utility policy, he created a dynamic system of production in which flow was the cardinal principle, flow of production from raw materials such as coal to the consumption of kilowatt hours by various consumers. Unlike European utility magnates, he stressed, in a democratic spirit, the supplying of electricity to masses of people in Chicago in the form of light, transportation, and home appliances. In Germany, by contrast, the Berlin utility stressed supply to large industrial enterprises and transportation, but was relatively indifferent to domestic supply to the lower-income groups. In London, utilities supplied at a high profit luxury light to hotels, public buildings, and wealthy consumers.[76] Fully aware that the cost of supplying electricity stemmed more from investment in equipment than from labor costs, Insull concentrated on spreading the equipment costs, or interest charges, over as many kilowatt hours, or units of production, as possible. Much as Ford later pushed the evolving Model T through his production plants as rapidly as possible, Insull processed energy as quickly as possible in his power plants. During the time in plant, the product was absorbing the cost of capital—interest charges. Smooth flow became a fixation for Ford, Insull, their managers and engineers. Flow was smoothest when the means of production were coordinated systematically.

Because electricity could not be economically stored, Insull and the electrical-utility managers felt especially keenly the pressure of maintaining flow of production and consumption. In the case of most manufacturing industries, the product could be stored, or stockpiled, when consumption fell, and products could be fed out from storage when consumption increased. In the case of electricity, customer demand had to be met instantly. When darkness fell on a cold December day while the factories and the transportation system were running at capacity, Commonwealth Edison had to respond. The company had to have elec-

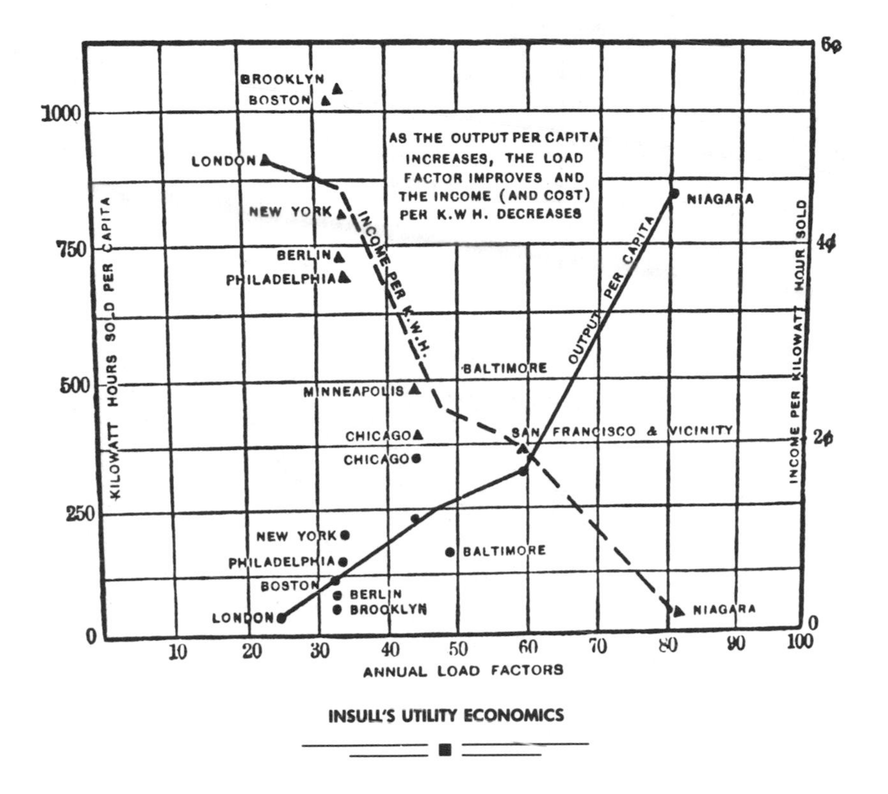

INSULL'S UTILITY ECONOMICS

tric generators with sufficient capacity to meet this peak, even though for the remainder of the twenty-four hours the consumption was lower—in the late evening and early morning hours, much lower. Production and consumption had to be coordinated.

It became customary in the utility industry to draw a curve showing the variation, throughout a twenty-four-hour period, of production of kilowatts of energy. Insull often showed these in his lectures and articles. The "load curve" graphically showed the valleys of low electricity demand and the peaks of high. More generally, Insull and others in the electrical industry realized that the load curve portrayed graphically one of the cardinal realities of a capitalistic society: the relationship between investment and the utilization of investment. In the case of an electrical utility, the curve usually displays a valley in the early morning, before the waking hours, and peaks in the early evening, when business and industry use power, homeowners turn on lights, and commuters increase

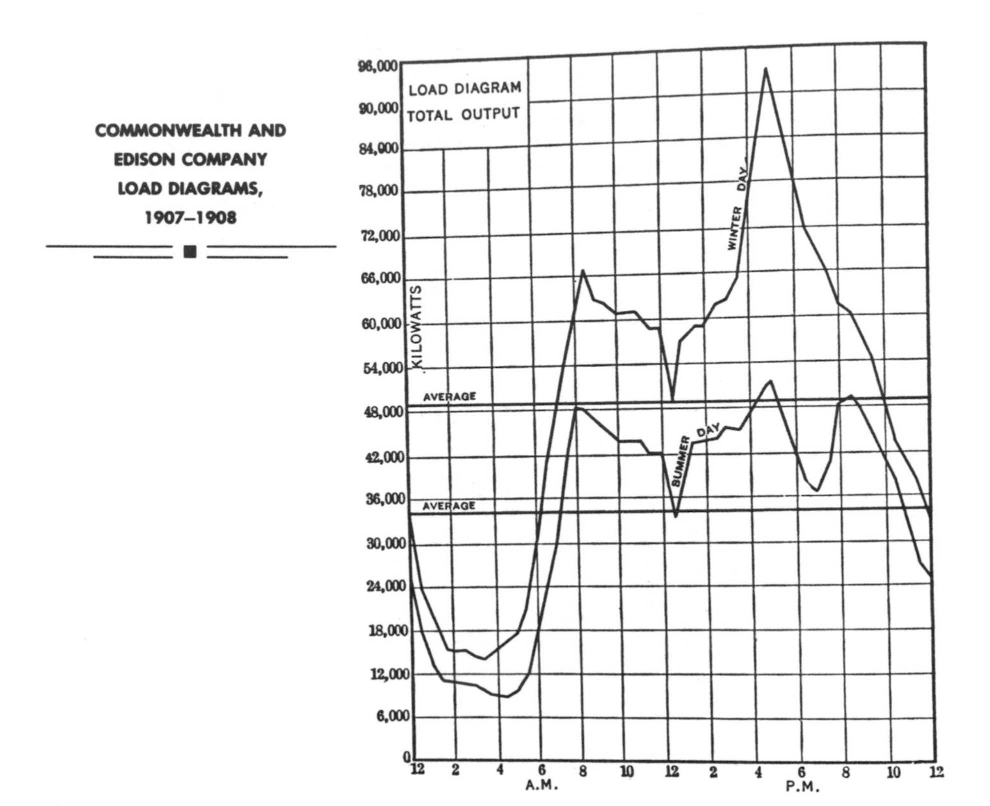

COMMONWEALTH AND EDISON COMPANY LOAD DIAGRAMS, 1907–1908

their use of electrified conveyances. Showing graphically the maximum capacity of the generator, power plant, or utility—which had to be greater than the highest demand—and tracing the load curve with its peaks and valleys, starkly revealed the utilization of capacity, or investment. With this information, Insull and his associates then did everything possible to fill the valleys. In this way the interest charges arising from the equipment installed to meet the peak load could be spread over many units and lower the cost of each kilowatt. The goal was simple and obvious, but the means to obtain it were complex, and the result for society momentous. As costs in other modern technological systems, such as the commercial airlines, computer networks, and communications, have become increasingly dependent on the cost of capital (interest), load-factor problems have also loomed larger and larger for them. Today the bewildering variety of rates devised to encourage us to fly and telephone during hours of low traffic are one indication of the pervasiveness of load-factor considerations and of load management. Utility managers realized this by the turn of the century and began to introduce rates according to time-of-day usage.[77] This is one more example of how we live our lives in a human-made time and space filled by the forces unleashed by technological systems.[78]

To fill the valleys in the load, or demand, curve, Insull resorted to a policy that came to be known as "load management." It was also social manipulation. Through expansion, variations in the charges for electricity, and advertising, Insull manipulated consumers of electricity. By expanding the service area of Commonwealth Edison and other progressive utilities in the United States, he exploited diversity of consumption. Expansion to achieve diversity is an infrequently recognized but major explanation of the inexorable growth of technological systems. Often the uninformed and suspicious simplistically attribute the expansion of systems only to greed and the drive for monopoly and control. All other circumstances being the same, a utility is more likely to find in a large area, rather than a small, a diversity of consumers, some of whom would use electricity during the valley—rather than the peak—hours of consumption. Then the utility attracts them by favorable rates. Chemical plants filled the valleys well, because their nearly labor-free processes could be carried on throughout a twenty-four-hour period. Through advertising, utility salesmen also pushed the sale of home appliances such as irons, fans, vacuum cleaners, refrigerators, and, later, air conditioners. After World War II the sale of air conditioners ironically created a new and undesirable peak of consumption on hot summer days. Insull pioneered in load management through appliances by establishing in 1909 an "Electric Shop." The ground floor emphasized domestic appliances; on the floor below, an Industrial Power Room displayed a wide variety of electric-motor-driven machines. Australians on an observation tour of North American utilities found the Electric Shop a remarkably effective marketing scheme. Chicagoans, they decided, did not hustle more than others elsewhere in the world, but their hustling was more organized. Insull was pleased. American consumers also seemed pleased by load management and manipulation when it meant shiny new gadgets.

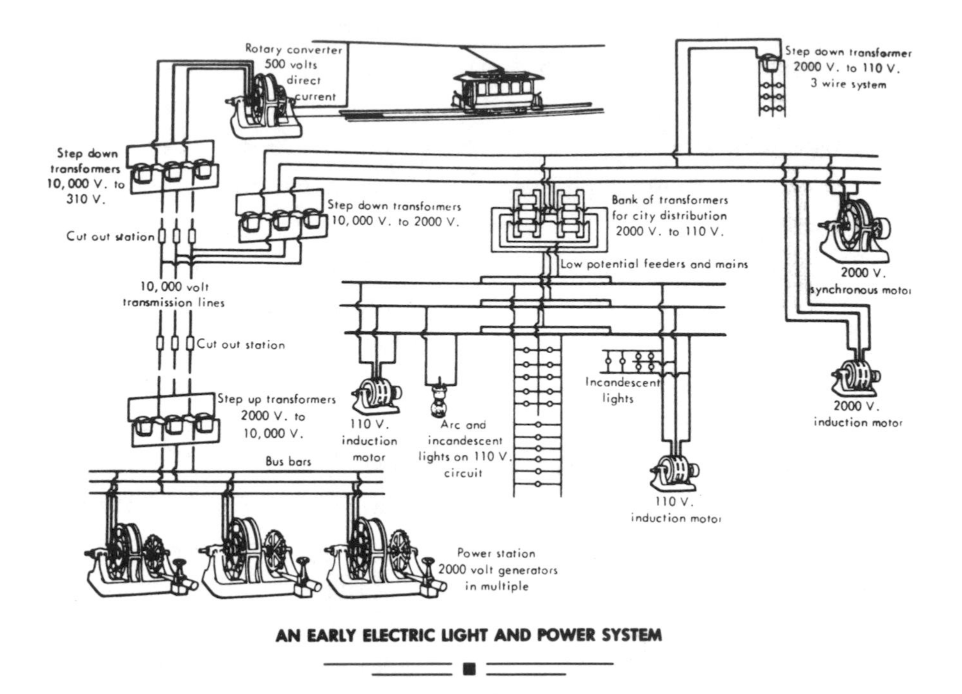

AN EARLY ELECTRIC LIGHT AND POWER SYSTEM

Behind the scenes, in places like load-dispatching centers and power plants about which American consumers were only dimly aware, Insull and other utility magnates and engineers pulled the strings of control even more deftly. They organized and systematized to achieve an "economic mix." Managing a mixture of interconnected power plants, some old, some new, some efficient, some inefficient, some using coal as fuel, others using waterpower, the utility engineers combined the output of the different plants to achieve the most efficient "economic mix." Sitting in a control center that was the esoteric high technology of the 1920s, the load dispatcher had before him a mass of indicator lights, diagrams, and switches that allowed him to keep the most efficient power plants in the system "on line" continuously and to switch in and out the less efficient plants to meet peaks of demand. To anticipate the rapid-fire switching necessary, the load dispatcher had information about the peaks and valleys on the same day of the previous year. Few better examples can be found of the direct applicability of history. The intellectual attraction—the elegant puzzle-solving aspect—that load factor, economic

A LOAD DISPATCHER, 1902

mix, and load management had for Insull the system builder and for the engineer-managers of rapidly expanding electric-power systems becomes understandable. This is not to deny the drive for power and profit, but to acknowledge human delight in "sweet" problem solving.

By the mid-1920s the far-flung Insull empire incorporated diversity and economic mix. To use the jargon of the day, it was a utility with a "good load curve." Not only were diverse customers interconnected by high-tension transmission lines carried on the great towers that symbolized modernity, but power plants in waterpower regions were connected to those in coal-rich regions hundreds of miles distant. These technical systems demanded organizational innovations. One of the most inventive organizational responses was the utility holding companies of the 1920s. Holding companies had a long history, extending back at least as far as those designed for railroad empires in the nineteenth century, but the

INSULL IN HIS PRIME

holistic subtlety of the electric-utility holding companies was unprecedented. Insull organized Middle West Utilities Company in 1912. Like others mushrooming in the United States about the same time, Middle West acquired the securities of far-flung small electrical utilities in exchange for its own stock and cash. The holding company also sold its stocks and bonds to the public. In this way the holding company gained control of a large number of utilities, and if these were physically contiguous they were often physically interconnected with high-voltage transmission lines, with the resultant benefits of diversity, economic mix, and higher load factors. Holding companies not only financed the technical and organizational improvements in the numerous companies under their control, but often constructed the new facilities and managed the small companies. Well-conceived and -managed holding companies efficiently integrated the financial, engineering, and management aspects. Others became the settings for financial chicanery and the watering, the ballooning, and the pyramiding of holding-company stock.

The heyday of the holding company came in a later stage of the evolution of electric light-and-power systems. In the opening stage, inventor-entrepreneurs like Edison solved the major technical problems. In the following stage, Insull and other manager-entrepreneurs presided over the organizational innovations facilitating growth. During the next stage, financiers took over the leading role. They had the capability to solve the critical problem of raising the immense amount of money needed by holding companies to form regional networks of power. During the 1920s Insull, essentially a manager-entrepreneur, began to fear that he would lose control of his utility empire to financiers, especially to the New York ones whom he as a Chicagoan found especially rapacious.

In 1928 he saw the major threat coming not from New York, however, but from Cyrus S. Eaton, a creative Cleveland capitalist who had also built utilities from the ground up. Despite this, he had a reputation as a "financial buccaneer" whom many saw as swimming sharklike in "the big leagues of utility operations."[79] Insull observed with concern that Eaton was quietly buying large blocks of stock in Commonwealth Edison, Middle West, and other Insull companies. Insull the manager-entrepreneur, like Elmer Sperry the inventor-entrepreneur, believed that he, like a shoemaker, should stick to his last. But he violated this rule to enter the unfamiliar world of high finance in an effort to thwart the raid he expected from Eaton. Insull decided to pyramid in order to transform his control of his utilities from that of manager to that of proprietor. In 1928 he established Insull Utility Investments (IUI), an investment trust, the controlling stock of which he and his friends acquired in exchange for the utility stock they owned. Utility Investments then raised money to acquire control of the various Insull utilities.[80] In 1930 Insull took the fateful step of removing the Eaton threat by having Insull Utility Investments and the Corporation Securities Company, another newly formed Insull investment trust known as "Corp," purchase for $56 million the Eaton holdings in the Insull empire. Because Chicago bankers could only partly finance the purchase, Insull had to borrow a substantial part of the money from New York financiers and give as collateral IUI and Corp stock. If the value of these stocks fell, Insull would have to increase the amount of stock correspondingly.

According to Forrest McDonald, "The New York financial club, abristle with excitement at the prospect of giving Insull his comeuppance, at last, and atremble at the prospect . . . of all those millions," moved in to take over from the once-independent and still-defiant Chicagoan.[81] It was 1931, the country was sunk in the Great Depression, and in September Britain abandoned the gold standard and stock prices plunged. Increasing amounts of IUI and Corp securities had to be put up as collateral against the Insull bank loans. By mid-December the combined portfolios of the two investment trusts fell into the hands of the bank creditors. "Coolly and with finesse, the Morgan group moved in for the kill. The kill took about six months, for though the House of Morgan could be devastatingly predatory, it was never impatient and it was never messy."[82]

Ostensibly seeking a solution to the plight of the Insull companies, the bankers called for an audit of Insull books and incidentally a report on any improper transactions. After the auditors changed the mode of calculating depreciation to an industrial one, rather than the one used by Insull and most utilities, they could then declare the Middle West holding company insolvent, a situation that the auditors maintained was concealed before by improper bookkeeping. The auditors also found a number of corporate indiscretions, so rumors began to circulate about the plight of the Insull empire, and hints were even dropped about fraud and embezzlement. Insull worked furiously to save Middle West, for the utilities it controlled were in good economic condition, but in April 1932, at an afternoon meeting in New York, the bankers told Insull that no one would put up more money for Middle West and that it was in receivership.

For the seventy-three-year-old Insull, the end then came quickly. In June, Stanley Field, a trusted friend of Insull's and like him a generous Chicago philanthropist, carried the message, not only from New York bankers but from Chicago bankers and business leaders as well, that they wanted his resignations from all of his companies. They argued that, because of the failure and suspicion now surrounding him, the remaining credit of the companies would vanish unless he complied. Insull uncharacteristically acquiesced without a struggle, and signed resignations from sixty-odd corporations. He and his wife, Gladys, then sailed for

Europe, to live in Paris in relative obscurity until the politicians decided that from the Insull collapse they could make political capital. In September, John Swanson, state's attorney for Cook County (Chicago), told Insull's son-in-law, "You know Sam Insull is the greatest man I've ever known. No one has ever done more for Chicago, and I know he has never taken a dishonest dollar. But Insull knows politics, and he will understand . . . I've got to do it."[83] Swanson then announced that he would launch an investigation of the scandals involved in the collapse of Insull's empire.[84] On the initiative of Swanson, a Cook County grand jury indicted Insull and his associates for embezzlement and larceny. Newspapers headlined the investigation, and local politicians throughout the country played on the resentments of stockholders in the far-flung Insull companies. Insull, persuaded that he would be subject to a political lynching by angry stockholders led by publicity-seeking politicians, sailed to Greece, where he believed he could avoid extradition until the tempest had passed. One anonymous note to his wife said, "You can get ready to buy a cemetery lot as the gang will send you your crooked boys head. you will pay as we have paid our good money. that has been stolen by the dirty yellow crooked Insulls Jews. . . ."[85] The Roosevelt administration succeeded in having him extradited, so in October 1934 Insull and his associates stood trial in Illinois for using the mails to defraud in the sale of the "worthless" stock of the corporation.

The prosecution rested its case on a mass of evidence taken from the records of the Insull companies. The defense, led by Floyd Thompson, a brilliant trial lawyer, succeeded in showing that critical prosecution arguments depended on the interpretations—not illegality—of accounting methods. For instance, a key prosecution witness testified that the Insull company had improperly treated certain expenses, but then, under cross-examination, had to admit that the system used by Insull was used by the government itself. The defense built its case on a sentimental account of Insull's life story that he had been persuaded to organize into autobiographical reminiscences while he was awaiting trial. On the stand, Insull told a story of the rise of a young immigrant to a position of wealth and power. He stressed his long association with the legendary Edison and the build-up of the utility industry through technical and organi-

INSULL RELEASED FROM JAIL, CHICAGO, 1934

zational changes. The jury was fascinated, and even the prosecuting attorney half-said, half-inquired, privately to Insull's son, "Say, you fellows were legitimate businessmen."[86] The jury, impressed by Insull's system building and persuaded that a crooked business would not have exposed all of its crimes in its books as, in effect, the prosecution was maintaining it had, quickly returned with a verdict of not guilty. Insull was also found not guilty on other charges in other trials. In 1938, having retired with his wife to Paris, Insull died of a heart attack in a subway. McDonald has pointed out that all of the operating electric and gas utilities Insull had managed survived him—and the Depression—and ultimately the personal collateral Insull gave his creditors in 1932 became worth $10–15 million more than the debts. Only about one-fifth of the Insull securities in public hands in 1932 were forfeited in any way, in contrast to a figure of close to forty percent for the securities of all American corporations.[87]

Insull's legacy, however, remains unclear. Despite his acquittal by the courts, a leading historian recently referred to him as a crook. At the bar of history, pronouncements about Insull's legacy will depend much on whether he is judged as a financier or as a system builder. As an organizer and a financier of holding companies he was, in comparison to a J. Pierpont Morgan, inept. As a system builder, he carried on in the tradition of Edison by bringing the urban utility system he invented and organized to the next-higher stage of development. Insull was an inventor, too, but his creations were coordinated organizations instead of integrated electrical circuits.

TENDERS OF TECHNOLOGICAL SYSTEMS

As electric light and power, automobile production and use, and numerous other production systems spread, not only system builders and financiers but also trained technicians and managers were needed. They provided a rank and file for operation and maintenance. For centuries expanding production systems had absorbed countless workers, but in the new era of system building, tens of thousands of persons, almost exclusively men with training in scientific engineering and management, were needed as well. As a result, there was an enormous increase in the number of university- or college-trained engineers who knew how to use engineering science and apply economic principles in solving the day-to-day problems of production. Earlier, the majority of engineers had constructed canals, roads, railroads, buildings, and bridges. Most learned engineering by apprenticing with leading engineers engaged in construction projects or in the supervision of machine shops associated with manufacturing plants. In 1862 the U.S. Congress, through the Morrill Act, appropriated funds to the states to support colleges of agriculture and mechanical arts. By 1917 there were 126 engineering schools on the college or university level. Between 1870 and the outbreak of World War I the number of engineering graduates leaped from 100 to 4300 annually. In 1900 there were 45,000 engineers; by 1930 there were 230,000. By 1928 the enrollment in electrical-engineering courses exceeded that of the older civil- and mechanical-engineering courses by fifty percent, and chemical engineering, the newest branch, was already half the size of mechanical engineering.[88] College training for managers followed in the wake of the boom in engineering education. Among the institutions establishing courses in management were the University of Pennsylvania, New York University, the Massachusetts Institute of Technology, Columbia University, Dartmouth College, and Purdue University. Engineering colleges also gave courses in scientific management.

Electrical- and chemical-engineering departments in engineering schools and the corresponding professional engineering societies did not exist in the United States until after 1880. From the beginning, most of the electrical and chemical engineers found employment as salaried employees with the rapidly increasing number of electrical and chemical manufacturers. No tradition of professional independence such as that known by the mechanical and civil, or construction, engineers in the earlier decades of the nineteenth century delayed the adaptation of the electricals and chemicals to corporate employers. In 1930 a study of the engineering profession reported that "the independent private practice of engineering [was] distinctly on the decrease and . . . engineers to a greater extent [were] going into the employ of corporate organizations, particularly of the large industries."[89]

During the first two decades of this century there was a short-lived and minor "revolt of the engineers," who in the name of professionalism insisted that the first responsibility of the engineer was to society. Morris Cooke, Taylor's disciple and a progressive engineer, wielded enough influence by 1919 in the American Society of Mechanical Engineers to have a society committee recommend a code of ethics stating that the first professional obligation of the engineer was to the standards of the profession, not to his employer. Yet the overwhelming pressures and the definition of problems, standards, and tasks continued to come from the corporations.[90] The engineers were the tenders of the technological systems.

The large corporations encouraged their engineer employees to join their respective professional organizations, such as the American Society of Mechanical Engineers (founded 1880), American Institute of Electrical Engineers (1884), and the American Institute of Chemical Engineers

(1908), and to take active roles in the Society for the Promotion of Engineering Education (1894). Through their employees in these organizations, the corporations could influence the formation of professional-society committees to define technical problems of common interest and to design engineering-school curricula. The corporations also greatly influenced the engineering schools by hiring faculty as part-time consultants, providing them equipment for their laboratories, and generally bringing them and their students to study and carry out experiments on problems of interest to industry. After 1900 cooperative plans between university and industry permitted students to work part-time for large manufacturers while pursuing their studies in engineering schools. The president of the American Society of Civil Engineers (formed 1852), with its longer tradition of professional independence, warned in 1909 that the engineer was becoming "the tool of those whose aim it is to control men and to profit by their knowledge."[91] In 1928 a professor of chemical engineering at the University of Michigan could write, in the *Transactions of the American Institute of Chemical Engineers*, that:

> There is some analogy between the college and the manufacturing plant which receives partially fabricated metal, shapes it and refines it somewhat, and turns it over to some other agency for further fabrication. The college receives raw material. . . . It must turn out a product which is saleable. . . . The type of curriculum is in the last analysis not set by the college but by the employer of the college graduate.[92]

The salable product brought to the employer an ability to solve routine technical problems using organized information about the natural sciences, management, and engineering practice. The corporations quickly heightened the young engineer's sensitivity to economics. Henry Towne, a leading American engineer whose ideas greatly influenced Taylor, wrote in 1886, "the symbol for our monetary unit, the dollar is almost as frequently conjoined to the figures of an engineer's calculations as are the symbols indicating feet, minutes, pounds, or gallons." In 1896 the president of the Stevens Institute Alumni Association told the students, "The financial side of engineering is always the most important. . . .

[The young engineer] must always be subservient to those who represent the money invested in the enterprise."[93]

Engineers entering the corporate world often aspired to move, after a stint of engineering work, into the ranks of corporate management. The career patterns of engineering graduates from 1884 to 1924 showed "a healthy progression through technical work toward the responsibilities of management."[94] Within fifteen years after finishing college, about two-thirds of the graduates had become managers. In the 1920s the chief executives of General Electric, Du Pont, General Motors, and Goodyear had been classmates at the Massachusetts Institute of Technology.[95] They had risen from the ranks of system tenders into the heady atmosphere of system builders.

VEBLEN'S SOVIET OF ENGINEERS

An inordinate appreciation of order, centralization, systematization, and control spread from the realms of the system builders, the scientific managers, and engineers throughout American society and culture. The engineering societies, the American Society of Mechanical Engineers among them, widely publicized the gospel of efficiency. The ASME even elected Taylor, an atypical member, president in 1906. As the technological spirit spread in the United States before World War I, it found a warm reception with the Progressives, an ill-defined political and social movement given impetus by the election of Theodore Roosevelt, a self-styled Progressive, to the presidency. Roosevelt ran for president in 1912 as a third-party Progressive. This party was not content that experts bring order, control, system, and efficiency only to resources and work; they wanted social scientists—also "scientific experts—to direct their reforming zeal to city, state, and federal government."[96] Those who applied the technological spirit to such diverse realms of society came to be known to the public as "efficiency experts." Among this group fell Taylor's scientific managers.[97] He encouraged the spread of his doctrines beyond industry when he wrote:

> . . . the same principles [of scientific management] can be applied with equal force to all social activities; to the management of our homes; the management of our farms; the management of the business of our tradesmen large and small; of our churches, our philanthropic institutions, our universities, and our government departments.[98]

In 1919 one of the country's most original and eccentric economists took the rise of the technological spirit, the Progressive movement, Taylorism, large technological systems, and wartime efforts to organize and plan the economy as signals that society was on the verge of a dramatic transformation. Because of an indifferent teaching style, a hostility toward bureaucratic authority, and amatory relations that college campuses considered scandalous at the time, Thorstein Veblen never found a permanent niche in the academic hierarchy.[99] His unorthodox books, *The Theory of the Leisure Class* (1899) and *The Theory of Business Enterprise* (1904), however, attracted attention. In 1919 he set down his views on the coming transformation in a series of articles, first published in a new radical journal, *The Dial*; in 1921 they appeared as a book, *The Engineers and the Price System*. In the meantime, the *Dial* essays were widely read, *The Theory of the Leisure Class* reissued, and an essay on Veblen by the widely read H. L. Mencken published in the magazine *Smart Set*. When *Vanity Fair*, the journal of the sophisticates, spoke approvingly of him, he became required reading among intellectuals.[100] "Veblenism was shining in full brilliance. There were Veblenists, Veblen clubs, Veblen remedies, for all the sorrows of the world."[101]

Veblen pursued the logic of the technological spirit and system building to its rational conclusion: the entire industrial system of the country should be under the systematic control of "industrial experts, skilled technologists, who may be called 'production engineers.' " He believed the nation's industry to be "a system of interlocking mechanical processes." In writing about the industrial system, he gave a fundamental definition of a system, "an inclusive organization of many and diverse interlocking mechanical processes, interdependent and balanced among themselves in such a way that the due working of any part of it is conditioned on the due working of all the rest." Veblen conceived of a national industrial system, a great productive machine, that would dwarf Ford's and Insull's. Veblen's interlocking system, or "network," of processes and exchanges of materials included "transport and communication; the production and industrial use of coal, oil, electricity and water power; the production of steel and other metals; of wood pulp, lumber, cement and other building materials; of textiles and rubber; as also grain-milling and much of the grain-growing, together with meat-packing and a good share of the stock-raising industry."[102]

Borrowing terminology from the Russian revolutionaries of 1917, Veblen called for soviets, or governing committees, of experts to take the management of the nation's industrial system away from parasitic financiers and inexpert entrepreneurs who were wasting the resources and manpower of the country through their counterproductive greed for profits and their competitive instincts. Veblen was of the opinion that, because of its highly technical nature, the interlocking industrial system had already drifted into the keeping of the corps of production specialists who would become members of the industrial soviets. He numbered inventors, designers, chemists, mineralogists, soil experts, production managers, and engineers as suitable members of organizing and controlling soviets to displace the "captains of finance" who had wastefully commercialized and exploited the experts.[103]

Veblen mistakenly believed engineers like Taylor, Gantt, Cooke, and the radical-minded engineers who followed Cooke in the American Society of Mechanical Engineers to be the tip of an iceberg. He erroneously assumed that a wartime commitment to the rational planning of the industrial economy would extend beyond the emergency into years of normalcy. He badly erred in not seeing that most of the engineers from whom he would draw the members of the soviets were salaried employees of the great industrial corporations, infused with corporate values, and content to move up the ladder of engineering and management in organizations often controlled by the "captains of finance."

TAYLORISMUS + FORDISMUS = AMERIKANISMUS

After World War I, Europeans and Russians wanted to know how the United States had become the most productive enterprise in the history of the world. This was especially true of liberals and radicals in defeated and despairing Germany, and among the Soviet leaders in a Russia prostrated by World War I, the revolutions of 1917, civil war from 1917 to 1921, famine, and disease. While middle-class Americans believed that the world was waiting to hear about their political system and their free enterprise, Germans and Russians were asking about Taylorism and Fordism. The Soviets believed that the American system could consolidate the Bolshevik revolution; the Germans believed that the American system could shore up the newly created Weimar Republic, successor to the fallen German empire. The United States had never enjoyed greater respect, or been more envied, than after World War I. Many foreign liberals and radicals perceived its examples as opening for their nations a path to the future. Their image of America was one of inventors, industrial scientists, and system builders. Other peoples were fascinated by, and derived hope from, the example of the creation of modern America.

LENIN, TAYLOR, AND FORD

To Vladimir Ilyich Lenin, Leon Trotsky, and the other Russian Bolsheviks who seized power during the November Revolution of 1917, hydroelectricity generation at Niagara Falls, steelmaking in Gary, Indiana, and, above all, Ford automobile manufacture in Detroit were the essence of modern American technology. They believed that, if such technology were developed in a Soviet context, American means of production could lead the way to the socialist future. By 1926 dreams of the "Americanization" of the means of production mesmerized engineers and managers.[1] Soviet planners believed that the wave of the future involved large systems of production on a regional scale, larger even than those in the United States. Unlike capitalism, the Bolsheviks argued, socialism would not be burdened by political and economic contradictions that constrained the full development of modern production technology. Lenin, who had read technical essays on Taylorism and listened closely to comrades who were engineers, understood that the second industrial revolution involved more than machines, processes, and devices—it involved order, centralization, control, and systems. The concepts that Ford, Insull, and their associates embodied at River Rouge and in Middle West Utilities enthralled the Soviet planners, managers, and engineers. Even the Russian engineers who had survived from tsarist days could resonate to such a shimmering vision of similar technological achievement in the Soviet Union. Driving the peasants and workers mercilessly to gather the grain and cut the wood and dig the minerals for export, the regime exchanged grain, wood, and other raw materials in prodigious quantities for foreign technology, especially American. In the Soviet Union to a far greater extent than in the United States, the Taylor dictum applied that in the past man had come first but in the future the system must. Stalin summarized the Soviet celebration of American technology and management in 1924, when he proclaimed American efficiency a cardinal Leninist doctrine:

> American efficiency is that indomitable force which neither knows nor recognizes obstacles; which continues on a task once started until it is finished, even if it is a minor task; and without which serious constructive work is inconceivable. . . . The combination of the Russian revolutionary sweep with American efficiency is the essence of Leninism. . . .[2]

The Soviet economy passed through several phases during the decades between the world wars. First, the period of war communism, 1917–21, witnessed the Bolsheviks' desperate and unsuccessful attempt to seize industry and turn it over to trade unions and committees of workers to administer. The country survived the foreign and civil wars of the period, but only after Soviet leadership began to restore managers and engineers, the experts, to their customary leadership roles. Lenin justified this by appeals to a new, Soviet form of Taylorism. In 1921, with the country near exhaustion and industrial production seemingly unable to recover, he called for a new phase, which he termed the New Economic Policy (NEP). It involved a temporary retreat from the goal of centralized government planning and control. After 1921 the regime tolerated a substantial amount of private and market enterprise while it retained control of the "commanding heights." These included heavy industry, transportation, and electricity supply. During the NEP period, the Soviet government embarked on the planning process that culminated in 1928 with the first Five-Year Plan and a drive to eliminate private enterprise in both industry and agriculture.

During the war-communism period the importation of Western technology and experts had been impossible; during the period of the New Economic Policy the Soviet Union tried granting concessions to Western manufacturers that allowed them to establish and operate plants in the Soviet Union. After the Soviet-German political and economic agreements reached at Rapallo in 1922, participation by German manufacturers and engineering experts in Soviet economic development was especially heavy. During the first Five-Year Plan, after 1928, the Soviets turned to the outright purchase and transfer of foreign-designed and

JURIJ PIMENOV, *BUILDING SOCIALISM*, 1927

NIKOLAJ DOLGORUKOV, *FIVE-YEAR PLAN*, 1931

-manufactured production plants. Foreign experts set these plants into operation and then turned them over to Soviet managers, engineers, and workers. U.S. manufacturers, industrial architects, and engineering construction and consulting firms played leading roles in this massive transfer of technology.[3] Throughout the decade of concessions and direct purchase after 1921, the Soviet leadership, always fearful of dependence on the capitalist world and of having to fight a war without allies, consistently maintained a policy of avoiding the import of manufactured goods but pursued instead the transfer of the means of producing these. This intensive and large-scale transfer of technology, historically unprecedented, should be recognized as one of the major chapters in Soviet history, but it has been virtually forgotten in both the Soviet Union and the United States. The former has denied its dependence on capitalism and the latter does not wish to boast of its contribution to the establishment of Soviet industrial power.

American engineers and management experts visiting the Soviet Union in the 1920s reported with some surprise the intense effort of the new regime to instill enthusiasm for technology in its engineers, workers, and ordinary citizens. One American engineer observed in 1930, "it is this amazing sign (something akin to religious fervor) that is of such interest to the foreign visitor."[4] American engineers found store windows on the major thoroughfares of the large cities filled with books and apparatus for the home study of physics and chemistry. The visitors observed, too, that on holidays countless workers and peasants took—or were taken on—guided tours of large steam-driven and hydroelectric power stations. The dramatic contrasts between the primitive and the modern also struck visiting Western experts. Excepting the pockets of intense technological transformation and highly urbanized Leningrad and Moscow, Russia, with its lingering reliance on humans and animals instead of machines, the use of wood instead of metals, and the predominance of agriculture, still resembled a preindustrial nation. Joseph Stalin later claimed that he found Russians using a wooden plow and left them with a nuclear reactor. This, however, was rhetoric rather than fact: railroad building, iron and steel production, textile manufacture, and foreign loans were rapidly industrializing many regions of Russia before the revolutions of 1917.[5]

In 1916, when Lenin the Russian revolutionary discovered Frederick W. Taylor the American revolutionary, there was a paradoxical meeting of the minds. Lenin, who made detailed notes on production techniques used in the capitalist world, found Taylorism profoundly impressive. Not satisfied with general principles, Lenin noted in detail Taylor's thousands of machine-tool experiments designed to find the steel with the best characteristics for metal cutting. Frank Gilbreth's views on scientific management also impressed Lenin, because Gilbreth did not seem so much intent on speeding up, or exploiting, the worker as he was determined to find the one best, energy-saving way of doing work. Lenin observed in the margins of a Gilbreth article that scientific management could provide a transition from capitalism to socialism. Taylor's and Gilbreth's claims that their doctrines were scientific greatly impressed Lenin, the student of scientific Marxism.

Lenin insisted that Taylorism would not be used in a socialist state to exploit the worker for the profits of greedy capitalists, but that the greatly increased production would be distributed among the workers and peasants. The lowered cost of production, the short installation time ("two to four years!!"), and the vastly increased reliance on engineering and management skills required by scientific management fascinated Lenin.[6] He believed that the detailed work instructions and planning done by engineers under the Taylor system would make possible the use of the large pool of unskilled peasant labor in industrially backward Russia. He also saw that Taylor's centralized control of the workplace, the work process, and the workers would allow politically reliable experts to monitor the industrial system closely during the period of transition from capitalism to socialism. It would also help to root out bourgeois saboteurs and other corrupted persons and attitudes. Lenin seemed confident that he could locate the required number of politically reliable industrial engineers. Lenin, with his need to control, shared many personality characteristics with Taylor and the Taylorites.

In the spring of 1918, as the country lay disorganized and demoralized, Lenin in one of his many speeches said:

> The task that the Soviet government must set the people in all its scope is—learn to work. The Taylor system, the last word of capitalism in this respect, like all capitalist progress, is a combination of the subtle brutality of bourgeois exploitation and a number of its greatest scientific achievements in the field of analyzing mechanical motions during work. The elimination of superfluous and awkward motions, the working out of correct methods of work, the introduction of the best system of accounting and control, etc. The Soviet Republic must at all costs adopt all that is valuable in the achievements of science and technology in this field. The possibility of building socialism will be determined precisely by our success in combining the Soviet government and the Soviet organization of administration with modern achievements of capitalism. We must organize in Russia the study and teaching of the Taylor system and systematically try it out and adapt it to our purposes.[7]

Lenin intended to bring American engineers to the Soviet Union to help install the Taylor system. Labor unions and some members of his own party opposed his call for Taylorism, predicting that its introduction in the Soviet Union would result in terrible mistakes, like those that labor unions abroad had been fighting. Lenin, however, after witnessing the chaos attendant on leaving work organization and control in the hands of labor unions inexperienced in management, preferred to have control and accounting in the hands of engineers and expert managers, even if, for a time, they were holdovers with bourgeois traits from the old regime, or foreigners. In violation of egalitarian sentiments, he was also prepared to pay higher wages to the experts and to the more productive workers. Increased production was a matter of survival for the nation, and Lenin's pragmatic approach prevailed. When conservative businessmen and engineers in the United States heard of his advocacy of Taylorism, they took it as proof that socialistic ideology was unworkable and that the American system was the one best way.

Not only Lenin, but Trotsky, commissar for war and the leader best known after Lenin, also espoused Taylorism. Intent upon the restoration of discipline and of leadership by experts, he tried to introduce Taylorism into the Red Army and into the decimated armaments industry. In his autobiography, Trotsky recalls his reliance on "Kili," an American engineer who came to the Soviet Union about 1918 of his own free will to Taylorize industry. Kili believed that a nationalized industry provided the Taylorite with an ideal opportunity for dramatic and encompassing reforms. He reported to Trotsky that soldiering in industry took about fifty percent of all productive time, a figure that made Taylor's earlier problems with soldiering at Midvale seem trivial. Trotsky, desperately concerned during the period of war communism that the Soviet Union faced a total industrial collapse, advocated a "militarization" of labor, the details of which corresponded to an extreme form of Taylorism on a national scale.[8]

Drawing up their first Five-Year Plan during the mid-1920s, the Soviet leaders used scientific-management experts, especially American, and "all of the multitude of techniques in the Taylorite armory."[9] In the Soviet Union, Taylorism shifted from the limited field of factory organization to the grand scale of the national economy. The Communist Party, wanting to persuade the entire nation that scientific management accorded fully with scientific Marxism, translated and published the Taylor book and other American books, articles, and countless commentaries on scientific management. The highest Soviet planning authorities brought Walter Polakov, a follower of Henry L. Gantt, who was one of Taylor's most fervent disciples, to the Soviet Union to provide a liaison between the scientific-management movements in the United States and the Soviet Union, and to draw up Gantt production charts for the entire first Five-Year Plan.[10]

Alexei Gastev, the Soviet "worker poet," the "Ovid of engineers, miners, and metal workers," and a trade-union leader, gave American Taylorism an exotic Soviet flavor. Written in Siberian exile and prison before 1914, his poetry fired the imagination of a generation of Soviet youth after the November Revolution. Shrill factory whistles, whirring machine tools, and the glowing eruptions of blast furnaces were the images in his poetry. He saw the industrial workers with "nerves of steel" and "muscles like iron rails" as extensions of the machines they tended.[11] Giving a poetic cast to Taylorite thoughts, he lauded the fusion of man and machine in his poem "We Grow Out of Iron": "I grow iron arms and shoulders—I coalesce with the iron form." He accepted machines as controllers of men.[12] Like Taylor, also a former machinist, he knew the workers and the workplace. Fascinated by what he had read about Taylor's and Gilbreth's approaches to the organization of work, Gastev as a trade-union official saw the opportunity after the revolution to create something new besides poetry. He became the bard of scientific management.

Lenin found Gastev's ideas, enthusiasm, and energy appealing, and supported what Gastev called his last artistic creation, the Central Labor Institute. In the 1920s the institute became the fountainhead for Soviet Taylorism, and time-and-motion studies became Gastev's *idée fixe*. His critics complained that the institute neglected the more complex aspects of scientific management, such as the organization of the workplace. Gastev's imagination, however, ranged even further than scientific management, for he dreamed of establishing a new worker's culture for the new technology that was coming into existence. He wrote:

> . . . the metallurgy of this new world, the motor car and aeroplane factories of America, and finally the arms industry of the whole world—here are the new, gigantic laboratories where the psychology of the proletariat is being created, where the culture of the proletariat is being manufactured. And whether we live in the age of super-imperialism or of world socialism, the structure of the new industry will, in essence, be one and the same.[13]

Following Taylor, he saw work becoming standardized and most workers as machine tenders. Gastev called the relationship of groups of workers with groups of machines a "mechanized collectivism." "Many find it repugnant that we want to deal with human beings as with a screw, a nut, a machine. But we must undertake this as fearlessly as we accept the growth of trees and the expansion of the railway network."[14] He predicted that Taylorism would usher in a new historical epoch in which society would be mechanized, in which social engineering would reign. The seat of creativity and of control in the workplace and in society would become the offices and planning rooms of the managers and engineers. In his pamphlets he sought to stimulate inventiveness in the sense that

"Taylor was an inventor, Gilbreth was an inventor, Ford was an inventor."[15] He longed for the American spirit; he advocated a "Soviet Americanism," and wanted to see Russia transformed into a "new, flowering America."[16] The Stalinist purge that decimated social scientists eliminated Gastev's institute and its directors in 1940. He died in 1941; at least one source reported that he was shot.[17]

The introduction of Taylorism into the Soviet Union brought countless problems. A system of scientific management developed in and for a highly industrialized nation with a highly complex and productive market economy was being tried in an industrially backward nation attempting to plan the economy. The American engineers and management experts who played a leading role in transferring technology to the Soviet Union in the 1920s came back with stories of enthusiastic, frantic, and harsh efforts to install the Taylor system, punctuated with disastrous failures. That peasants-become-industrial-workers did not arrive at work on time because they had no clocks in their homes suggests the magnitude and depth of the problem of technology transfer. In addition, party members with the responsibility to monitor and push the engineers, managers, and workers often had a poor grasp of the systematic character of scientific management. They demanded speed-ups in one part of an integrated factory system while they neglected other parts of the system. As a result there were monumental reverse salients, bottlenecks, and logjams. Pressed by unrealistic production norms and threatened with the loss of meager income, workers drove newly imported machinery destructively in order to meet quotas and, after cutting corners, turned out shoddy products. Engineers and managers who made mistakes in their efforts to find new ways of planning and administering production became pathologically cautious when they found that honest mistakes might be labeled criminal sabotage by high-level functionaries not willing to take responsibility for production lapses. Ford's irrational outburst in Detroit would have seemed inconsequential to a Soviet engineer accused of high crimes. Irrational and unsystematic speed-ups and production-quota rises became a lasting characteristic of Soviet Taylorism and Soviet technology.

Lenin also decided that electrification on a national scale would open the way to constructing a modern Russia. Deeply influenced by Karl

Marx's philosophy of history, in which technological changes bring social ones, Lenin reasoned by analogy. He thought, as Marx had argued, that, since steam power and the factory system had brought industrial capitalism and the dominance of the industrial middle class, electrification plus large, then regional, systems of production would bring the next great social change, the formation of socialist society. Lenin's adviser, the engineer G. M. Krzhizhanovsky, persuaded Lenin that electrification could not be fully developed in a capitalistic shell where competition prevailed, but only in a socialist context. Collective enterprise and co-operation, instead of competition, would facilitate a nationwide system of energy production, a national grid that would function like a single great machine. Insull's vision paled by comparison. "The rise of a machine culture on an electrical basis can be achieved," Krzhizhanovsky wrote, "in the most perfect and unfolded form only in conditions of socialist economy."[18]

Lenin's interest in electrification can be traced to the 1890s and his association with Krzhizhanovsky, with whom he shared exile in Siberia.[19] Like so many contemporary social reformers and intellectuals in the industrial West, Lenin saw electrification as bringing an ideal industrial society: "the electrification of all factories and railways" he believed would "accelerate the transformation of dirty, repulsive workshops into clean bright laboratories worthy of human beings, and electric light and heating of every home would ease the life of millions of 'domestic slaves.' " At times he seemed more enthusiastic than technically informed, as when he called for installation of electric lights in all rural districts within one year, the manufacture of highly complex insulators for high-voltage transmission lines in local and small ceramic factories, and the acquisition of copper for wiring by collecting it in rural districts ("a little hint . . . church bells, etc.").[20] H. G. Wells, the British novelist and social reformer, visiting him in 1920, concluded that "Lenin, who like a good orthodox Marxist denounces all 'Utopians,' has succumbed at last to a Utopia, the Utopia of the electricians."[21] Lenin attempted to popularize the call for rapid electrification. He believed that it would overcome the backwardness and darkness of the countryside. He instructed that a copy of the electrification plan go to all schools. Illiterate peasants should learn to read, using the plan as their basic book. He asked that a railroad car outfitted to show movies about electrification be sent around the countryside. Even though he laid major stress on electrification for industrialization, Lenin also proposed electrification of the countryside for purposes of irrigation, soil electrolysis, soil heating, night lighting, and the manufacture of fertilizers.[22] Perhaps the ephemeral nature of electricity appealed to this intellectual who felt so at home with abstractions. Lenin, though no engineer, understood that the essence of electrification was large power stations using the most economical regional energy joined by transmissions lines to form an extended regional, or even national, system. On this he and Samuel Insull saw eye to eye.

Lenin and Krzhizhanovsky in 1920–21 embarked on a campaign to persuade the Communist Party leadership that, failing the possibility of immediately introducing a nationwide, nationalized, and planned economy, the best alternative was a national plan for electrification, and especially a national grid. This, they argued, would provide the foundation, or structure, for the eventual planned economy. In February 1920 the Soviet leadership approved the formation of the State Commission for the Electrification of Russia (GOELRO), which would have the responsibility for extensive information gathering, planning, organization, and administration. On a historic occasion depicted in a heroic painting of Lenin announcing the electrification plan, he told the Eighth Congress of the Soviets, assembled in the Bolshoi Theater in Moscow, that electrification and modern large-scale production would bring the ultimate victory of communism over capitalism. He predicted that "if Russia is covered with a dense network of electric power stations . . . our communist economic development will become a model for a future socialist Europe and Asia." Believing it necessary to win over the masses as he had through his speeches and decisive actions during the November Revolution of 1917, he called for the widest possible propaganda and the conversion of "every electric power station we build into a stronghold of enlightenment to be used to make the masses electricity conscious."[23]

In industrially backward Russia, electrification proceeded during the years of the New Economic Policy, 1921–28. The construction of electric-power networks, part of the state-controlled "commanding

ELECTRIFICATION OF RUSSIA, C. 1930

ELECTRIFICATION OF AMERICA AS PORTRAYED IN
LOUIS LOZOWICK'S LITHOGRAPH *SENTINEL*, 1930

heights," depended on the massive transfer of technology. The only other precedent for such a visible and trauma-inducing influx of Western technical experts and technology had been the dramatic effort of Peter the Great to Westernize Russia in the late seventeenth and early eighteenth centuries. In the twentieth century the Soviet government resorted, on a prodigious scale, to the tried and tested ways of transferring technology, including the translation of technical and scientific books, the hiring of foreign managers, engineers, and skilled workers, and the purchase of machines and processes. The Soviet effort, however, took a new turn in its concentration on the importation of entire systems of production and incorporating them in hydroelectric regional complexes. Observers who labeled this simply "gigantomania" missed Lenin's point that to be modern was to be large-scale and to be large-scale was to introduce the material conditions for the transformation from capitalism to communism. Albert Kahn, the American industrial architect who designed Ford's Highland Park and River Rouge plants, observed:

> It is indeed difficult to understand the Russian psychology which dictates the erection of such huge establishments. We in this country would begin with a smaller lay-out, so arranged as to make expansion easily possible. . . . Not so in Russia . . . where they say, "We haven't time to learn to run, we must fly."[24]

In constructing such large power stations, however, the Soviets were assuming that capacity would be well utilized and that load curves would be favorable. Samuel Insull might have cautioned them that this was often not the case unless demand was being carefully cultivated as supply was being created.

The falls of the Dnieper River, once dominated by a fortress of Ukrainian Cossacks, became the site of a mammoth Soviet hydroelectric project, the most ambitious of the Soviets' new construction schemes.[25] It became the showpiece of the new Soviet technology policy and an epic chapter in the history of technology transfer. The head of the electrification section of the Soviet planning agency said: "The whole of Europe is watching the construction. . . . This is our examination in technology, and we must create the best possible atmosphere, so that we can pass it."[26] Often compared to the Muscle Shoals project, which became the first unit in the Tennessee Valley Authority system, the Dnieper project (Dnieprostroy) was done in an American style. The Soviets named Hugh Cooper, an American, the chief consulting engineer for the Dnieper project. I. Aleksandrov, a Soviet engineer, headed the project. American companies supplied equipment and engineers to supervise installation. International General Electric built five of the giant generators needed, while the Soviet Electrosila Plant in Leningrad built the other four under American supervision. The Newport News Shipbuilding and Drydock Company constructed the nine 85,000-horsepower turbines, the world's largest. German and Swedish firms assumed responsibility for other major items at the dam and power station but, in sum, about seventy percent of the hydroelectric equipment was American. Steam shovels, hoists, locomotives, rock drills, and construction steel also came from the United States. "Such a forest of equipment . . . could not be seen elsewhere on any single construction site in the world."[27] When American photographer Margaret Bourke-White visited the construction, she found a notable international exchange when she observed four soft-spoken Virginians in charge of the barefoot Russians installing the turbines.[28] Construction began in 1927. Tens of thousands of workers were involved. On 1 May 1932 the Soviets dedicated the power station, "V. I. Lenin," and placed in operation the largest hydroelectric station in the world. The project had schooled countless Soviet engineers and workers in Western technology, and they carried their experiences to other projects throughout the country. While the Soviets used foreign engineers, they relied on native workers even though their lack of skill resulted in countless mistakes and accidents. In a similar way, in the early nineteenth century, the Erie Canal had been a training school for Americans. Hugh Cooper believed that the learning experience of constructing Dnieprostroy, coupled with an abundance of human and natural resources, would place the USSR in a position to take a commanding position in the world.[29]

In accord with Lenin's insistence that Soviet projects had to be large-scale, planners envisaged the construction of an entire industrial region based on cheap electric power from the dam, a power system similar to the one that had mushroomed around Niagara Falls. The Soviets wanted the hydroelectric station to constitute the core of "a unified industrial complex economically and technically inter-connected."[30] They projected a nitrogen-fixation plant, a cement works, an aluminum production plant, and a steel-producing complex knit together by high-voltage power lines and an electrified railroad system. The construction of a complex of canals around the falls and the dam, to make possible unbroken navigation of the Dnieper from northern Russia to the Black Sea, coincidentally fulfilled a dream of Catherine the Great. They also planned high-voltage transmission lines to carry power from the dam to industry in the Don basin, almost two hundred miles away.[31] As customary for them, the Soviets also projected modern housing and a new city for 150,000 workers in the heart of the electro-industrial complex. Planners predicted a population growth in the industrial region of from one to eight million.

During the ceremonies dedicating the hydroelectric station in 1932, the government awarded Cooper the Order of the Red Star, its highest award. He was the first foreigner to be so honored.[32] Born in Sheldon, Minnesota, in 1865, Cooper had built hydroelectric projects throughout the world. He apprenticed with the Chicago, Milwaukee, and Saint Paul Railroad. All his college and university degrees were honorary. After rising to the rank of chief engineer with the Chicago Bridge and Iron Works, he entered the booming field of hydroelectricity in the 1890s, and in 1901 established Hugh L. Cooper & Co., his own consulting firm. Over the next decades, he designed and constructed hydroelectric projects with more than two-million-horsepower capacity, among them the mile-long Keokuk dam and power plant on the Mississippi and the U.S. government installation at Muscle Shoals, Tennessee. He also helped plan the heightening of the Aswan dam in Egypt. Under contract from 1927 to 1932 with the Soviet government, he spent one or two months each year at the Dnieper site. A leading Soviet engineer recalled that "the dry and cautious American expert Cooper said that, since he had got to know about the Dnieper project, he had turned into a poet."[33] He and his American staff lived in a *sloboda*, or foreign quarter, with

NIKOLAJ DORMIDONTOV, *DNIEPROSTROY*, 1931

comfortable housing, excellent provisions, and access to a swimming pool and a golf course.

Before naming Cooper the principal engineer for the project, the Soviets considered the German firm of Siemens Bauunion. The German firm figured prominently in the construction of the dam.[34] Its meticulously theoretical approach and a papier-mâché model of the project impressed the Soviets, but Cooper's use of the standard Russian gauge for all construction railways, his resort to simplified techniques that the inexperienced Russian laborers could learn, and other highly practical and adaptable recommendations secured him the principal consulting contract. He received the highest fee ever paid a technical expert by the Soviet Union. The government imported American food for Cooper and

CONSTRUCTION OF DNIEPROSTROY DAM

his staff and housed them in brick cottages that would have graced "an American garden city development."[35]

Cooper said that he did not accept any "isms" except good, old-fashioned, American common-sense-ism, but that he found all the Soviet leaders with whom he had business dealings and long conversations—and these included Joseph Stalin—to be men of great intellectual ability. They were, he believed, committed to improving the conditions of the people through technology. Stalin he found "kindly minded, but firm and confident that their economic plans are correct."[36] He approved of their forthright business dealings and was gratified by the absence of corruption. Cooper also liked the Russian workers and found them eager participants in the dramatic construction project. Trying to instruct the laborers, many of them peasants, how to use complex equipment was heartbreakingly frustrating, but, because of their joint determination, he made headway. The Soviet engineers' and managers' authority over the

HUGH COOPER AND JOSEPH STALIN (LEFT AND MIDDLE)

workers and their use of incentive, piecework wages for the workers pleased Cooper. In the United States he sponsored Soviet-American relations before his country formally accorded diplomatic recognition to the Soviet Union. He headed an American-Soviet Chamber of Commerce, whose directors came from leading American firms eager for trade. Among these in 1932 were International General Electric, Westinghouse Electric International Company, Thomas A. Edison, Inc., General Motors Corporation, W. Averill Harriman & Co., Chase National Bank, and American Locomotive Company.[37]

By 1928, when the Soviets inaugurated the first Five-Year Plan, Henry Ford had become a greater hero of production to the Soviets than Taylor. An emotional cult grew up around the methods and even the person of Ford. By 1925 his 1922 autobiography, *My Life and Work*, had had four printings in the Soviet Union. He was read with a zeal usually reserved

COOPER (LEFT) AND MAKSIM LITVINOV, COMMISSAR FOR FOREIGN AFFAIRS

for the study of Lenin. An American observer reported that even managers studied Ford with as much enthusiasm as others studied Lenin, and that villages adopted the name of the Fordson tractor. Peasants who had never heard of Stalin knew about the man who manufactured the "iron horse."[38] Walter Duranty wrote in 1928 that "Ford means America and all that America had accomplished to make her a model and an ideal for this vast and backward country. . . . Cheap mass production is a Soviet goal, more precious from the practical standpoint than world revolution. . . ."[39] The Soviets used a massive Ford-designed plant in Russia along with the great Dnieper hydroelectric project as symbols of modern Soviet technology. As in Weimar Germany, Ford's social philosophy of mass production and mass consumption fired as much enthusiasm in the Soviet Union as did the machinery and layout of his Highland Park and River Rouge plants. Fordism conjured up a vision of humming factories supplying an abundance of consumer goods to once-oppressed and -depressed workers and peasants become wholeheartedly committed citizens of a communist society. In 1919 a Soviet delegation asked for a meeting with Ford, adding, "We believe we could make you understand that Soviet Russia is inaugurating methods of industrial efficiency compatible with the interests of humanity."[40] Ford's role as Soviet hero and provider of technology must have caused at least a minor identity crisis, for in *My Life and Work* he wrote, "Nature has vetoed the whole Soviet Republic. For it sought to deny Nature. It denied above all else the right to the fruits of labour. . . . The fact is that poor Russia is at work, but her work counts for nothing. It is not free work."[41] He added, "The same influence that drove the brains, experience, and ability out of Russia is busily engaged in raising prejudice here. We must not suffer the stranger, the destroyer, the hater of happy humanity, to divide our people."[42]

Ford's views on the Soviet regime did not penetrate Soviet consciousness as deeply as the import of Fordson tractors did in the early 1920s. By 1926 the Soviets had ordered 24,600 Fordsons and 10,000 had been delivered. The Ford Company in 1927 boasted that eighty-five percent of the trucks and tractors in Russia were Ford-built.[43] In 1924 only about 1,000 tractors had operated in the vast Russian countryside, but by 1934 the number had increased to 200,000, most of which were of U.S. manufacture—Ford, International Harvester, and others. Some were made in Soviet factories from American designs.[44] A number of American agricultural experts had come to the Soviet Union to help the Russian peasants learn to use the tractor and to help them introduce an appropriate agriculture. Trotsky said that "the most popular word among our forward-looking peasantry is Fordson." The peasants celebrated Fordson days and Fordson festivals in their villages.[45] Superb as a symbol, the Fordsons served less well as tractors. The Russians found them often too light to plow Russian soils deeply enough.[46] Low-priced and economical to operate, the lightweight Fordsons suffered breakdowns that could be promptly attended to in the United States, where service was a part of the Ford system. This was not the case in the Soviet Union.[47] The Fordson there in the early 1920s turned out to be an inappropriate technology, because it burned gasoline, a fuel in short supply in the Soviet Union. The Russians needed naphtha-burning engines. After 1928 the

UNLOADING FORDSON TRACTORS, NOVOROSSISK, 1923

Soviets imported larger and sturdier tractors than the Fordson from other U.S. companies, including International Harvester, John Deere, Case, and Allis-Chalmers. In 1931, when the Russian import of tractors peaked, ninety-nine percent came from the United States. After that imports dropped sharply, as the Soviets increased their own tractor production—mostly in plants of American design.

At Stalingrad the Soviets built an immense plant designed by Albert Kahn. Construction was under the supervision of John K. Calder of Detroit.[48] International Harvester Company provided the technical advisers and the design of the tractor to be manufactured. Approximately 380 American engineers and foremen helped with construction at Stalingrad; then the foremen remained in Russia until the Russians were capable of operating the plant. In 1930 the plant began producing tractors, but Stalingrad soon became known for poor quality, late delivery, and gross mishandling of machinery by workers, many of whom had never before seen an electric light or operated machinery.[49] John Calder also

FORDSON TRACTORS LEAVING PORT, 1923

supervised the construction of a tractor plant at Chelyabinsk, the assembly building that the Soviets boasted would be the largest building in the world. They confidently predicted an annual production of fifty thousand of these "Stalinets" tractors. In 1933 production began with a replica of a Caterpillar crawler. ("As usual, the Russians paid no royalties to the American patent holder."[50]) The Soviets specified that Chelyabinsk could be quickly changed over to tank production. Leon A. Swajian, who had supervised construction at Ford's River Rouge, also presided over the expansion of a small tractor plant that the Russians had at Putilov in Leningrad and the building of another plant at Kharkov to produce a replica of an International Harvester model.

Frustrations experienced by American managers, engineers, and foremen trying to bring the tractor plants into production while dealing with Soviet functionaries and workers hardly exceeded those encountered by Americans helping the peasants use the machines. The American Harold

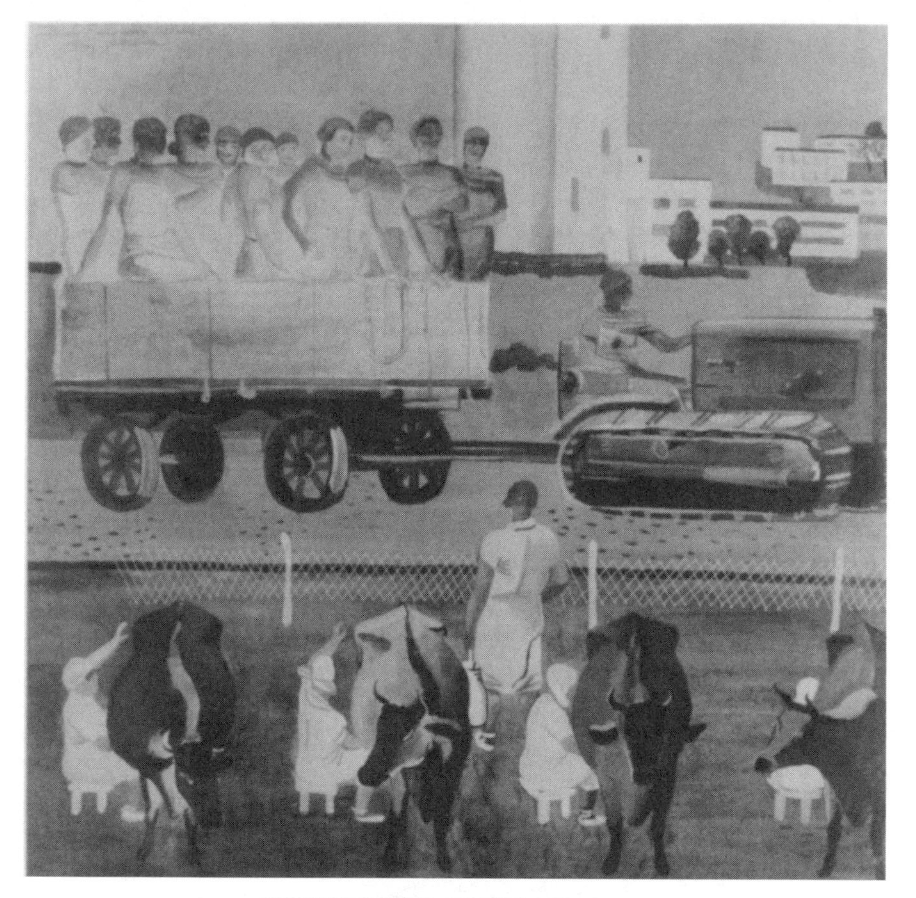

ALEKSANDR DEJNEKA, *WOMEN'S BRIGADE*, 1931

M. Ware, his wife, and eight American farmers traveled to Russia in 1922 with some tractors to teach the peasants how to operate them. The government assigned them a fourteen-thousand-acre farm to demonstrate techniques. Trotsky greeted the American farmers, happily noting that the big, strapping fellows were nearly all first-generation Americans of Scandinavian descent. "So," he said, "in one generation you make Scandinavian peasants into American farmers and American tractor experts. Well, we can make Russian peasants over like that too."[51] The one generation proved a long and hard one in the Soviet Union.

In 1924 Ware brought over more American farmers. Earlier, in 1922, Joseph Rosen brought over a number of young American farmers and tractors under the auspices of the Jewish Joint Distribution Service. Ware and Rosen helped establish tractor farming in the Soviet Union and received high praise from the Soviet leaders for their contributions. Other Americans followed in their wake, but there were also countless stories of heartbreaking setbacks. The increasing number of tractors made in the Soviet Union after 1930 had leaking radiators, poorly cast cylinder heads, loose bearings, and broken valve springs. One returning American instructor wrote, "I can't begin to tell how the Russians mistreat their machinery. . . . Tractors good for ten years hard work will last through three seasons there. . . . He [the Russian worker] does not care whether it runs or not. In fact, if it doesn't run he has more time to sleep, and sleep is one thing he loves."[52] Another observer reported that fleets of disabled tractors dotted the landscape. Spare parts were short and repair work poor. Frantic efforts to meet production quotas broke down many machines. One Soviet farm manager received Americans assigned to his farm with a revolver on his desk as he recited their duties. A five-man delegation from the Ford company making a six-thousand-mile tour of tractor and other facilities in 1926 found the Russians entirely mad on the subject of charts, diagrams, and colored tables of figures, but learned that these meant nothing. After being shown a chart identifying in detail a large number of tractor repair shops in the Ukraine, the Ford delegation was unable to locate a single repair facility there. This delegation found the machine tools modern and layouts logical, but factories dirty, supervision by worker committees poor, and workers lazy. In their confidential report to the Ford company they expressed their shock that political considerations took precedence over technical ones.[53] Yet there were other reports of peasants, initially skeptical and recalcitrant about using the machinery, being won over and demanding more tractors. In sum, it seems, despite setbacks and disappointments, that by World War II the American experts and the predominantly American-designed tractors had substantially facilitated the collectivization of agriculture in the Soviet Union.[54]

Besides tractors, the Russians craved the famous Ford automobile and truck. Following the Soviet policy of importing the means of production rather than the products of the capitalist world, a delegation from Amtorg, the Soviet-American Trading Corporation, and from the Moscow Automobile Trust visited Detroit in 1928, the year after Ford had changed over from the Model T to the Model A. The Ford company had shown interest in negotiations with the Soviets as sales of the Model T waned throughout the world and Soviet orders for Fordson tractors sharply declined.[55] The Soviet delegation announced that it wanted a Ford automobile factory in Russia. In May 1929, a year after the inauguration of the first Five-Year Plan, the Ford Motor Company signed a contract with the Supreme Economic Council of the Soviet Union. Ford agreed to furnish detailed plans for construction and to equip assembly and production plants able eventually to produce annually a combination of a hundred thousand Ford Model A automobiles and Ford Model AA trucks. Ford would also equip the plants, but the Austin Company of Cleveland, Ohio, the engineering-and-consulting firm, was to supervise erection of the production plant, the assembly plant, and a model city for the workers at Nizhni Novgorod (renamed Gorky in 1932). Albert Kahn directed construction of a smaller assembly plant in Moscow. The assembly plants were to assemble imported parts until the Soviet production plant was in operation. An exchange of several hundred Russian and Ford engineers and foremen facilitated the placing of the Ford plants into operation. The Soviets had agreed to purchase seventy-two thousand unassembled Ford Model A cars and Model AA trucks, and parts in any proportion it preferred over the initial four-year period. During this time production at the Soviet plant would increase and then, when production had reached a hundred thousand annually, the importation of Ford units was to stop.[56]

Ford's attitude toward the Soviets had changed since he had written *My Life and Work*. "Russia is beginning to build," he reputedly said, adding, "I have long been convinced that we shall never be able to build a balanced economic order in the world until every people has become as self-supporting as possible. . . . [Only stupid greed] can think of the world as continuously dependent on us. . . . No, the nations will do as Russia is doing."[57] Believing that the Soviets would acquire a half-century of experience through the Ford contract, he was content to think that

FORD AUTOMOBILE ASSEMBLY LINE IN MOSCOW, 1930. FIRST MODEL A ASSEMBLED IN THE SOVIET UNION

industrialization meant prosperity and that prosperity would bring world peace. Those who recalled Ford's ambitious World War I venture in commissioning a peace ship to sail to Europe to persuade the belligerents to lay down their arms could hardly doubt his sincerity—and his determination to sell Ford automobiles.

When the Nizhni Novgorod assembly plant began production in February 1930, the townspeople celebrated wildly, visiting the factory in droves. Serious professional men asked to work at the assembly line for a time just to take part in the momentous process. At a celebratory banquet, the Russians tossed first the Ford supervisor in a blanket, then a much-decorated general present for the occasion. Shrewdly, the Soviet officials insisted that the Ford supervisor take a luxurious vacation on the Black Sea for several months until they were sure the assembly plant in

Nizhni Novgorod and Moscow would run without his presence. In January 1932 the large production plant began operations. River Rouge was established in Nizhni Novgorod on the Volga; Muskovites could sport Model A's. Ford financial records do not make clear whether his company made a profit or suffered a loss from the contract, for the Soviets purchased less than half the vehicles for which they had contracted. Since it was changing over to the V-8 engines, however, the Ford company shipped to Nizhni Novgorod for the production plant as much as $3 million in production equipment that would otherwise have been discarded.[58]

MAGNITOGORSK

The forcers of a technological revolution also dared the prodigious feat of constructing a steel complex based on the one at Gary, Indiana. The complex lay behind the Urals at Magnitogorsk, a village on the Kirgiz Steppe in Russian Siberia. Near the village were two small mountains rich in magnetized iron. The mountains of iron ore situated on the eastern slope of the Urals near the Ural River, had aroused superstition and interest ever since early explorers had noted that compass needles were deflected near the mountains. In the eighteenth century this remarkably rich source of magnetic ore was mined, but the great distances from behind the Urals to markets in western Russia and the primitive means of transport kept the output of iron small. On the eve of World War I, foreign entrepreneurs funded modest operations with the cooperation of the tsarist regime but, until the Soviet first Five-Year Plan, exploitation of ore was limited. The Soviet leaders projected an ambitious plan calling for the construction of an integrated regional system of production. They wanted nothing less than the world's most modern and largest iron-producing facility. Plans specified facilities for magnetic separation, concentration, and sintering of ore, eight giant 1,500-ton blast furnaces, twenty-eight (later forty-two) 150-ton open-hearth furnaces, three 25-ton Bessemer converters, forty-five coke ovens with related by-product chemical works, and three rolling mills.[59] The iron- and steel-making facilities were to be part of a larger regional complex including gold, platinum, silver, copper, nickel, lead, and aluminum mines; machine-building and armament-producing plants at Sverdlovsk; a tractor factory at Chelyabinsk; machine-building units at Orsk, Ufa, and Perm; nonferrous metallurgical installations near Orsk; railway-car works in nearby Nizhni Tagil; oil fields and oil refineries at Ishembayevo; and transportation connections with fourteen-hundred-mile-distant coal resources at Kuznetsk in Siberia. Another source of coal for the complex was distant Karaganda. The plan also called for socialist working communities to be built in the barren region. The scheme, a creation of grandiose regional planning, was called the Ural-Kuznetsk-Kombinat.[60] The Soviets intended to show that they could learn from and transcend the capitalistic system of production.

Peasants from Russia and nomads from Siberia streamed into the project at Magnitogorsk, seeking higher wages and better living standards, some having escaped from the newly collectivized farms. Engineers and technicians from abroad and from the Soviet Union soon appeared at the construction site. Arthur G. McKee & Company of Cleveland, Ohio, was the principal foreign contractor at Magnitogorsk. Freyn Engineering Company, another U.S. consulting-and-construction firm, shared with Soviet engineers the responsibility for design and construction of the Kuznetsk ironworks, which were systematically linked to the facility at Magnitogorsk. Since 1927 Freyn had had fourteen of its engineers advising the Soviets on general plans for the development of their metallurgical industries.[61] Originally McKee was to design and supervise construction of all of the Magnitogorsk iron and steel facilities but, because of the company's inability to furnish sufficient credit, other firms received subcontracts. The contract for construction of the rolling mill went to the German consulting-and-contracting firm of Demag and Klein. Responsibility for the coke plant was taken over by the U.S. firm of Koppers and Company. McKee retained responsibility for the blast furnaces and mining operations. The various Soviet engineering organizations supervised construction of the open-hearth furnaces, transportation system, water supply, and other facilities.

John Scott, a young American, has left an account of his experiences as an American worker in Russia's city of steel. (After World War II, he became senior editor of *Time* magazine.) Working there for five years, he saw much sweat and blood, but also "a magnificent plant built."[62]

Scott left the University of Wisconsin in 1931, apprenticed as a welder at General Electric, and brought this skill to the Soviet Union. He recalled that he had been sadly disillusioned by the Depression in the United States, had found little use for his energy and enthusiasm as a twenty-year-old, and decided to "lend a hand in the construction of a society which seemed to be at least one step ahead of the American."[63] Scott, who lived with the workers assigned to the blast-furnace construction, believed their living conditions to be better than those in nonindustrializing areas. Nevertheless, he saw many workers die or suffer terribly from cold, hunger, fatigue, and the industrial accidents that often resulted from the inexperience of workers and foremen. Tens of thousands of political prisoners, dispossessed Kulaks, and others also worked at Magnitogorsk under the surveillance of the secret police. Scott did not think that their lives were much different from those of the free workers, except that they had little opportunity to learn skills. Using "political" tactics, special political troubleshooters, representatives of the Communist Party, came from Moscow to enforce schedules and quotas. Scott believed them sources of initiative and energy, able to force the work forward despite their unnecessary intriguing and heresy hunting. He also wrote, in 1942, perhaps stimulated by the Soviet Union's stand against Hitler's Germany, that "Stalin's indomitable will and his ruthless tenacity were responsible for the construction of Magnitogorsk and the entire Ural and Western Siberial industrial areas."[64] But he also recalled the many casualties from cold, hunger, fatigue, and industrial accidents.[65]

The ranking managerial personnel usually possessed histories of party political activity and subsequent training in technological institutes. Because of their special skills, engineers of the old regime, though suspect in the eyes of the new regime, were used and, in order to maintain their morale, treated fairly well. Young communist engineers fresh from Soviet engineering schools were enthusiastic, but not particularly well trained. The work foremen often achieved their positions by attending evening school after a day's exhausting labor. Evening schools for workers at the construction sites offered free political-indoctrination courses and technical courses, but enrollment in the technical courses required a more substantial educational background. The schools had a high priority among the project's leaders. Successful completion of courses meant consequent improvement in status, better living and working conditions. Persons who had successfully completed the highest level in the Soviet technical schools and obtained a degree comparable to that of engineer received wages six to eight times those of the unskilled workers at Magnitogorsk. Scott, however, was not particularly impressed with the academic level in the engineering schools: a Russian friend of his could barely design a truss after two years of evening civil-engineering work. Upper-level managers made twenty or thirty times as much as laborers. U.S. and other foreign engineers had their own living quarters and the highest living standards.

The Soviet planners brought Ernst May, one of Europe's foremost avant-garde architects, to plan a modern city for the workers at Magnitogorsk. In the 1920s, May had used factory techniques to build modern housing settlements in Frankfurt am Main. The presence of May shows how determined Soviet leadership was to establish modern industry and the forms, symbols, and way of life thought of as modern. In 1930 he and his staff paid their first visit to the site of the Soviet city they were planning. On the railroad trip to Magnitogorsk, the coach for May and his staff was provisioned with crates of caviar, chocolate, cigarettes, and sausages. At almost every station stop, peasant women brought fresh eggs, milk, butter, and poultry. John Scott reported, by contrast, that the Magnitogorsk workers usually had thin, watery soup and black bread. During their four-day stay at the site, the May contingent drove Ford cars and used the small, clean, excellently provisioned settlement for American experts. May found all of the engineers and managers of the American consultant firm energetic and clearheaded.[66]

After it was built, the new settlement was not "really a very good example of a Socialist city."[67] Consisting of some fifty large, balconied apartment houses three, four, and five stories high, painted various colors that stood out attractively against the winter snow, and surrounded by open squares, fountains, flower gardens, and playgrounds, the apartments were desperately overcrowded in 1937, with as many four or five people per room. Apartments were equipped with electricity, central heating, running water, and bathtubs. The latter were usually used for storage, because the Russians preferred community bathhouses.

Despite the presence of foreign consultants, advisers, and equipment, despite efforts at training and education, despite the enthusiasm of young Soviet engineers, and despite the drive of party representatives, Magnitogorsk experienced unending frustrations and small-gauge failures in carrying out the bold projects of a Five-Year Plan. Unskilled labor used imported machinery ineffectively, and unrealistic schedules bred distorted bureaucratic progress reports and corner-cutting. R. W. Stuck, an American engineer employed by McKee, reported that the Communist Party functionaries insisted that the first blast furnace start up in January 1932, even though it was only three-quarters complete. He also recalled that the Soviet authorities had the furnace's great stacks built first because they made impressive publicity pictures. Stuck, not mindful of politics and power, found such considerations absurd. Inadequacy of transportation and materials-handling facilities, bureaucratic overhead, disastrous and unexpected shortages of supplies, and poor planning, allocating, and coordination also plagued the project. Projected plans were unfulfilled by the end of the first Five-Year Plan. The operation was too poorly coordinated to produce millions of tons of iron and steel. The project was revised and schedules were moved forward to the end of the second Five-Year Plan. In 1937 the great purge struck the project. "No group, no organization, was spared." The methods used in the investigations "were indefensible according to most civilized standards."[68] Thousands were arrested, incarcerated for months, and exiled.

Around 1934–35 emphasis had begun to shift from construction of the means of production, to production. The transfer of responsibility from foreign engineers and specialists to personnel took place in 1936 and 1937. Foreign engineers who had been pampered by the Soviet Union began to be accused of various obstructionist tactics. The young Soviet engineers replacing the foreigners were more experienced than they had been earlier and began to enjoy the prestige and respect that had formerly been granted to the foreign specialists. Russian workers also began to prove themselves. A blooming mill, which had been shut down frequently

MARGARET BOURKE-WHITE, *BRICKLAYER, MAGNITOGORSK*

in 1934 because of inexperienced personnel, lack of fuel, electricity, and malfunctioning of poorly maintained electrical equipment, began processing all the steel the open-hearth furnaces could produce by 1935. In 1934 the personnel could not efficiently operate a rolling mill equipped with the best machinery available, but the mill ran at capacity after January 1935. It was in this period of growing achievement that Scott believed the presence of party members among the workers had a good effect upon morale and work achievement. He was not impressed, however, by the Stakhanov movement, named for the miner who had achieved a fantastic work output and who was intended to be a model for all Soviet workers. Scott found that the quota demands of the movement resulted in workers' overusing equipment and failing to give proper time and attention to maintenance.

By 1938, despite the favorable trend, Scott believed the plan had been fulfilled only to about forty-five percent of the ultimate goal.[69] Nevertheless, he was impressed by the miracle that had been wrought in the desolate wilderness. Despite the failure to meet projected goals, the Soviet Union could boast that Magnitogorsk was producing more pig iron than all the plants in Czechoslovakia, Italy, or Poland. Yet, as another observer noted, the full capacity of Magnitogorsk for the foreseeable future could not even produce the steel rails the Soviet Union needed to build the railroad system required by the giant regional complex.

WEIMAR AND THE AMERICAN MODEL

Prostrate from war and revolutions, the Soviets looked up to the United States for a way to produce a new society. Defeated in an exhausting and demoralizing war, Germans also looked to the New World as a model for industrial and technological development. During the interwar years the Germans made a second discovery of America. Countless books and articles explained to them the transformation of a wilderness into the world's most productive nation. In 1924 an American, Charles Dawes, later U.S. vice-president, proposed a plan to facilitate German payment of reparations. This spurred Germany's economic recovery and stimulated large U.S. investments. German envy of and admiration for American economic and technological power dramatically increased. Germans looked to the source of the American dollars shoring up their newly constituted and fragile Weimar Republic. They found there the mightiest of modern nations, a mode of production pouring forth forty-nine percent of the world's finished products. Julius Hirsch, a professor at the University of Cologne and later a member of the German Ministry of Economics, wrote, in *Das amerikanische Wirtschaftswunder* (1926), of a prosperity built on a new form of industrial organization.[70] German fascination with the New World in the 1920s compares to—even exceeds—American fascination with Japan in the 1970s.

During the brief period of post–World War I stabilization, many German supporters of the Weimar Republic took American technological civilization as a model to imitate. They admired not only America's highly rationalized production system but its high standard of living and its functioning democracy. Many admirers decided that the interaction between technology and democracy defined the essence of America. The liberals concluded that a German adaptation of the combination could strengthen and preserve the shaky Weimar Republic. Revisionist socialists saw American large-scale, centralized-production technology as a step toward the socialist society. Weimar Germans believed they had discovered a modern land in which there was peace between labor and capital, steadily rising wages, rapidly increasing consumption, and generally a hitherto undreamed-of prosperity. Many thought the United States to be the first classless society, where everyone had an equal opportunity to—and many did—own his or her own house, car, and, last—and perhaps not least—to enjoy personal freedom. The *sine qua non*, these Germans assumed, was the organized, centralized, large-scale system of production. "Taylorismus" (scientific management) and "Fordismus" (mass production) constituted "Amerikanismus." Many Americans mistakenly assumed that foreigners looked to colonial Philadelphia as the cradle of the republic. Instead, foreigners believed that modern Pittsburgh and Detroit had forged the new nation.

Even before 1914 Germans had seized on the Taylorite doctrines as a key to understanding the remarkable productiveness and power of human-made America. In so doing, they were greatly reducing the deeply complex nature of America's technological achievement, but Taylorismus in Germany, as in the case of Fordismus during the interwar years, became a political program and an agenda for social change, not just a mode of raising worker efficiency and industrial productivity. German proponents of Taylorismus extended Taylor's doctrines to the far horizons that he himself had envisaged in his later years. By the outbreak of World War I, Taylorismus was a household word and could be read about in almost every newspaper.[71] Taylor's *Shop Management* and *Principles of Scientific Management* appeared in Germany soon after publication in the United States. Gustav Winter, journalist, editor, publicist, and self-anointed disciple of Taylorismus in Germany, boasted in 1920 that his booklet on *How to Introduce the Taylor System in Germany* had sold a hundred thousand copies.[72] In 1920 Winter promised that Taylorismus, properly adapted and applied to German conditions, would place the "deathly ill" German economy back on its feet. Other enthusiastic Germans transformed Taylorismus, as they would Fordismus, into a social and political philosophy.[73] The Taylor enthusiasts came from a broad spectrum of economic interests and political persuasions. Liberal supporters of the new parliamentary Weimar Republic, conservative business leaders, trade-union leaders, revisionist socialists, and even reactionary intellectuals found reasons to support the scientific-management doctrines of Taylor, at least as each variously interpreted them.

On the other hand, countless German workers, like the Americans, opposed Taylorism when they saw it as giving management more control over the workplace and work routine, as de-skilling workers, and as a possible cause of unemployment. Defenders of Taylorism insisted that Germans could avoid the American mistakes.

The common goal of those advocating Taylorism's diverse elements was efficiency and increased productivity. The supporters of the new republic believed that increased productivity would bring higher living standards for workers and thereby woo them away from the radical left and revolution. Many business leaders wanted increased efficiency and production for larger profits, and some believed that the increase of both profits and wages made possible by Taylorismus would greatly reduce the strife endemic in Europe among classes and between labor and management. Reactionary intellectuals believed that traditional German culture could survive in the modern world only if reinforced by the technological power to which Taylorismus provided access. Furthermore, Taylor's passion for planning and order resonated well with traditional German, especially Prussian, attitudes. Many Germans also took him as one of their own, because he was born in the Germantown section of Philadelphia, which they assumed meant he was of German background.[74]

During the 1920s Ford supplanted Taylor as the American messiah who could show Weimar the way to mass production and social harmony. Publication in 1923 of a translation of Ford's *My Life and Work* (*Mein Leben und Werk*)[75] prepared the German readers for a rash of books on Fordismus. The autobiography quickly became a best-seller, with sales

WORK AND SAVINGS IN ALL PROFESSIONS THROUGH TAYLORISM

of more than two hundred thousand copies. In 1925 Paul Rieppel, in a book about Ford management and Ford methods that was characteristic of the enthusiasm for Ford in Germany,[76] insisted that Henry Ford had an unprecedented and incomparable genius for creative activity and that he was first among those promoting the welfare of his country. Rieppel noted ironically that "the good old Prussian ideal of service to the people, which contemporary Germany had virtually lost, appears suddenly in America, from which we had only expected the most crass materialism."[77] The spirit and practice of Fordismus would, he predicted, repair the breach between labor and management in Germany. Instilled with Fordismus, the workers would display once again the craftsmanship and work ethic for which old Germany had become known. Believing it to be a high compliment, Rieppel called Ford "the greatest Prussian in America" and a fulfillment of the ideals of Prussian socialism as defined by Oswald Spengler, author of the immensely influential *Decline of the West* (1918–22).[78]

In 1924 Friedrich von Gottl-Ottlilienfeld, an influential professor of political economy at the University of Berlin, along with the economic historian Werner Sombart and the sociologist Max Weber, published *Fordismus: Über Industrie und technische Vernunft* (*Fordism: Concerning Industry and Technical Rationality*).[79] He carefully laid out the ingredients of the Ford system: standardization of product (the Model T); vertical integration of production processes from raw-material production to assembly line; mass production to lower unit costs; constant flow of production from raw material through the dealer's showroom; and, above all, organization in accord with the principles of "rationalization," often a German synonym for "efficiency."[80] Carl Friedrich von Siemens, a member of the family that established and presided over the electrical-manufacturing giant, one of Germany's greatest industrial enterprises, provided the foreword for another of Gottl-Ottlilienfeld's books, *Vom Sinn der Rationalisierung* (1929), in which he described the Ford method as the epitome of rational systematization. Later a National Socialist, Gottl-Ottlilienfeld argued that Fordismus represented a higher stage of capitalism that had absorbed socialistic elements and was dedicated more to the common good than to profit making. Ford, he continued, was no reactionary, but a progressive who saw the danger of red socialism and proposed as an alternative a white socialism filled with the spirit of creativity and technical rationality.[81]

Germans like Gottl-Ottlilienfeld and Rieppel were heartened by their belief that Fordismus would soothe the restless workers without bringing drastic social change or revolution. A socialist, Jakob Walcher, however, took the unpopular stance of attacking Fordismus in *Ford oder Marx* (*Ford or Marx*), published in 1925.[82] Walcher saw as illusory the promise of harmony between worker and management under Ford's brand of capitalism. He rejected the supposed contrast stressed by the Ford cult between exploitative capitalism and Ford's so-called creative capitalism. He called attention to the authoritarian characteristics of the Ford empire, such as the close surveillance of the workers' private lives and Ford's hostile attitude toward unions and strikes. He predicted that Ford production would oversupply and exhaust the market, despite the regular lowering of automobile prices, and he foresaw unemployment. He believed that the practices of the American industrial giants would cause another imperialistic world war. Fordismus, he concluded, bore witness to the need for communism.[83] Walcher disdained Ford's flagrant anti-Semitism. Relatively few Germans in the heyday of Weimar, however, shared Walcher's view of Ford as the "false messiah."[84] Few foresaw, either, that the Nazis would find spiritual sustenance in aspects of Ford's philosophy and admit him to their Valhalla as a creative entrepreneur.

The far more influential and popular Gottl-Ottlilienfeld boasted that he had introduced the term "Fordismus" and spread the benign and progressive concepts associated with it. By "Fordismus," he and other Germans had in mind far more than the Ford mode of production; they took Fordism as a practical economic and social philosophy, one appropriate for Weimar Germany. Ford, they pointed out, had shown that higher wages led to higher consumption, higher consumption to greater production and higher wages, and so on, in a great spiral. The enthusiasts accepted Ford's self-portrayal as an entrepreneur motivated primarily by an ethic of social service, not one of profit. In his 1927 book, *The Great Today and the Greater Tomorrow* (*Das grosse Heute und das noch grössere Morgen*), Ford assured his German readers that his social system, with its higher wages, would end the attractiveness of communism and revolution for the worker. Ford promised that his followers could transform "the wasteland of industry into a blooming garden." His books became a canonical statement of the Weimar stabilization period.[85]

Idealizing labor relations at Ford in a way curiously like the way in which Americans idealize Japanese labor-management relations today, German disciples of Fordismus wanted the peace, harmony, discipline, and morale of Ford plants spread through their entire society. In support of Fordismus, liberals allied with elite trade-unionists against feudal or agrarian interests in Germany.[86] Trade-union leaders and revisionist socialists saw Fordismus as a peaceful transition to white socialism. They were also impressed that the Soviet leaders believed that the American mode of production, when adapted by socialism, would provide the route to the new society.[87] Fordismus appealed not only to liberals and to the left, but to segments of the political right. Conservatives of anti-Semitic tendency associated Ford with healthy, Germanic creativity and production; they relegated Jews to the sphere of soulless commercial transactions.[88] Early Nazi leaders, especially the engineer and party ideologue Gottfried Feder, praised Ford as a productive, capitalistic entrepreneur. They contrasted his positive virtues with the negative ones of the capitalistic, parasitic, Jewish financiers, whom they found more international than German. Ford's anti-Semitism made him no less acceptable to the Nazis.[89]

With the influx of Taylorismus, Fordismus, and U.S. money and business and financial influence into Germany, German engineers, factory managers, journalists, and academicians traveled to the New World to see the source for themselves. They voyaged to the New World in order to see the future. Their enthusiastic preconceptions usually colored their reports, their *Wunschbilder* (wishful images).[90] Otto Moog, a German engineer, reported on his voyage of discovery in a book, *Drüben steht Amerika* (*Over There Stands America*).[91] As was the case with so many other Germans, a visit to Detroit and Ford's Highland Park plant was the high point of the tour for him, a chance to pay homage at the shrine of mass production and to Ford, the "*Automobilkönig.*" The ceaseless to-and-fro of the four hundred thousand or so Detroit automobiles, the workers' small but freestanding houses with tiny green lawns, and the omnipresent spirit of Ford fired Moog's imagination. Of the Highland Park plant he wrote: "No symphony, no *Eroica*, compares in depth, content, and power to the music that threatened and hammered away at us as we wandered through Ford's workplaces, wanderers overwhelmed by a daring expression of the human spirit."[92] Whirring and flashing machine tools, great presses sounding like artillery as they stamped out automobile parts, and the turning, climbing, and descending conveyors left indelible impressions. Afterward he saw the foundry and blast furnaces at River Rouge, the "colossal" spaces filled with immense numbers of workers so concentrated on their tasks that one could not exchange a word with another. Patrolling foremen stepped in to caution him if any worker missed a beat in the rhythm of production. Observant and curious, Moog asked about the rows of clothes hung among the machinery and learned that they were the street clothes of the workers, which they donned, with no dressing-and-washing room as in Germany, to wear in their automobiles during the ride home. Not impressed by the Spartan working conditions Ford provided, Moog took the high wages—and the worker-owned automobiles—to be a compensation. He met a Ford worker who earned 85¢ an hour and who in eleven years had saved $25,000, in part through speculation, "which almost all Americans do, especially in a booming . . . [city] like Detroit."[93] Recently that worker had traded in his Ford for an elegant $1,100 Nash that now stood before the $2,800 house he had built. Moog found the self-sufficient Detroit workers full of spirit, unlike the sullen German workers with their downcast eyes. Organized plant tours and selected interviews probably persuaded him that the Ford workers were satisfied by improvement in income and fulfilled by the realization that they were a part, albeit small, of a historical transformation of the process of production.

Franz Westermann, another German engineer, also felt compelled to make the pilgrimage and write a book about "Amerika," even though scarcely a week passed in Germany without newspapers, journals, books, or lectures telling something old or new about the United States.[94] He believed it his responsibility to impress on his countrymen an understanding of the "marvelous" economic life, the perfected factory system, and the pervasive efficiency and productiveness of America. Not given to understatement, Westermann began his journey full of anticipation. America was for him the land of "Onkel Tom and Winnetou" (the heroic Indian of German popular literature), a country that filled the engineer with longing, a land of skyscrapers and automobiles, "of Taylor and Ford." After visiting Manhattan, the Brooklyn Bridge, New York subways, and Niagara Falls, he felt Detroit exert its irresistible attraction. Chicago's stinking stockyards with their disassembly lines only briefly detained him. He could encompass all that he had seen until he got to Detroit; then impressions overwhelmed him. The General Motors office building, with its elevators for twenty-four persons, the body works of the Fisher Company, the giant spread of the Ford plants, as many automobiles as pedestrians, and his hotel with a thousand rooms—these filled him with awe. (A half-century later Americans find it difficult to imagine the enchantment of technological America, or to call Detroit to mind without grim thoughts of murder statistics and a decaying city center.) With Dante-like undertones, Westermann recalled that a hellish concert of the sounds of mass manufacture alarmed him as he entered the Highland Park plant, but that the untroubled, even smiling faces of the workers emboldened him to proceed. Taking the ever-popular tour, Westermann learned from his German-American guide that Highland Park was a systematically linked maze of individual factories, each concentrating on the manufacture of a single automobile part, and each populated by an anthill of diligent workers. Westermann wrote that he had always been moved, like so many of his countrymen, by the beauties and romance of nature. He had seen the shimmering surface of woodland lakes on nights of the full moon; he had felt the power of the endless ocean while standing on the tossing deck of a steamer in a storm; and he had been deeply moved by the sight of snow-covered Alpine peaks and dark mysterious valleys. Yet "the most powerful and memorable experience of my life came from the visit to the Ford plants, where the hand of man had created in a short time a gigantic production complex, which not only through its size and technical characteristics made a staggering impression, but also filled the viewer with the powerful organizing spirit of its creators."[95] At every turn a new machinescape, "a Bacchanal of work," stimulated the engineer.

Not all Weimar Germans shared such enthusiasm for the United States, Taylorismus, and Fordismus. Ambivalent attitudes toward technology were far more characteristic of Weimar than of America. Conservative and reactionary Germans, unlike the Weimar liberals, associated the United States with materialism and formless, soulless, chaotic liberalism and capitalism. At the same time they grudgingly acknowledged the prodigious achievement of its production technology and believed that America's economic and political power derived from it. So they presumed that the Old World could and should take American technology, purge it of soulless, materialist capitalism, and infuse it with aesthetic, philosophical, and spiritual values to establish a superior German culture—as contrasted with America's modern civilization. Those of such mind have been called "reactionary modernists."[96] Unlike most Americans, the reactionary modernists in Germany believed that modern technology and capitalism were distinguishable. In *The Decline of the West*, Oswald Spengler popularized the distinction between a material and formless civilization and a coherent, organic culture.[97] The reactionary modernists aspired for a German culture with both soul and American technological muscle. Many other Germans of the right, however, rejected Amerikanismus, technology and society alike. *Amerika und der Amerikanismus* (1927) by Adolf Halfeld became for them the centerpiece of the resistance to all that America represented. Halfeld's chapter titles expressed the prevailing clichés and prejudices: "The Business State," "Chains of the Spirit," "The Omnipotence of the Idea of Success," and so on. American materialism, rationalization, and mechanization of life posed deadly threats, many felt, to German culture, a culture for which many Germans believed they had fought during World War I.[98]

After the National Socialists came to power in 1933, their ideological leaders adopted an attitude toward technology similar to that of the reactionary modernists. Fritz Todt, a Nazi and an engineer, whom Albert Speer succeeded as minister of armaments and war production in 1942, when Todt died in an airplane crash, articulated an ideology for German engineers. He drew heavily from Adolf Hitler's views on technology and culture as expressed in *Mein Kampf.* Todt expressed the Nazi reaction against Weimar—and, therefore, American—modernism, which the Nazis characterized as soulless and materialistic in its single-minded commitment to the production of consumer goods. The Nazis wanted the inventors and engineers, the German equivalents of Edison and Ford, to solve problems, but they wanted these problems to be defined by the Nazi politicians, the bearers and creators of culture. In Nazi Germany, Todt insisted in his speeches to the engineers, technology would be a means for fulfilling the work ethic, satisfying heroic and creative instincts, making a German habitat, and establishing the Aryans in their rightful role as the master race. Todt said, "Who would solve material problems by material means alone will be dominated by the material. Mastery will come through the spirit. We idealists will master dead matter through the Nazi spirit of combat and will." Hitler wrote in *Mein Kampf,* "everything we admire on the earth today—science and art, technology and inventions—is only the creative product of a few peoples and originally perhaps of one race."[99] In essence the Nazis were not rejecting modern technology, but the modern, technology-based culture that was being articulated by the avant-garde in Europe and the United States, a culture which by Nazi standards was soulless, material—and prodigiously productive.

7 THE SECOND DISCOVERY OF AMERICA

During the opening decades of this century Europeans, including the Russians, began the second discovery of America. The first discovery had been that of the virgin land, nature's nation; the second was of technology's nation, America as artifact. Some foreigners continued to associate the United States with the frontier, but others realized that for more than a century the nation had been the world's most active construction site. Americans, too, realized that the nation was undergoing a rapid industrial transformation, but, with the perspective of distance, Europeans perceived the transformation to be more than a technological and an industrial revolution, that it bore the seeds of a cultural mutation as well. European intellectuals, architects, and artists making the second discovery believed that the United States was leading the world into a uniquely modern era. With their deep cultural sensitivity, Europeans felt that they could best articulate the modern culture that modern technology was inaugurating.

In the United States, the definers of technological and social change, accustomed to celebrating their nation's material advances but modest about its cultural achievements, focused on what they believed to be a second industrial revolution rather than a cultural transformation. Celebrating industrial revolutions was, and is now, an American pastime.

Since World War II, enthusiasts for various technologies have announced the beginnings of diverse technological, or industrial, revolutions. They have celebrated an industrial revolution brought, they believed, by the advent of nuclear power. More recently there has been talk of the computer, or information, revolution. Other enthusiasts have told us that space flight, or exploration of "the new ocean," would also inaugurate a new era. Earlier, prophets of a "second industrial revolution" anticipated that it would bring even greater social transformation than the first Industrial Revolution, which had occurred originally in Britain during the eighteenth century and then in the United States and Western Europe in the nineteenth. This revolution introduced the technology of steam, iron, and textile factories. Drawing a lesson from the laissez-faire attitudes during this revolution and the consequent social disorders, prophets of the second insisted that they must plan technological and related social change during the new industrial revolution. They dated the origins of the second industrial revolution variously, usually around 1870 or 1880, the era of the rise of the electrical industry. But the early 1920s was the time of greatest excitement among contemporaries who believed they were experiencing such a revolution. On a superficial level, they associated the new industrial revolution primarily with electric light and power, with the internal-combustion engine and its use in automobiles, and with airplanes, wireless communications, and synthetic, or human-made, chemicals. On a deeper level, social critics stressed the social changes coinciding with the appearance of new machines, devices, and processes. They were aware that inventors and industrial scientists had organized the process of invention and discovery, and that there was a greater possibility than ever before that human beings could create and control their material environment. Reflective American commentators pointed out that in the modern world the environment was becoming less and less natural and more and more a created artifact. They believed that, like the British Industrial Revolution, the newer one involved social, institutional, and political changes. The new wave of industrialization, the so-called second industrial revolution with its unique set of characteristics, was particularly American. Only the newly united German empire among the industrial nations rivaled, and in some instances surpassed, the United States after 1870 in its development of peculiarly modern machines, devices, and processes, and in the cultivation of modern institutions such as giant industrial corporations, industrial laboratories, engineering schools, and research universities.

Architects and artists, first in Europe and then in the United States, sought forms and symbols to express the emerging technological culture. Lewis Mumford in 1921 defined style as "the reasoned expression, in some particular work, of the complex of social and technological experience that grows out of a community's life." He asked: "How is it that the modern style has been so slow to realize itself—is still so timid, so partial, so inadequate?"[1] A year later, however, he wrote: "With the beginning of the second decade of this century, there is some evidence of an attempt to make a genuine culture out of industrialization."[2] By then both Americans and Europeans were beginning to see themselves as uniquely modern, and the age in which they were living as the first modern one. They perceived the early twentieth century as the beginning of a modern age extending into a limitless future. Their belief in the power of modern technology to bring a second creation of the world was perhaps the major reason for their belief that the modern age had arrived. First German, then American, artists, architects, and social critics wished to transform a materialistic technical civilization into a modern culture. Their use of "culture" transcended art and architecture, and referred to an embracing array of human thought and activity. They wanted to find the artistic and institutional forms to express the values and meanings of modern technological culture. They believed that technology, thought of as a means of transforming the material world, had been brought to an unprecedentedly high state of development, and that it was the characteristic modern achievement. No wonder those who believed that a second industrial revolution was under way and that a new culture based on technology was being defined could encourage an interwar generation to believe that it was living in the modern age. Robert Hughes has perceptively written of modernism, "between 1880 and 1930, one of the supreme cultural experiments in the history of the world was enacted in Europe and America . . . a historical space that we can enter, look at, but no longer be a part of."[3]

SECOND INDUSTRIAL REVOLUTION OR NEOTECHNIC AGE?

The coming of steam and iron, the rise of the industrial middle and worker classes, the physical transformation into "black country" of the Midlands and northern England, and the reinforcement of political, laissez-faire liberalism had characterized the earlier Industrial Revolution. The second brought the perfection of mass-production technology, the giant industrial corporation, the mushrooming of the industrial city, the rise of higher technical education in universities and colleges, and the growth of a class of professional experts, especially engineers and industrial scientists. Progressive social reformers between the two world wars argued that, because the inhabited world was undergoing a second creation, the process should be planned, that it should embody benign and aesthetically satisfying values. They believed that the British Industrial Revolution had laid waste labor and the land because of the domination of heedless capitalists and narrow-minded engineers. Social planners, or social engineers, they argued in the 1920s, should play leading roles in the second industrial revolution.

Progressives and social reformers of the early twentieth century saw, as historians have demonstrated recently, that during the so-called second industrial revolution the changes were not simply technical but also managerial.[4] The spread of Taylorism and Fordism and sharp increases in productivity made this clear, as did the increase in centralized control of large production systems by such private corporations as Insull's utility empire, and by investment houses, like those that financed the spread of communication and power networks. Centralized management and control were bringing the coordination and integration of the means of production, distribution, and marketing. Progressives and social reformers asked why management and control—and planning—should not be in the hands of government rather than private enterprise.

The increased flow of goods during the period after 1870, or the second industrial revolution, often caused crises of control. Managers were frequently unable to control the heightened flow through the production, distribution, and marketing system. Not only was the invisible hand of the market functioning poorly, but the visible hand of the managers often failed as well. The response to these crises has been called a "control revolution."[5] The organization of bureaucracies, the rationalization of these to simplify procedures and increase efficiency, the development of communication systems, such as the telegraph, postal system, telephone, and wireless, and the invention and development of means of acquiring, storing, and retrieving information were among the responses.

Between the world wars, progressive and radical reformers committed to the perfection of technology believed that the creative power of organized invention and research, the new techniques of management and control, and the power of electricity were providing a broad avenue leading to a new and thoroughly modern society. The ethereal nature of electricity, combined with its great potency, appealed especially to those who believed that modern technology and culture were something new under the sun. Like Vladimir Lenin in the Soviet Union, American progressives believed that innovative government plus electrification would bring revolutionary social changes. Governments during World War I had introduced technological and scientific planning and control on an unprecedented scale. This reinforced the commitment to government initiative and control. Shortages of energy, especially coal and electricity, stimulated the belligerent nations to fund, among other measures, the building of large hydroelectric and coal-fired power plants and the forced interconnection of electric utilities to improve load factor. This increased the output of power from a given amount of capital equipment.[6]

In the Soviet Union, Lenin said, "Communism is the Soviet power plus the electrification of the whole country, for without electrification progress in industry is impossible."[7] In the United Kingdom, proponents of a national scheme for electrification insisted that it would transform Britain into a modern industrial power, and a nation once again in a position of world industrial leadership.[8] In Germany, supporters of the newly formed and fragile Weimar Republic believed that formation and nationalization of an all-German electrical network, or grid, could contribute greatly to saving the republic.[9] In the United States, a government-funded study of 1921 recommended the establishment of "Superpower" in the Northeast, or Boston-Washington, industrial region. The plan called for the construction of privately funded Superpower plants of 60,000-to-300,000-kilowatt capacity to be interconnected by transmission lines of 110,000 to 220,000 volts.[10] Progressive reformers saw Superpower and similar giant electrification schemes as much more than a way of increasing production: they believed that the technology of the second industrial revolution would bring a new society and a new culture.

Mumford was among the most articulate and influential of those envisioning a new industrial era based on new forms of electrical and political power. An independent writer, Mumford was an intellectual who wrote for the general reader about the pressing social, architectural, and technological issues of his day. In 1934 he published a seminal social history of technology, *Technics and Civilization*,[11] in which he delineated three ages of history: the eotechnic, the paleotechnic, and the neotechnic. The neotechnic was his designation for a period that others associated with the second industrial revolution. The neotechnic era for Mumford promised to be one of electric generators, water turbines, aluminum, new alloys, rare earths, and synthetics such as celluloid, Bakelite, and synthetic resins. In so periodizing history, Mumford was influenced by a similar scheme of Patrick Geddes, the Scottish sociologist and regional planner whom he greatly admired. Henry Adams, American historian and social critic, also caused Mumford to think of a modern transition from a "mechanical phase" to an "electrical phase."[12]

Mumford and Geddes characterized the ages of history in accordance with the prevailing energy and material technology of the times. In the paleotechnic age, for instance, the primary energy was steam, the material iron. Earlier, the eotechnic period had been one of wind, waterpower, and wood. Mumford, however, was no simplistic, technological determinist: he realized that social and psychological conditions had to favor the reception of a new technology. "The gains in technics," he wrote, "are never registered automatically in society: they require equally adroit inventions and adaptations in politics. . . ."[13] His grand schema for history also introduced layers of complexity, for each of his ages had its prevailing political and economic power structure closely tied to the technologies being used. Yet he did not slip into simplistic Marxism with its determination of societal and cultural superstructures by the means of production. Mumford did not believe that the factory owners, financiers, miners, and militarists of the paleotechnic age of iron and steam were simply creatures of technological forces, but that they were both shaped by and shaped the prevailing sources of energy, materials, and means of production. During the neotechnic or modern age he expected new political and social forces to emerge. He hoped that these would be benign and motivated by values other than economic gain and military aggression.

Mumford thought that electric power, the principal generator of change in the neotechnic period, would make possible the elimination of many of the evils of the coal-driven paleotechnic era, more commonly designated the British Industrial Revolution. In the neotechnic period coal would be converted at the mine mouth into electric power, or white coal. At the mine mouth, coal would not be burned wastefully, but first distilled, so as to drive off, refine, and utilize its most volatile ingredients as dyes, medicines, oils, and fabrics, leaving coke to be burned in the electric-power plant. With these technically advanced processes at the mine mouth and electrical transmission to manufacturing sites, urban factories belching smoke and cities smothered in smog would become archaic.

Mumford coupled the new energy sources and modes of transportation and communication with regional planning and development. Electrical power, with the automobile, the radio, and the telephone, could transform an industrial society of congested cities into one of economically and demographically balanced regions. No longer would the industrial and commercial population heap up at the mines, at railroad lines along valley bottoms, and at ports, but it could spread to upland and rural areas. Hydroelectric power could transform isolated mountain areas now suited only for forestry into industrial areas carefully planned to avoid the congestion and ugliness of paleotechnic industrial centers.[14]

The coming of electricity, then, especially hydroelectric power, persuaded Mumford that the West had raised the curtain on a new age. Whether civilization moved onto this new stage of history depended, of course, on our abandoning the economics and politics of the paleotechnic period. Profit- and power-seeking financiers, militarists, and miners, Mumford believed—and clearly hoped—would give way to scientific engineers and scientists as the administrators of the new world.[15] He cited as harbingers of the new era Michael Faraday, the selfless British physicist of the early nineteenth century; Joseph Henry, his American contemporary in physics; and Louis Pasteur, the French chemist and bacteriologist interested in the organic, not simply the mechanical.

The displacement of coal and iron by hydroelectricity, alloys, and aluminum would have profound geopolitical consequences, he predicted. Writing in the early thirties, Mumford anticipated that the industrial dominance of Western Europe and the United States during the coal-and-iron regime would give way as waterpower-rich Asia, Africa, and South America exploited the technics of the new era. Even within Europe and the United States, the industrial center of gravity was shifting to "Italy, France, Norway, Switzerland and Sweden . . . [and] to the two great spinal mountain-systems of the United States."[16] Electricity would also change the character of work and the workplace, and would redirect industrial organization. Workers would become technical supervisers of automated production processes, and these processes would disperse, rather than concentrate in industrial cities. Mumford, like so many prophets of the computer age today, anticipated that transmission of power and information would permit the utilization of small units of production by large units of administration. The large factory would be displaced by the small one, because machines need no longer be grouped to run from a belt drive but could be individually driven by electric motors. The widespread distribution of electric power and the standardization of manufacturing techniques would make possible manufacture in villages, "possibly even in the farm-home. . . . The little pieces for lawn mowers need no longer be made by men who live in the crowded homes of Philadelphia. . . ."[17] Also anticipating the visions of computer enthusiasts today, Mumford spoke of the displacement of the proletariat, by which he meant the end of assembly-line labor in massive factories and industrial cities. Small, automated-production plants would be supervised by highly trained worker-technicians, for during the neotechnic era power transmission lines and efficient small motors would permit efficient, small-scale production in far-flung sites characterized by pleasant living conditions.

Others of his day were even more enthusiastically committed to the concept of a new industrial revolution made possible by electric power. One of them was the governor of Pennsylvania, Gifford Pinchot. In 1925 he proposed for his state the Giant Power plan that specified giant, three-hundred-thousand-kilowatt mine-mouth power plants in the coal regions of western Pennsylvania and transmission lines of one hundred thousand volts or more to carry the power as far as the heavily populated and industrialized area of eastern Pennsylvania, two hundred miles distant. Distribution lines branching off from the cross-state transmission lines were intended to supply the small cities and rural communities along the way. Pinchot attributed the idea of Giant Power and the articulation of the plan to the reforming, progressive engineer, Morris Cooke, who as a federal adviser during World War I had seen and approved of the government planning and control of power resources.[18] Because of the strict and detailed regulation of the proposed Giant Power system by the state, this plan encountered the opposition of various interests and was not carried out, but the publicity and controversy surrounding it reveal how thoroughly committed its proponents were to the proposition that such power developments would bring a social and industrial revolution, and how determinedly they tried to persuade the general public of this.[19] Governor of one of the most industrialized and heavily populated states, Pinchot enthusiastically envisaged the spread of electric-power transmission bringing an industrial revolution as great in its effects as those of the British Industrial Revolution. He displayed a technological enthusiasm equal to the most fervent we hear today among the prophets of the information age.

Before World War I, as a leader of the Progressive movement and an ardent supporter of Theodore Roosevelt, Pinchot invented the term "conservation" to describe the scientific management of the environment. He believed that society, in not providing for the negative effects of the steam-powered Industrial Revolution, suffered from a long train of disturbances and struggle between capital on one hand and labor and agriculture on the other. He declared that the power of steam had shaped, for better and for worse, the centralized industrial order and the civilization of his day. Steam had brought such dramatic social changes as the prodigious increase of production, the rise of the urban complex, the decline of rural life, the decay of small communities, and the weakening of family ties. Because steam "might well say of electricity, 'One mightier than I cometh, the latchet of whose shoes I am not worthy to unloose,' " Pinchot warned that society must prepare for a greater technological and industrial revolution as well as provide against negative consequences and for opportunities of far greater magnitude than those of the steam era.[20]

The governor promised the Pennsylvania General Assembly that electricity could bring to the housewife the comforts of electric lighting, cooking, and other appliances, to the farmer the safety and convenience of electric lighting and power for milking, feed cutting, wood sawing, and countless other tasks. Pinchot went on to say that electricity could bring to every worker a higher standard of living, more leisure, and better pay. The electric-power revolution promised "to shower upon us gifts of unimaginable beauty and worth," to form the basis "for a civilization safer, happier, freer and fuller of opportunity than any the world has ever known."[21] "The day is coming," he prophesied, "when from morning to night, from the cradle to the grave, electric service will enter at every moment and from every direction into the daily life of every man, woman and child in America."[22]

In his optimism, Pinchot also predicted that electric power would reverse the trend that steam had brought toward industrial concentration, mass factory labor, and noisome, smelly, slum cities. Pinchot, however, focused the attention of his audience, the state legislature, on the political action needed if the electrical age were to fulfill its great promise of becoming "incomparably the greatest material blessing in human history."[23] Legislation was needed, he believed, to thwart an "evil spider . . . hastening to spread his web over the whole of the United States and to control and live upon the life of our people."[24] Pinchot made reference to an electric-power monopoly being established by the owners, especially large investment bankers, and managers of privately operated utilities. Surely Samuel Insull was one of those whom the governor had in mind. Pinchot, like Mumford, believed that the technological revolution must be accompanied by changes in the political and economic power structure.

Pinchot reasoned that electric-power interconnections would continue and that electric-power grids, or networks, would spread across state boundaries. He even expected an interconnection so vast in the United States that it would in effect be a national grid. The physical interconnections of the grid would bring with them the interconnections of operation and finance. These interconnections, he warned, were those of privately owned utilities and were being masterminded into a great private-enterprise monopoly that soon would transcend control and regulation by state governments. He compared the monopoly to one in which all sources of steam power in the country should be under the control of a single monster corporation. The primary goal of this gigantic monopoly would be profit making, not the public welfare. He doubted if even the federal government were arming itself to exercise regulation against the day of the national grid. So in his legislative package the governor proposed means for the state to regulate the spread of electric power. Private utility companies could participate in the great electric-power grid, but only if they provided vastly better service and vastly lower rates to the consumers, especially the farmer, the homeowner, and the small-workshop owner.

Joseph K. Hart, a university professor and an associate editor of the influential *Survey*, an American periodical that conveyed progressive social-science views to a broad readership, heralded with even more enthusiasm than Pinchot the opening of a new era of electric power.[25] Hart's views represented those of a community of liberal and progressive reformers, especially social scientists. Writing in the 1920s, Hart surveyed centuries of history, from the Greeks to the modern era.[26] Equating power with culture, he found man throughout history subduing or dominating nature to bring order out of chaos. Before the twentieth century, however, the exploitation of power had demanded a heavy price of society. The Greeks had enslaved and degraded nine-tenths of their population to create a system of services and production permitting the minority to live a serene life and cultivate philosophy, arts, and the sciences. The British Industrial Revolution brought steam power and a remarkable capacity to organize the world as a system of production. The price to be paid, however, was again terribly high. Hart blamed steam, "the great centralizer," for tearing people from their old roots and gathering them into polyglot centers. The outpouring of goods and services in the steam age also charmed men and women into elevating machines to the status of the means and ends of life. Furthermore, Hart wrote, "steam has torn us free from old standards of workmanship, taste and culture."[27] Thus, using the machine to master nature, man created a new master, the steam-driven machine, more orderly but less sustaining of the human spirit than nature, once the fearsome environment. Having used the abundance of steam power in the conquest of nature, modern man replaced nature with an artificial, urban environment; in turn, that environment has made human beings artificial. Humans need once again, Hart insisted, to hear the breaking of waves, to view silent snow fields, to explore the wilderness, and to walk along streams in green valleys. Hart, like many other Americans living in a period of rapid urbanization and industrialization, felt at the core of his being the loss of the land and rural life. Electric power, Hart continued, would end servitude to steam. Using electricity as a means of transmitting energy where needed, "men may feel the thrill of control and freedom once again."[28] This remarkable social and psychological transformation would come about because industry, with electric power, could be decentralized. With it the future "lies open before man, as it did in the day when Joshua said to Israel: 'Behold, I have set before you life and good, death and evil: choose ye this day which ye will serve.' "[29] Hart then projected a softly focused vision of small communities using the electric genii to realize intellectual freedom and "the culture of the spirit. . . . Such decentralization of living will tend to regenerate our culture by releasing it from the city's hot-houses, where it attains a superficial brilliance, and restoring it to its native rootage in reality."[30]

Not all of those sharing a vision of a new society made possible by electric power were political liberals and progressives inclined to displace the power of capital. Henry Ford, the master of River Rouge and Highland Park, also envisioned hydroelectric power as a way of decentralizing industry, deurbanizing the city, and revitalizing the countryside. After

World War I, Ford saw the possibility of physically decentralizing industrial production while maintaining closely centralized control. His intentions along these lines anticipated by a half-century present-day prophets who enthusiastically predict a new postindustrial era of decentralized, or neo-cottage, production. Today the spread of computer networks stimulates such visions. Waterpower networks caught Ford's imagination.

In wanting to "break up cities, narrow farming down to a twenty-five day year, and make producing villages, with their clustered homes and plants, the foundation stones of a new structure for American life and labor," Ford appeared not only an industrial but also a social reformer.[31] The mainsprings of his reforming zeal, in this case, seem to have come from his rural background and his late-in-life sentimentalization of the nineteenth-century agrarian Midwest. He fondly recalled his boyhood on his parents' farm at Dearborn, on the Rouge River, near Detroit, and especially building a primitive dam and mill on a small stream. On a small scale, young Ford was replicating the flour and other mills that still stood along the Rouge. When he was older, his painstaking reconstruction of nineteenth-century village life at the museum he founded at Dearborn testified to his wish to recapture this past. Paradoxically, Ford, who contributed so greatly to the urbanization and industrialization of Detroit, spoke longingly of a past when there were no slums, "or any of the other unnatural ways of living."[32]

Ford's vision in the early 1920s included electric power from dams, but he did not call for large electrical grids or networks. For the transport of energy he wanted to use rivers instead of transmission lines. He argued that "Streams are better transmission lines than wires. There is a good deal more loss in wires. Water goes over a dam and is power still. . . ."[33] Instead of transmitting power from a large, central dam site, he preferred each hydroelectric plant along the rivers to be independent of the others. Ford the ecologist also added that damming would create ponds, the evaporation from which would encourage more regular rainfall in the area. In his tidy, orderly, and circular world, use would increase availability. He also said he was pleased because workers and farmers could live along the pleasant shores of the ponds.

By 1924 he had made a substantial start along the Rouge River. At his home and large farm in Dearborn, he installed "Dam No. 1" to supply lights for the residences and farm buildings. Farther upstream, he converted the Nankin flour mill into a hydroelectric plant and factory where seventeen men used electrically driven machine tools to manufacture screws and Ford carburetor parts. At another converted mill at Plymouth, water turbines generated twenty-five horsepower, and twenty-five skilled mechanics worked at taps and machine dies. At Phoenix, 150 women made Ford generator cutouts in a village factory. Not far away, at Northville, 350 men made valves for the Model T motor, and at Waterford, another village, a new Ford dam provided 280 horsepower for a small Ford factory making gauges. Ford was dispersing some of the five hundred manufacturing departments at the Highland Park plant to these village waterpower sites. The parts made in these village plants and factories fed into the great Ford production and assembly systems. Eventually, he prophesied, only one or two major manufacturing processes would be concentrated at Highland Park. "A thousand or five hundred men ought to be enough in a single factory," he added.[34]

By the mid-1920s, in a similar decentralization move, the Ford company had located assembly branches in ten countries and assembly or service branches in thirty-four American cities. In so doing Ford had reversed the earlier trend to concentrate production under a single factory roof, but in this case the move was not limited to villages and small towns. The Ford company also constructed hydroelectric stations. At Saint Paul, Minnesota, for instance, a former federal dam provided a Ford factory with twenty-eight thousand horsepower.[35] Technical rationalization and economic efficiency could explain this large-scale decentralization, but, with his village mills and shops, Ford embarked on a far more visionary scheme, for which he gave additional reasons neither economic nor technical.

In small villages of ten or eleven houses like Nankin Mill, Ford pointed with pride to his workers, tidying up and painting their houses, and to the splendid drinking water. In another village, Ford happily reported, grocers began placing fresh goods on their shelves. So that the workers could balance industrial work with farming, Ford let the men off from the plants at planting and harvesting seasons. He once remarked, "I am a farmer. . . . I want to see every acre of the earth's surface covered with little farms, with happy, contented people living on them."[36] Ford also said that his costs were lower in the villages than at Highland Park, but, if unused property and projects unfinished had been taken into account, the village industries would presumably have shown a loss.[37] On the other hand, he probably found the smaller number of workers less prone to discontent and labor agitation. His motives were certainly overdetermined and mixed, but the results differed from those we associate with the Ford of the massive Highland Park and River Rouge plants.

L'ESPRIT NOUVEAU

As we have observed, Europeans looked to Henry Ford and the United States for modern technology, but not for modern high culture—architecture, art, and literature. European, not American, architects first sought ways to use in their buildings both modern production technology and an aesthetic vocabulary to express modern technological values, such as efficiency, precision, control, and system. In architecture, the Americans Frank Lloyd Wright and Louis Sullivan imaginatively used advanced construction techniques and materials, but the forms of their buildings did not become the prevailing modern architectural style of the first half of the twentieth century. Sullivan used steel framing in his skyscrapers and helped set the pattern of construction for the century. Wright urged architects to use reinforced concrete, glass, and standardized components imaginatively. Both men spoke admiringly of technological progress, and Sullivan identified bridge-building engineers as heroic figures. Yet both used biological and nature metaphors to describe their work, while the European pioneers of the modern or international style preferred machine and industrial metaphors to express their aesthetic commitment. Social Darwinism influenced Wright and Sullivan, whereas European avant-garde figures, namely Walter Gropius and Le Corbusier, were less obviously affected by Darwinism but deeply influenced by the production philosophies of the Americans Frederick W. Taylor and Henry Ford. Sullivan did write that form must follow function, but he had in mind the functioning of a Darwinian-like organism in a complex economic, social, and technical environment, not simply shapes arising from engineering considerations. Social and worker unrest deeply concerned the pioneers of the international style of architecture, especially the Germans Ernst May and Martin Wagner, city architects respectively of Frankfurt am Main and Berlin. Sullivan, on the other hand, is remembered for Chicago commercial high-rise buildings and Wright for upper-middle-class suburban homes in Chicago and elsewhere. The two Americans sought a national, even regional, style, while the Europeans took pride in the international character of their work. Considering all of these contrasts, one cannot place the Americans in the mainstream of the development of the international, modern style that prevailed until the rise of postmodern architecture. Le Corbusier wrote of American architects in 1928, "And you construct your skyscrapers in the manner of students of the École des Beaux Arts building a private house. I repeat: a hundred years of new materials and new methods have made no change whatsoever in your architectural viewpoint." "The Americans," he continued, "are the people who, having done most for progress, remain for the most part timidly chained to dead traditions."[38]

In their search for modern forms to express an aesthetic inherent in modern technology, several leading early-twentieth-century German architects anticipated the international-style architects of the interwar years. Peter Behrens believed that engineers expressed in their industrial and engineering designs the principles of modern technology, such as efficiency, flow, and system.[39] Yet their machines and structures lacked the aesthetic aspect that only an artist or architect could express. Behrens, as chief architect and industrial designer for Allgemeine Elektrizitäts-Gesellschaft, Germany's largest electrical manufacturer, designed factories, electrical appliances, exhibitions, and graphics that expressed the spirit of large-scale production and business organization. In his effort to achieve "those forms that derive directly from and which correspond to the machine and machine construction," Behrens anticipated Gropius, Ludwig Mies van der Rohe, and Le Corbusier, all of whom worked in Behrens's atelier, and all of whom became pre-eminent modern architects.[40] Among his most notable buildings was the turbine-manufacturing

PETER BEHRENS: AEG TURBINE FACTORY

factory in Berlin. In its design he expressed the concentrated power, order, precision, and regularity of the production methods and the products, including the turbines, made in the building. Behrens used, for instance, a series of pillars resting on hinges along an exterior wall to give the impression that these could be multiplied, with a rhythmic cadence, to any length. The visual metaphor suggested a sequence of powerful and precise mechanical movements. A contemporary observed: "Is it not strange, how such a logical and unified building becomes a symbol of that which it encloses?"[41] Trying to make architecture an expression of modern technological culture, Behrens wrote, "Whether or not technology can succeed in becoming a means to and an expression of culture rather than remaining an end in itself is therefore a question of great historical importance."[42]

Hermann Muthesius, another influential German architectural theorist and critic before World War I, also lastingly influenced European avant-garde architects in their search for the modern form. As a leading member of the German Werkbund, an organization of industrialists, architects, and artists dedicated to promoting industrial design,[43] Muthesius argued that products in a machine era had to be designed for machine production. The shapes of traditional handicraft were not suited to being efficiently used by machines.[44] As a homely example, Muthesius referred to the modern design of clothing, simply cut and sewn for machine manufacture. Plane surfaces, right angles, cubes, spheres, and other geometric forms suited manufacture with most production machines better than the undulating and convoluted organic shapes often favored by the craftsperson and the traditional architect. Technical and economic efficiency also called for the stripping of decoration from design. Along with these design principles, modern architects also took into account the availability of inexpensive human-made materials such as glass, cinder or concrete block, and structural steel. The European pioneers of modern design and architecture were not simply inspired by the forms of modern bridges, railway stations, and the products of mass production; they were strongly influenced by the methods used in constructing and manufacturing these.

GROPIUS, TAYLORISM, AND FORDISM

When the design philosophies of Muthesius and Behrens merged in interwar Germany with the production methodology of Ford and Taylor, one important branch of modern architecture, later known to many as the "International Style," began to take shape. Gropius, who worked in Behrens's atelier from 1907 to 1910 and was a member of the Werkbund, became an advocate of Taylor and Ford methods in architectural construction. In his essays and architecture, he articulated both the principles and forms of modern architecture. Founder and director from 1919 to 1928 of the widely and lastingly influential Bauhaus, a school of art, crafts, and architecture located first at Weimar and later at Dessau, Gropius left National Socialist Germany in 1934. In 1937 he joined those who were bringing the International Style to the United States. He taught at Harvard, established an architectural practice, and designed major and

WALTER GROPIUS
(1883–1969)

widely emulated buildings in the modern style.[45] Having drawn on American production technology to define modern architecture in the 1920s, he brought to the United States aspects of an American technological style that had been transformed into a European architectural style.

Gropius argued that architects had to incorporate aesthetic considerations and values into engineering designs. Without architects, engineers would never have introduced modern design into industrial products or buildings. An engineer's bridge, for example, showed the calculations and rationality of engineering thinking but failed to include the texture, tones, and light and shadow that only the architect as artist could introduce. Gropius was not proposing decoration; he wished to transcend the functionalism and materialism of engineering design and express more fully the principles and values of modern technology and, therefore, modern culture. By the mid-1920s, avant-garde architects like Gropius were consciously and explicitly defining themselves and their creations as uniquely modern. By this they did not mean, as we might today, that their work was contemporary: they insisted that they were defining a style they designated as "modern," as earlier styles had similarly been designated either Gothic, or Renaissance, or Baroque.

Even before World War I, Gropius envisaged the need for a new architecture for his age. He felt that the artistic genius should always express the greatest ideas and themes of his or her times. A constituent fact of his age, he believed, was the revolution in the way work was done. No longer did individual handwork prevail, but specialization, organization, and centralization.[46] Architects, he assumed, always had to come to terms with the technology of their era and, since his own was experiencing a new technology, a new architecture was needed.[47] In 1911 Gropius believed that factories and other industrial buildings would best express the spirit of his age, as cathedrals had expressed the culture of the Middle Ages. In the 1920s he decided that housing settlements for workers could also become a major expression of the spirit of the modern industrial age. He believed, as did so many political and social reformers of post–World War I Europe, that unless housing were built for the workers, their unrest might bring a "social catastrophe."[48]

American technological values greatly impressed the young Gropius. Like Le Corbusier later, he found them expressed especially well in late-nineteenth- and early-twentieth-century American concrete silos. In a set of illustrations prepared for a lecture he presented in 1911 on "Monumental Art and Industrial Building," he showed silos in Baltimore, Ohio; Buffalo, New York; and Minneapolis, Minnesota.[49] Like that of Egyptian pyramids and classical temples, the form of the silos expressed for Gropius monumentality and power, but the monumentality and power of American technology and civilization. Gropius aspired to ennoble the technical forms of America with European culture. Adolph Behne, architectural critic and associate of Gropius, spoke of this as "aesthetic filtering of American directness."[50]

Like many of the American professional inventors, Gropius was also an entrepreneur. Not satisfied with only designing (inventing) buildings, he was also interested in the technical, economic, and organizational aspects of constructing (developing) buildings. As was true of the American independent inventor-entrepreneurs, his heart was in inventive design, but he realized that he had to preside over the introduction of his designs into use. In his holistic approach to architecture, he also shared the essential characteristic of the American system builders; thus he gravitated to Taylorismus and Fordismus. He marched to a different drummer from most of his tradition- and discipline-bound architectural contemporaries. Like so many other truly innovative persons, he distanced himself from the ongoing institutions and ideas of his field and sought applicable analogies in other realms of human activity.

In the mid-1920s, when he headed the Bauhaus, Gropius behaved like a technological enthusiast.[51] The move of the Bauhaus from Weimar to industrial Dessau in 1925, the end of inflation and beginning of the period of stabilization in Germany, the growing expenditure of socialist city governments on public, especially worker, housing, and contacts with other avant-garde German architects who shared his interest in modern technology, like Bruno Taut and Ernst May, all reinforced Gropius's interest in the interactions of architecture, technology, and industry. His letters, memoranda, and publications testify to his technological enthusiasm and his familiarity with the means of modern production, especially Ford's and Taylor's methods.[52] His references to Ford and Taylor were explicit. He advocated in design and construction the standardization of precision component parts, the use of capital-intensive, labor-saving, special-purpose machinery, and the division of labor. He and his architectural staff used cost-accounting techniques and emphasized the lowering of unit costs through maximum utilization of capacity. Not only were these the signs of his technological awareness, but he even wrote of invention and patents on construction techniques, research and development in a laboratory setting, and the factory production of housing. After the Bauhaus moved to Dessau, Gropius described it as an experimental laboratory for the housing industry.[53]

In 1924 he predicted that the time had arrived to mass-produce the machine-made house. At hand, he insisted, were the means to fulfill the age-old dream of inexpensive, attractive, and healthful homes for the masses. Only inertia and sentiment, he believed, held back the production of houses in a fashion similar to the way Ford produced his inexpensive

MASTERS OF THE MODERN AT THE BAUHAUS: WASSILY KANDINSKY, NINA KANDINSKY, GEORG MUCHE, PAUL KLEE, AND GROPIUS (LEFT TO RIGHT)

Model T. Gropius envisioned a "Wohnford" (house Ford).[54] In fact, he was more flexible than Ford, for he wanted factories to produce standardized, interchangeable house components that could be assembled rationally into various combinations. In automobile parlance, Gropius wanted different models of houses, but he would limit variation from standardized components to a few types.[55] In 1924 he entered into an agreement with the architects Martin Wagner and May to design and build experimental houses to serve as models for large housing projects. They agreed to design model houses suited for large-scale, systematized, and rationalized production. Labor-saving, full-plant utilization, economical materials handling, and vertical integration of production facilities were their explicit goals, ones not usually associated with modern architects but more often with Fordismus and Taylorismus.[56] In an essay, "How Do We Build Cheaper, Better, More Attractive Dwellings?," Gropius summarized his approach: the use of mass-production methods, factorylike assembly-line processes, flow charts, and other control techniques.[57]

From the determination of Gropius and other avant-garde European architects to use modern production methods and to express modern values emerged a fully articulated modern form, design, or style. They displayed it in 1927 in the various buildings designed for the Weissenhof settlement, with its houses designed by him, Le Corbusier, Behrens, Mies, and others of the avant-garde for the Werkbund exposition in Stuttgart, Germany.[58] Henry-Russell Hitchcock, an architectural historian, and the architect Philip Johnson introduced this European avant-garde style to a U.S. audience with the publication in 1932 of *The International Style* and with an exhibition of the works of Gropius, Le Corbusier, Mies, and others at the Museum of Modern Art in the same year.[59]

A major explanation for the paradox that a German architect, Gropius, and not American architects, should have pioneered in the development of a style that featured both the construction techniques and the principles of modern American technology lies in the rash of public-housing projects built in Weimar Germany. Funded mostly by socialist city governments, the housing projects were designed by gifted avant-garde architects, most of whom shared idealistic and leftist values. These architects saw American Taylorismus and Fordismus as the way to mass-produce economical housing. In 1926 the city of Dessau commissioned Gropius to plan and supervise the construction of a housing settlement at Törten, a suburb of the city. Gropius seized this opportunity to employ on a large scale the Ford and Taylor methods and to design buildings using forms expressing modern technological values such as efficiency, regularity, ingenious technical solutions, standardization, and system. For Dessau-Törten, Gropius and his associates designed several basic house types and built more than three hundred of them. The two-story, attached houses stood in long rows of eight or more units. Their austere, geometrical

HOUSING SETTLEMENT, DESSAU-TÖRTEN

design featured plane surfaces, cubic volumes, flush windows, and flat roofs. The appearance shocked Germans accustomed to the steep roofs, inset windows, curvilinear forms, and applied decoration then common to great and small houses alike. In the interior, Gropius emphasized an efficient kitchen, small, spare rooms, specially designed furniture, and, in the rear, a small kitchen garden.

In construction he used composite building blocks and reinforced-concrete beams fabricated on the site. He and his associates took inordinate pride in the systematic layout of the construction site to achieve, like Ford, the smooth flow of materials and work. Because the houses stood in rows, rails could be laid parallel to them along which traveling cranes and building materials in trolleys could move. Storage bins for concrete, sand, cinders, and gravel were conveniently and systematically located. Gropius enthusiastically utilized various labor-saving machines, among them concrete mixers, stone crushers, building-block makers, and reinforced-concrete beam fabricators. He explained that the large scale of the project made feasible the investment in machinery. Following Taylor methods, the planners and builders of Dessau-Törten wrote out detailed schedules and instructions for the work process and used detailed cost-accounting methods. Planners broke down construction into precisely defined steps. Individual workers performed the same task on each of the standardized houses. Aerial views of the project reveal the systematic, assembly-line-like layout of the project. Gropius at Dessau-Törten pursued his goal of a product designed for, and made by, mass-production methods—a "Wohnford."

LE CORBUSIER

Charles-Edouard Jeanneret, the Swiss-born architect who has come to be known by his adopted name, Le Corbusier, contributed as much to articulating the modern architecture as Gropius, and even more to defining modern technological culture. Like Gropius, he was influenced by Fordism and Taylorism. But he was an even greater technological enthusiast than his German contemporary. For five months in 1911, as a draftsman, he, too, worked in Peter Behrens's atelier in Berlin. Earlier, he had had the good fortune to secure part-time work in the Paris architectural firm of Auguste and Gustave Perret. The firm was then making revolutionary architectural statements with their use of concrete-frame buildings: an innovation based on an understanding of the engineering science of statics and informed by mathematics. Le Corbusier settled in wartime Paris in 1916, then found work as a consulting architect for a construction firm. Among other projects, his design for a slaughterhouse provides notable evidence of his more than passing interest in modern production engineering. In this case, Le Corbusier turned the Ford system of assembling automobiles on its head. He used a conveyor-belt network to disassemble the animals. In 1917 he established a small company, the Société d'Études Industrielles et Techniques, that operated a brickmaking factory near Paris. The contemporary spirit of mass production (Fordism) and industrial organization (Taylorism) greatly stimulated Le Corbusier.[60] References to Taylorism appeared in almost all of Le Cor-

RATIONALIZED CONSTRUCTION, DESSAU-TÖRTEN

RAILS AND CRANE, DESSAU-TÖRTEN

RATIONALIZED LAYOUT OF CONSTRUCTION SITE, DESSAU-TÖRTEN

busier's books published between 1918 and 1935.[61] Just as Gropius had, he dreamed of producing houses the way Ford produced automobiles.[62] He also saw Taylorist production methods as an answer to problems of reconstruction after World War I.[63] By 1920 he was calling himself an industrialist as well as an architect.

He believed his age to be essentially technological and industrial, and himself, by commitment and aspiration, to be in the professional and intellectual mainstream. Fired by his technological enthusiasm, his commitment to a new architecture, and his vision of a technological culture, Le Corbusier and his close friend, the painter Amédée Ozenfant, in 1920 began publishing *L'Esprit nouveau* (*The New Spirit*). Through its pages, he intended to publicize the spirit of the modern era as he defined it.[64] Besides the essays on architecture in the journal, its twenty-eight issues (1920–25) contained articles on engineering and engineers, sociology, music, theater, beaux arts, sports, cinema, industrial science, and the high-technology achievements of the time, including ammonia synthesis. The obvious diversity of subject notwithstanding, *L'Esprit nouveau*'s persistent themes included order, rationality, engineering, and technology—or, in sum, the spirit of the new age. Seeing *L'Esprit nouveau* primarily as an architectural manifesto, which those interested mostly in his architecture sometimes do, fails to capture the intent of Le Corbusier and the journal. In his view, the new architecture did not bring the new spirit. Instead, the new spirit, *l'esprit nouveau*, brought the new architecture.[65]

For Le Corbusier, the clarity and rationality of twentieth-century engineering and classical architecture epitomized *l'esprit nouveau*. He found contemporary architects obfuscating the spirit of the age with their dreary nineteenth-century derivations and tired neoclassicism. By contrast, he found engineers, especially the American, with their mathematical calculation, their scientific deductions, and their bold empirical experiments, discovering the form and content of modern culture. American architects, he believed, had no more feel for the spirit of their times than their European counterparts. Among modern engineers, the American was for Le Corbusier the *primus inter pares*. "Listen to the counsels of the American engineers," he wrote, "but let us beware of American architects." They, too, had abandoned the purity of ancient classical, geometric forms for the decorative excesses of nineteenth-century historicism.[66] He believed that a new generation of architects, such as himself, would bring the aesthetic dimension to the purely rational designs of the engineers. Le Corbusier, like Gropius, was not a simplistic functionalist arguing that function determined form. He enthusiastically admired modern grain silos, airplanes, steamships, automobiles, and other works of the engineers, but he insisted that only the creative architect could transform such structures into architecture. One of his most memorable images was the juxtaposition of the Parthenon and a 1921 Delage sports car. Both, he believed, expressed with finesse the laws of proportion and harmony underlying universal beauty.

We remember Le Corbusier for his private houses, including Villa Savoye in Poissy (1929–31), his apartment complexes such as Unité d'Habitation in Marseilles (1952), his public buildings at Chandigarh, India (master plan, 1951), and his church at Ronchamp (1955), but he, like Gropius in the 1920s, longed to mass-produce housing. He feared that, unless the workers were well housed, there would be social revolution, but he also enthusiastically embraced mass production as essentially modern. He urged that buildings be modeled on large-scale mass-production techniques. World War I had stirred his imagination, as it had that of so many European industrialists and engineers. He reasoned that, if factories could mass-produce airplanes, trucks, and artillery, houses could be built similarly. He believed that automobile and airplane manufacturers could adapt their techniques to house building.[67] In 1921 he published sketches of a house, Citrohan, that drew by analogy on the techniques and style of the French mass-produced Citroën, an automobile whose manufacturer had been directly inspired, by the Ford precedent, to use the assembly line.[68] The mass-produced house should be as beautiful and functional, he believed, as the most elegant working tools and machines of modern technology. Among his proposals were houses that could be completed in three days, made of concrete poured into forms. Natural materials should be replaced by artificial ones, such as reinforced concrete, whose strength could be mathematically calculated so that it could be used most efficiently and economically. Within two decades, he prophesied in the 1920s, big industry would have standardized materials and rationalized construction of components in specialized factories. Transformed architectural schools and large investment houses would become involved.[69] In 1924 an eccentric Bordeaux industrialist commissioned Le Corbusier to mass-produce housing at Pessac for his workers, but the project had to be abandoned before completion because labor and contractors would not cooperate.[70]

Until the late 1930s the modern style of Le Corbusier, Gropius, Mies, and other avant-garde European architects remained mostly a European movement, despite its American technological ancestry. Exceptions were the California houses of Viennese architect Richard Neutra, which showed the influence of both the international and the Wright organicist styles, and the Philadelphia Savings Fund Society building of George Howe and William Lescaze, which combined international and U.S. skyscraper styles.[71] When the National Socialists seized power in Germany, they fanned a reaction against the international style, public interest in which had been high in the late 1920s. Supporters of the Weimar Republic had offered international-style architecture to the world as an example of the new German culture. Conservative architects and disgruntled craftspersons of the building trades, irked by the new construction techniques, launched a counterattack that Nazi ideologists soon joined. Paul Schultze-Naumburg, a conservative German architect well known for his use of historical references in his traditional buildings, characterized the international style as materialistic in its commitment to the spirit and forms of modern production technology and un-German in its claim to international rather than national character. He wrote that the houses of the new architecture looked like stationary sleeping cars designed by nomads of the metropolis. (During the 1920s archconservatives believed Berliners to be rootless followers of international fads in politics and culture.) Curiously enough, this bitter exchange with racist overtones focused for a time on the virtues of sloped (German) versus flat (non-German) roofs. A roofers' newsletter compared flat roofs with flat heads and said that using them was swinish.[72]

The hostility among Nazi leaders toward the international style, and several unsuccessful attempts to obtain commissions from the new regime, prompted Gropius to emigrate in 1937 to Britain, and then to the United States. After the government shut down the Bauhaus in 1933, Mies, who in 1930 succeeded Gropius and Hannes Meyer as head of the Bauhaus, also emigrated to America in 1937. After World War II Gropius, architects trained by him at Harvard, and Mies established the international style in the United States with major high-rise buildings in Chicago and New York City. The social concern and austere, imaginative, machine aesthetic expressed in the mass-produced housing of the 1920s, however, gave way to corporate and luxury-apartment high-rise buildings, private homes for the wealthy, and banal buildings designed with primary concern for cost savings made possible by simplistic, geometric structures, prefabrication, and standardization.

MODERN ART AND MACHINE TECHNOLOGY

Not only avant-garde architects but painters as well tried to express the values of an era of technological enthusiasm. Again, the Europeans, not the Americans, initially sought to express the values and forms of modern technology in nonrepresentational art. For centuries, in both Europe and America, artists had portrayed industrial landscapes, but in the twentieth century some sought through subject matter and style to convey their conviction that the modern culture was essentially a technological one. In doing this, they took an approach like that of the International Style architects. The artists believed that technology embodied a confident spirit of invention and discovery. For them, technology was infused with a belief that humans could create, order, and control their material environment. Technology involved a commitment to such values as order, precision, power, motion, and change. In their paintings they tried to symbolize this spirit. The subject matter of these artists was not simply industrial scenes, but mechanical, electrical, and chemical technical forms, such as pistons and cylinders, electric dynamos and transmission lines, and tanks and tubes. Some of them tried to capture the essence of the motion of fast-moving automobiles and locomotives, and others drew and painted with the precision and cool objectivity of engineers and architects. They also tried to convey the harmony, balance, and interaction of smoothly functioning technological systems.

Before World War I the Italian Futurists led those aspiring to articulate a style appropriate for a technological culture. From a less industrialized and more agrarian country than many of their European counterparts, the Futurists called boldly for technology and art to act as agents of radical social change. Italians who visited the United States, Germany, France, and Britain saw the social, political, economic, and military contrasts between their own country and those more heavily industrialized. They attributed this development to a commitment to use technology to force social change. Filippo Tommaso Marinetti, the Futurist founder, sponsor, leader, and prolific manifesto writer, found fascinating the drama and power of the new technology. He embraced the social transformation it brought. A writer of novels and poems, he gathered about him architects, painters, sculptors, photographers, dramatists, and writers who, between 1909, when Marinetti published the founding Futurist manifesto, and the end of World War I, attracted considerable attention in European artistic and intellectual circles. Marinetti's fiery enthusiasm for new technology and his disdain for the old order found expression in an anecdote included in his 1909 manifesto. In it he contrasted the trams that passed in front of his Milan house with the Renaissance canal (for which Leonardo had reputedly designed the locks) that flowed behind it:

> Suddenly we jumped, hearing the mighty noise of the huge double-decker trams that rumbled by outside, ablaze with coloured lights. . . . Then the silence deepened. But, as we listened to the old canal muttering its feeble prayers and the creaking bones of sickly palaces above their damp green beards, under the windows we suddenly heard the famished roar of automobiles. . . .
>
> "Let's go!" I said. "Friends, away! Let's go . . . !"
>
> We went up to three snorting beasts, to lay amorous hands on their torrid breasts. I stretched out on my car like a corpse on its bier, but revived at once under the steering wheel, a guillotine blade that threatened my stomach.[73]

Automobiles symbolized the Futurists' ideas about technology and modernity. Marinetti said that a roaring motorcar was more beautiful than the *Victory of Samothrace*. The Futurists found the essence of the new age in its exhilarating speed and power and its liberation of the spirit. Dynamic images of autos appeared in Marinetti's and other Futurists' paintings; in more than a hundred works, Giacomo Balla painted a speeding automobile as the primary subject.[74] Marinetti also celebrated arsenals and shipyards blazing at night, railway stations filled with "deep-chested locomotives," factories spewing smoke, airplanes flying, and cities crowded with people who were filled with radical ideas. On the other hand, the Futurists held in contempt, as symbols spreading the influence of an agrarian, moribund culture, the treasured Italian Renaissance art, architecture, libraries, and academies. For the Futurists, the new technology and art would create a material and spiritual environment conducive to the birth of a dynamic, strong, and aggressive Italy. Umberto Boccioni, one of the group, said that they wanted to Americanize themselves by hurling themselves into the all-consuming vortex of modernity through its automobiles, crowds, and relentless competition.[75]

The modern city caught their imaginations, too. Antonio Sant'Elia, a Futurist architect, presented his concepts of the new technological city at a Futurist "New Trends" exhibition in 1914. Hydroelectric dams, power plants, and factories were a part of his city concept but, clearly influenced by American urban vitality, Sant'Elia let skyscrapers dominate his designs, at a time when other European architects rarely resorted to urban verticality. He also had a systematic approach that integrated skyscrapers and other structures into a master urban plan. His ideas subsequently influenced the architects of the 1920s who defined the modern or international style.[76] Marinetti believed new cities would nurture the Futurist spirit. In a speech to the Venetians in 1910, he shouted that "it is time for electric lamps with a thousand points of light to brutally cut and tear your mysterious, enchanting and seductive shadows!"[77] He wanted the canals filled, streetcar lines laid, and bustling crowds making purchases in a manufacturing and mercantile metropolis. The economic determinism of tourism seems to have been lost on him.

Artists other than Italians found the Futurist approach expressive of the momentous technological and urban changes taking place around them. Marinetti's Futurist manifesto of 1909 was translated into several languages, and in February 1912 the Bernheim-Jeune Gallery in Paris mounted an exhibition of Futurist painters. Among those influenced by the exhibition was the Italian-born American painter Joseph Stella, who had been traveling in Europe since 1909. When he returned to the United States in 1912, he chose to portray motion, technological forms, and modern urban and industrial scenes by using themes and techniques like those of the Italian Futurists. He exhibited works in the Armory Show, or International Exhibition of Modern Art, first held in New York City in 1913 at the Sixty-ninth Street armory. The exhibition introduced contemporary art, including the works of American artists Morton Schamberg and Charles Sheeler, and of European Cubists, to an American audience. Among the best known of Stella's works are *The Gas Tank* (1918) and his *Brooklyn Bridge* (1920–22). He painted a number of scenes of Manhattan in which his assimilation of Futurist ideas clearly emerges.[78]

Alfred Stieglitz, American photographer and innovative art-gallery director, also prepared the ground for the rise of an American art expressive of modern technology. Born in Hoboken, New Jersey, in 1864, he studied engineering at the College of the City of New York and at the Berlin Polytechnic. Afterward he followed his interest in photography and became editor of several photography journals that also covered art. Stieglitz insisted that photography was a form of creative expression. Before World War I he made repeated visits to Berlin, Dresden, Munich, London, Paris, and other European cities. After seeing the stark contrast between European cities and New York, he began intensively photographing the machines and urban industrial landscapes of the New World. He focused on tugboats, trains, airplanes, and the city, always under construction. "The raw vitality of Manhattan and its dynamic growth served as a 'non-art' subject, one that was uniquely American as well." For him the skyscrapers under construction were an inspiration. After photographing the Flatiron Building in 1903, he wrote, "It appeared to be moving toward me like the bow of a monster ocean steamer—a picture of new America still in the making." He added, "The Flat Iron is to the United States what the Parthenon was to Greece."[79] Stieglitz called the American

ALFRED STIEGLITZ,
O'KEEFFE HAND AND FORD WHEEL

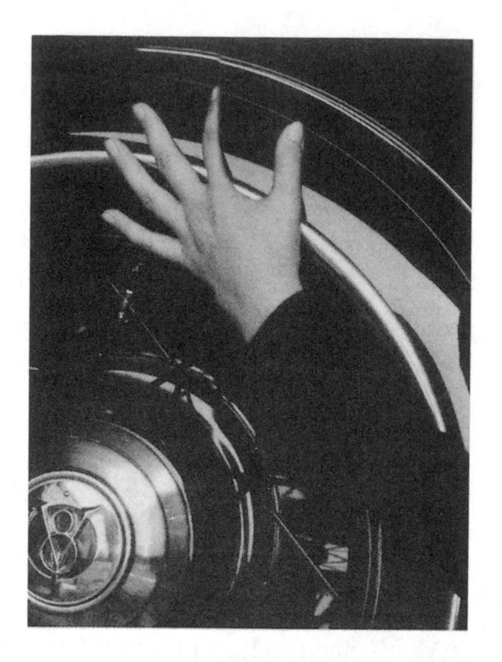

skyscraper a "giant machine." He "married Man to Machinery and he obtained issue."[80] He opened his "291" Gallery at 291 Fifth Avenue, New York, in 1905, published the magazine *Camera Work* from 1903 to 1917, and in 1915 and 1916 published the magazine *291*.[81] Through a number of exhibitions of photography and painting, Stieglitz introduced Americans to modern art, first European and then American. Later he showed the works of young American modern artists interested in the emerging urban, technological culture, among them Demuth, Arthur Dove, and Georgia O'Keeffe whom he married in 1924. O'Keeffe exhibited in Stieglitz's Intimate Gallery in the 1920s.

INVASION OF 1915

The slow, tentative movement in the United States toward an art expressive of a mass-producing, technological society that had transformed a wilderness into a construction site accelerated after Marcel Duchamp, Francis Picabia, and several other leading French painters arrived in America in 1915. Before their arrival they had all been seeking new modes of expression in order to distance themselves from Cubism. Duchamp, influenced by a photographic analysis of the moving figure by French physiologist Etienne-Jules Marey, painted *Nude Descending a Staircase*. Picabia, having met Duchamp in 1911 and joined a group working in the Duchamp brothers' house, also made several paintings expressing motion. Speeding automobiles fascinated Picabia, as they had Marinetti. With shapes and color he tried to communicate the thrill of speed. In their search for the spirit of modernity, Duchamp and Picabia found in their visit to the United States in 1915 a confirmation of their propensity to associate art with technology. They conveyed the enthusiasm of this commitment to a small coterie of American artists, who absorbed more of the positive than the skeptical in the two artists' attitudes toward technology. When Duchamp and Picabia arrived, the American press received them as art-world celebrities. Both had already challenged conventional American opinions about European art with their works displayed at the armory show of 1913 which included Duchamp's *Nude Descending a Staircase*.[82] They won over Americans in 1915 because, unlike most other European intellectuals and artists, they announced that they found much to learn and observe in the New World. The New York *Tribune* reported:

> . . . for the first time Europe seeks America in matters of art, for the first time European artists journey to our shores to find that vital force necessary to a living and forward-pushing art. . . . They in turn will pay us with what is perhaps equally necessary, the courage to break from the tradition of Europe.[83]

Duchamp said that he gloried in the vibrant electricity of the wholly new, young, and strong force in the world.

In 1915 Duchamp and Picabia were part of a mini-invasion of French artists whom the admiring *Tribune* reporter described as "modernist French artists." Among them were the sculptor Frederick MacMonnies, the Cubist painter Albert Gleizes, the poet and painter Juliette Roche Gleizes, and the artists Ivonne Crotti and Jean Crotti. All despaired about a Europe at war and enthusiastically approved the culture they believed to be taking shape in free, peaceful, vital, technological America. Europe, they considered, had become repressive before the war, and during the war even Paris had lost the free and inspiring atmosphere vital to stimulating art. Duchamp assured Americans that the Quartier Latin was gloomy. Omnipresent death and mutilation brought even the artists who were not at the front to lay down their brushes. "I adore New York," Duchamp added. "There is much about it which is like the Paris of the old days."[84]

Duchamp, Picabia, and others were making the second discovery of America, of the technological, or man-made world. This America revitalized Picabia. He predicted that in the future art would attain a "gorgeous florescence" as the bold, dashing, boundless spirit of America spread. In the New World, art and life would become one. His visit to America revolutionized his painting. Before, he had been a landscape painter, then a Cubist, and just before leaving Europe he became engrossed in presenting psychological insights through the forms he created. "Almost immediately upon coming to America it flashed on me," he reported, "that the genius of the modern world is machinery, and that through machinery art ought to find a most vivid expression." He continued:

> I have been profoundly impressed by the vast mechanical development in America. The machine has become more than a mere adjunct of human life. It is really a part of human life—perhaps the very soul. In seeking forms through which to interpret ideas or by which to expose human characteristics I have come at length upon the form which appears most brilliantly plastic and fraught with symbolism. I have enlisted the machinery of the modern world, and introduced it into my studio. . . . I intend to work on until I attain the pinnacle of mechanical symbolism.[85]

Picabia predicted that "modernist" art had found its future home in America.

Picabia believed that in discovering the machine he discovered America, and that the spirit of America was similar to modern art—in its indifference to tradition and history and in its commitment to action.

Marius De Zayas, a Mexican caricaturist, an art critic, a contributor to Stieglitz's journal *291*, and the owner of the Modern Gallery in New York, said of Picabia, "Of all those who have come to conquer America, Picabia is the only one who has done as did Cortez: he has burned his ship behind him."[86] Having served eight months in the French army, Picabia was sent by the French military to the United States to purchase Cuban molasses, a mission he characterized as secret.[87] He did not burn his ship but returned to Europe and conveyed his excitement about machine culture to the Dadaists, an avant-garde art movement that originated in Zurich in 1916. He did, however, prolong his stay in New York City into 1916, and contributed a set of machine drawings for the journal *291*. Engineering, or mechanical, drawing greatly influenced Picabia. In these and later works, he took from industrial catalogues machine drawings such as spark plugs, pistons, and cylinders, modified them freely, then titled his creations with references to women and men and with quotations from dictionaries, especially Latin. Thus he combined ancient wisdom, biology, and modern technology.[88] For him, machines presented a model for his own behavior, freedom from conventional restrictions and responsibilities. He found machines highly attractive, because they had no morals. In painting them, he revealed his own attitudes of skepticism, irony, and hedonism.[89]

American artists gathered in the New York apartment of Walter and Louise Arensberg to hear and absorb the ideas of Duchamp and Picabia. Wealthy intellectuals, art collectors, and patrons, the Arensbergs had moved to New York City in 1914. The 1913 New York armory show had a profound impact on them, and they bought there their first modern painting, a work by Jacques Villon, older brother of Duchamp. Subsequently the painter Walter Pach, one of the organizers of the armory show, guided them in their purchases of avant-garde art. They also befriended De Zayas, who encouraged the importation and native growth of modernist art in the United States. Pach introduced the Arensbergs to Duchamp when he came to New York in 1915, and they became close friends of the painter.[90] Duchamp rented a studio in the same New York building in which the Arensbergs had an enormous apartment, and they regularly gathered a circle of avant-garde artists around him. The

FRANCIS PICABIA,
ICI, C'EST ICI STIEGLITZ FOI ET AMOUR

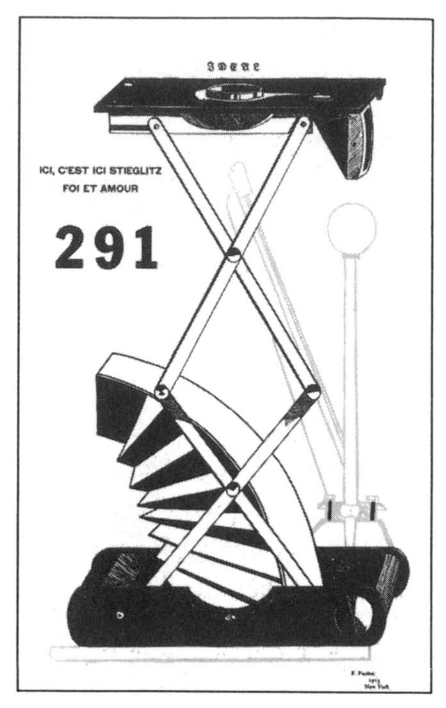

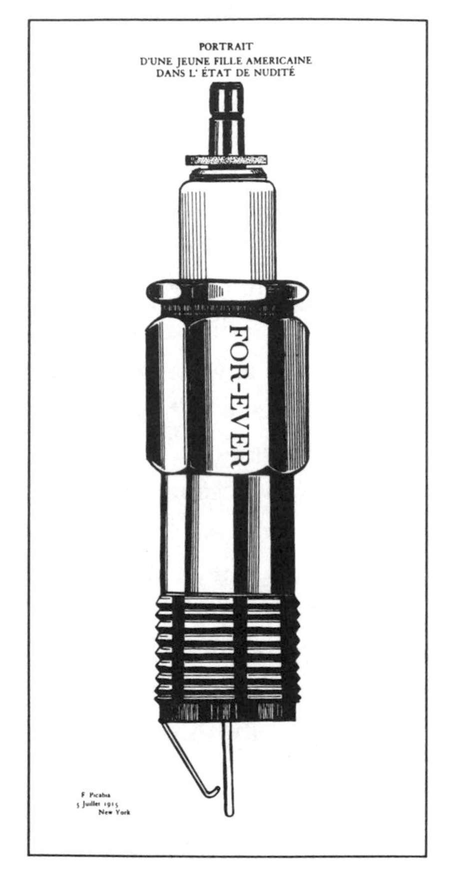

PICABIA, *PORTRAIT D'UNE JEUNE FILLE AMÉRICAINE DANS L'ÉTAT DE NUDITÉ*

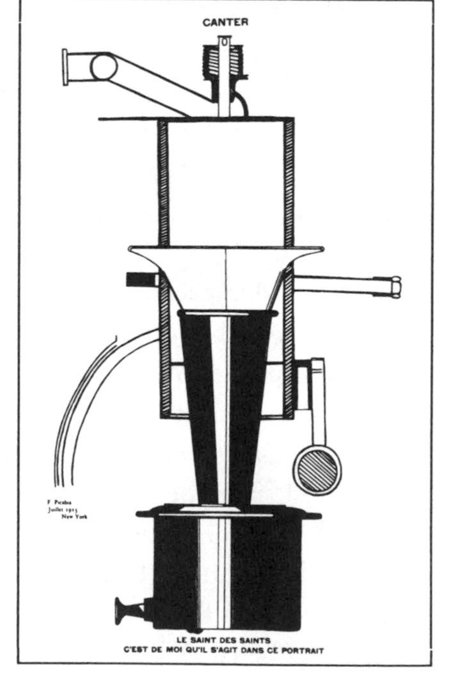

PICABIA, *LE SAINT DES SAINTS*

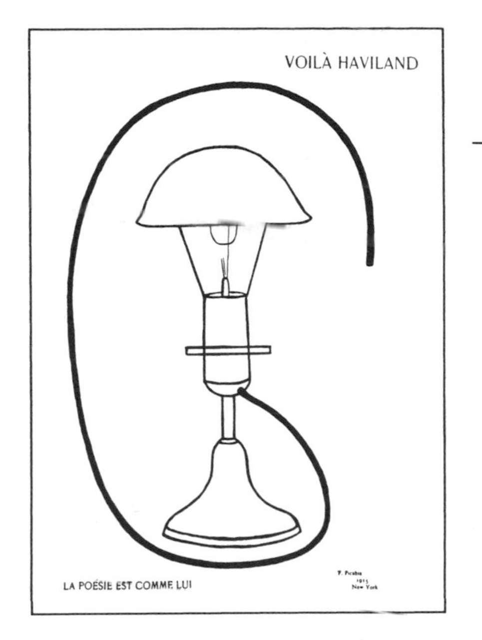

PICABIA, *VOILÀ HAVILAND*

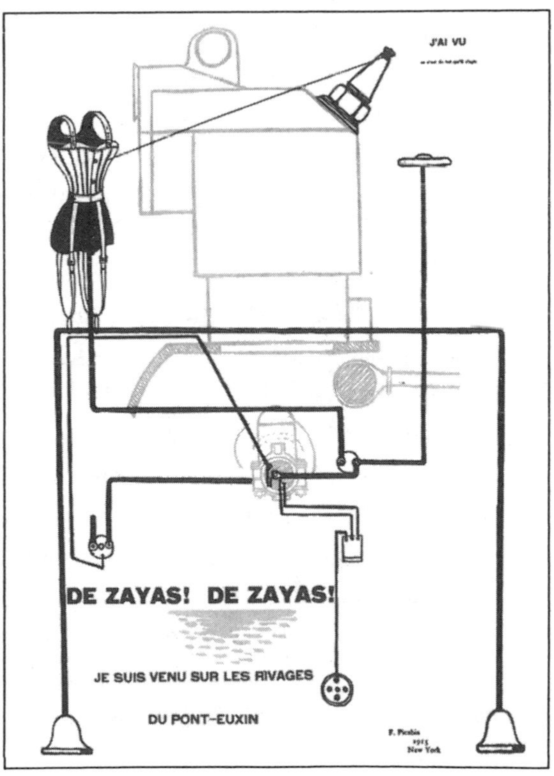

PICABIA, *DE ZAYAS! DE ZAYAS!*

Arensberg apartment became an outpost of French modernists and American artists and intellectuals committed to articulating the new American culture. The Americans included Stella, Marsden Hartley, Isadora Duncan, Katharine Dreier, Amy Lowell, William Carlos Williams, Sheeler, and Demuth. Their Sixty-seventh Street salon throbbed with genial confusion, intrigues, and long-staying guests. In 1922 the exhausted Arensbergs fled to California.[91]

Remaining in New York City, Duchamp also helped instill in a few American artists the confidence to turn away from Europe as mentor and to develop their own artistic approaches by reflecting on their own strong technological civilization. He showed them how to elevate commonplace machine products to the level of objects worthy of aesthetic consideration. Duchamp selected, signed, and submitted a common porcelain urinal that he named *Fountain*, a "ready made," to an American exhibition, only to have it rejected by a shocked jury. Duchamp argued that his perception of the urinal as an object of interest and one expressive of the machine age qualified it for exhibition. In calling attention to "ready mades," he made the New York art world aware of the omnipresence of mass-produced objects as a principal way in which American society defined itself. He also encouraged American artists to discover the symbolic character of machine artifacts and to present them as cultural expressions. Following Duchamp's lead, Schamberg fixed a plumbing trap for a sink in a miter box and titled it *God*. Man Ray attached tacks to the surface of a common flatiron and called it *Cadeau*.[92] In 1915, after his arrival in the United States, Duchamp began his most renowned work, a painting on glass entitled *The Bride Stripped Bare by Her Bachelors, Even*, which he left unfinished in 1923. Always enigmatic, Duchamp responded "no" when asked if it was a sketch for a mobile. He said it was like the hood of a car. All of the parts in the painting, some of which suggest machine elements and chemical apparatus, relate to one another in a functionally absurd way. He did acknowledge that *The Large Glass*, as it is also known, reflected an attitude toward machines. An incomprehensible mix of the biological, the erotic, and the technological also characterized the painting and its title. Picabia struck the same note when he referred to machines as daughters created by men and born without mothers.[93]

In the early 1920s Duchamp became a focal point for a New York Dada movement that survived until the middle years of the decade. Founded in February 1916 in the cabaret Voltaire in Zurich by the Rumanian-born poet Tristan Tzara, the French sculptor and painter Jean Arp, and the Germans Hugo Ball and Richard Huelsenbeck, the Dada movement spread to Berlin, Hanover, Cologne, and Paris. Prior to Picabia's arrival in Zurich in 1918 bearing his news from America, the Dadaists had not been particularly interested in mechanical technology. Proclaiming that they were anti-art, the European Dadaists had already rejected both the established subject matter of and approach to painting—that of the Cubists, for example. Denying their artistic heritage, they were delighting in the absurd and unpredictable in art. Once they absorbed Picabia's fascination with technology, however, some Dadists sought stimulation in what they saw as the new culture of technology. Finding the fullest development of modern production technology in the United States, they found the raw material of modern culture there as well. Following the lead of Duchamp and Picabia, and not that of the Futurists or the Constructivists, the Dadaists, Max Ernst among them, became known for their acerbic, irreverent, and surrealistic expression of modern technological forms and themes. They delighted in absurd juxtapositions of the mechanical and organic, of life and of the machine. Whereas leading Futurists and Soviet Constructivists were technological enthusiasts, the Dadaists were irreverent, skeptical, and iconoclastic in their views toward technology.[94]

Several editors and writers associated with avant-garde little magazines also encouraged American artists seeking to develop a mechanistic style and to present technological themes. Robert J. Coady, editor of *The Soil*, attempted the Herculean task of launching an American art magazine intended to fire the national consciousness with an awareness of the possibility of a peculiarly American art of high quality and sensibility. Beginning publication in December 1916, his magazine attracted a surprising amount of attention, selling more than five thousand copies of one issue at the high price of 25¢. Yet the funds gave out after four issues. He envisioned an American art that would arise from "a compost of vegetation discarded by various racial types."[95] Enthusiastically he wrote:

> An Englishman invented the Bessemer process and we built our sky-scraper. A Dane married a Spanish-African, and Bert Williams sings, "Nobody." We've dug into the soil and developed the Steam Shovel. We play and we box. . . . Our art is, as yet, outside our art world. It's in the spirit of the Panama Canal. It's in the East River and the Battery. It's in Pittsburgh and Duluth. It's coming from the ball-field, the stadium, and the ring. Already we've made our beginnings, scattered here and there, but beginnings with enormous possibilities. Where they will lead, who knows? Today is the day of moving pictures, it is also the day of moving sculpture. . . ."[96]

By "moving sculpture" Coady meant machinery such as the "Sellers Ten Ton Swinging Jib Crane, Locomotive No. 4000, the Industrial Works 120 Ton Crane, the Chambersburg Hammer, and the Erie Hammer," all of which he illustrated in *The Soil* (January 1917).[97] These machines, an admirer of Coady's wrote, expressed for him the greatness of the American nation, much as the pantheon of the gods had expressed for the Athenians the golden age of Greece. The product of a "chain of social-economic phenomena, vitalized by the assimilated hordes of immigrants, these colossal mechanical forms," Coady felt, "must, whether obvious or not to the academicians, be looked to as the most representative material for our national art."[98] The American artist accepted the exciting challenge of translating the rational and elegant designs of machines into artistic images. Artistic sensibility might alter a machine designer's handle here or a bolt there, straighten a line or modulate a curve, vary slightly the intersection of planes, but the machine itself remained the inspiration for mighty art. Coady anticipated Sheeler, Schamberg, Demuth, and other American artists of the interwar years when he found the rudiments of a truly American art in the "grim earnestness of industry and the childlike earnestness of sport and humor."[99]

Matthew Josephson was another of the little-magazine editors and contributors who urged American artists and writers to define a unique American culture neither derivative of the European nor shaped primarily by the great American wilderness and frontier. In France in 1921–23 and 1927–28, Josephson probably felt the influence of the art and ideas of Le Corbusier and his friend Amédée Ozenfant, who celebrated the classical harmony and purity of elegant machine forms. He also became aware of the enthusiasm of Dadaists for ordinary American things like advertising, dime novels, vaudeville, and movies as well as American business methods and new materials and artifacts, such as skyscrapers. A friend of Sheeler, Josephson articulated a romantic attitude toward the machine that influenced the painter and photographer.[100] He believed that the United States was essentially a nation shaped by machine technology. He championed an American Dada with its adventuresome exploration of the modern technological world. In the magazine *Broom*, he announced the age of the machine and a new art and new literature stemming from the technological civilization. Josephson stressed the new "American flora and fauna" generated by technology.[101] In a 1923 advertisement seeking new subscribers, *Broom* announced that "the Age of the Machine in America is an age of spiritual change and growth as well as one of economic ascendancy." "A new art and new literature," it continued, "spring sturdily from the machine civilization."[102] Surveying the French literary scene after the war in *Broom*, Josephson approved Apollinaire's dictum that young writers should be at least as daring as the mechanical geniuses of the age, that they should not condemn the machine for "flattening us out" or "crushing us," but should see it as a magnificent slave. "We are a new and hardier race," Josephson added, "friend to the sky-scraper and the subterranean railway as well." He did not want young American artists and writers to surrender their prerogatives in order to follow the French or Europeans, "for the fundamental attitude of aggression, humor, unequivocal affirmation which they pose, comes most naturally from America. The high speed and tension of American life may have been exported in quantity to Europe. But we are still the richest in material."[103]

THE PRECISIONISTS

Between World War I and the onset of the Great Depression, a loosely interacting group of American artists, influenced by the subject matter and the dynamism of the Italian Futurists, by the Cubist tendency to reduce objects to their basic geometric shapes, by the Dadaist inclinations of Duchamp, by the little magazines' celebration of the machine, and primarily by the remarkable technological world evolving around them, contributed to the rise of a style of painting variously called "Precisionist," "Machine Art," "Cubist Realism," and the "School of the Immaculates." Their subjects included technological forms, industrial and urban scenes, especially New York and its skyscrapers and Pittsburgh and its steel mills. Demuth, Sheeler, and Schamberg were prominent among the first wave of these artists. Elsie Driggs, noted for her "precise line" and especially for her 1928 paintings of Pittsburgh steel mills, and Georgia O'Keeffe, whose art has been described as having the precision of finely made machines, moved among and exhibited with the Precisionists.[104]

Sheeler, Demuth, and Schamberg all studied at the Pennsylvania Academy of Art in Philadelphia, then one of the world's leading industrial cities. Sheeler shared a studio with Schamberg there. All three, drawn into the Arensberg circle, fell under the influence of Duchamp and Picabia.[105] In 1916, when Schamberg completed eight canvases with a central machine motif, he was probably influenced by Picabia's machine drawings done for *291* and his paintings on display in the Modern Gallery the same year.[106]

Born in Lancaster, Pennsylvania, Demuth made several trips to Europe after finishing art school, including a five-month stay to visit Paris, Berlin, and London in 1907–8. After a longer period of study in Paris, 1912–14, he became acquainted with Duchamp and Picabia. He visited Berlin the last year before the war broke out. Later he expressed some regret at not having settled in Paris as a painter, but he acknowledged that in the United States and New York, "awful as most of it is," he found the root of all that was modern.[107] Converging interests in the modern technological culture and temperament brought together Demuth and Duchamp in New York. Both cosmopolitan, they enjoyed drinking together in Harlem. Demuth celebrated Duchamp's *Large Glass*, with its diagrammatic quality, as the great picture of their time. His approach to technology, however, differed from Duchamp's in that, like Sheeler, he preferred to paint industrial architecture and the industrial city, not only individual machines. His technique was both representational and abstract. In *Machinery*, a 1920 painting dedicated to his friend, the poet William Carlos Williams, the influence of Duchamp and Picabia becomes clear. Demuth depicted a large factory-ventilating system that in its tubing, bulky tank, and planes was at the same time fantastic, erotic, and sinuous.[108] In *My Egypt* (1927) he struck a familiar chord among modern artists and architects. The subject was an American grain silo, the same artifact that had fired the imagination of such European architects as Gropius and Le Corbusier. The title suggested the analogue between the monumental buildings that expressed the spirit of Egypt and the ones that captured the essence of the emerging American technological culture. Gropius and other reflective European avant-garde architects had insisted that grain silos, railway stations, great bridges, and factories were comparable in their monumentality and cultural expressiveness, if not in their form, to medieval cathedrals.

Sheeler, too, saw similarities between factories and Gothic cathedrals. He believed that, "In a period such as ours, when only a few isolated individuals give evidence of a religious content, some form other than that of the Gothic Cathedral must be found for our authentic expression. Since industry predominantly concerns the greatest numbers, finding an expression for it concerns the artist."[109] He vacillated between the representational and the abstract. Far less ironic than Duchamp and Picabia in his portrayal of modern technology, he believed that he could perceive in its structures an essential, pervasive, and rational world order. By careful and repeated observation, he composed an image in his mind's eye of a machine or an industrial landscape before he began to paint it. He believed that, much as an engineer or an architect had to design a machine or a building so that all the components interacted harmoniously

CHARLES DEMUTH, *INCENSE OF A NEW CHURCH*, 1921

DEMUTH, *END OF THE PARADE, COATESVILLE, PA.*, 1920

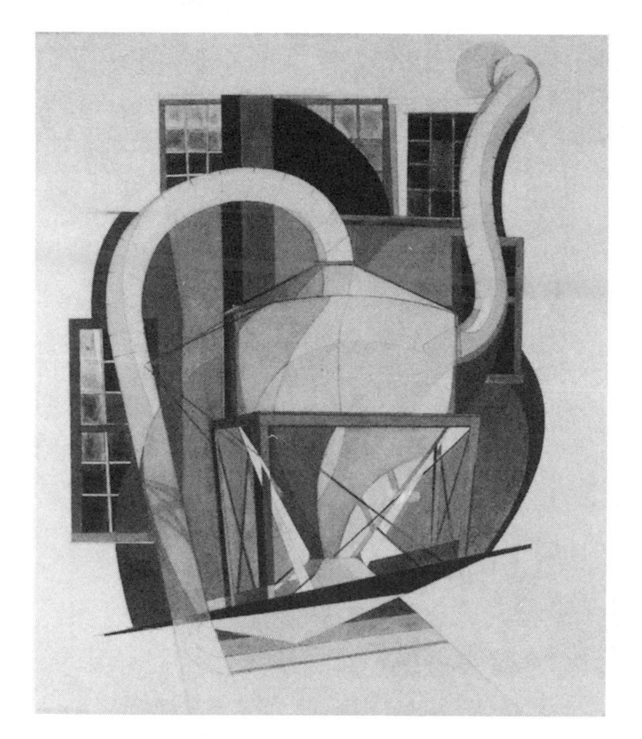

DEMUTH, *MACHINERY*, 1920

and functioned efficiently, so an artist such as he should try to relate all of the parts in a painting in order to achieve an integrated and total effect. He greatly appreciated architectural and machine artifacts created to interact and function efficiently. Like a system builder, such as a Ford, he also saw production technology as an embracing system. He graphically captured this in his *City Interior*, a 1936 painting of the River Rouge plant. Also a photographer, Sheeler often made photographic studies of scenes he later painted. The Precisionists believed the use of cameras to be particularly appropriate in a machine culture. In 1927 Henry Ford commissioned him to do a series of photographs of the River Rouge plant, several of which Sheeler used as the basis for painting the Ford complex. In the 1930s he turned increasingly in his paintings and photographs to the urban and technological landscape. He shared with Demuth the dedication to lucid and precise presentation of detail. Both omitted people from their technological landscapes.

In 1927 Jean Heap, editor of *The Little Review*, organized the Machine-

CHARLES SHEELER, *INSTALLATION*, 1939

SHEELER, *SUSPENDED POWER*, 1939

Age Exposition, which opened in May. She was influenced by Soviet Constructivism and by the enthusiastic attitude of Le Corbusier toward technology and the spirit of the modern age. The Soviet Constructivists, who flourished after the 1917 revolution, saw art and architecture as ways to construct a material environment that embodied the values of the new regime and modern technology. The environment, they believed, would help determine the character of the new Soviet citizen. In her journal, Heap had earlier urged the American artist, who could not expect government support similar to that provided by the Soviet Union, to "affiliate with the creative artist in the other arts and with the constructive men of his epoch: engineers, scientists, etc."[110] Her exhibition was a response to Le Corbusier's Pavillon de L'Esprit Nouveau at the 1925 Exposition des Arts Décoratifs in Paris.[111] In Steinway Hall, a commercial building in midtown Manhattan, Heap exhibited actual machines, including a Crane Company gate valve, a Studebaker crankshaft, a Curtiss airplane engine, and a Hyde Windlass propeller. She also exhibited photographs and drawings of skyscrapers, the Bauhaus in Dessau, Soviet industrial buildings, architecture by Gropius, and paintings, drawings, sculpture, and "inventions" by modern artists. The Russian-born American artist Louis Lozowick, whom the Soviet Constructivists also influenced, exhibited his American-cities series, and Demuth a canvas entitled *Business*. The exhibition also stressed the decorative arts. Sheeler, Demuth, Lozowick, and Duchamp were members of the organizing committee. Heap avoided having engineering creations in one section and the works of artists in another, as had been the case at the great world's fairs, such as the Paris Fair of 1900. She admired engineers, whom she called a great new race of men, but was disturbed that they worked in isolation, "practically ignorant of all aesthetic laws," so she displayed engineering works and art together. The engineers, she believed, should learn from leading artists who were "organizing and transforming the realities of our age into a dynamic beauty."[112] The exhibition received limited attention, and in 1929 *The Little Review* ceased publication. Jean Heap lost faith in collective activity by artists and engineers.[113]

Lozowick summarized the exhibition's confidence, as well as his own, in the technological society and its emerging culture in a statement

SHEELER, *AMERICAN LANDSCAPE*, 1930

SHEELER, *RIVER ROUGE PLANT*, 1932

SHEELER, *CITY INTERIOR*, 1936

prepared for Jean Heap's 1927 exhibition. Like Heap, Lozowick had been influenced by the optimism of the Soviet Constructivists and their belief that the artist, with the engineer and the scientist, in a politically radicalized society could create a new world. Lozowick, however, saw the future emerging in America. In his essay "The Americanization of Art," he wrote:

> The history of America is a history of gigantic engineering feats and colossal mechanical construction.
>
> The skyscrapers of New York, the grain elevators of Minneapolis, the steel mills of Pittsburgh, the oil wells of Oklahoma, the copper mines of Butte, the lumber yards of Seattle give the American industrial epic in its diapason. . . .

> Every epoch conditions the artist's attitude and the manner of his expression very subtly and in devious way. He observes and absorbs environmental facts, social currents, philosophic speculation and then chooses the elements for his work in such fashion and focuses attention on such aspects of the environment as will reveal his own esthetic vision as well as the essential character of the environment which conditioned it. . . .
>
> The dominant trend in America of today, beneath all the apparent chaos and confusion is towards order and organization which find their outward sign and symbol in the rigid geometry of the American city: in the verticals of its smoke stacks, in the parallels of its car tracks, the squares of its streets, the cubes of its factories, the arc of its bridges, the cylinders of its gas tanks.[114]

NEUE SACHLICHKEIT

American artists' paintings of machines and industrial landscapes, especially those of Sheeler, suggested engineering drawings and architectural models, as did the work of a contemporary group of German painters who came in the 1920s to be associated with a "New Objectivity" or "Neue Sachlichkeit." As we have seen, liberals and radicals in the Weimar Republic expected American mass-production technology to reinforce the new government. It is not surprising, therefore, that not only an architecture but also an art dedicated to the definition of a technological culture should have flourished there. Yet there remained among intellectuals and artists an ambivalent attitude that reflected not only enthusiasm for technology, but a deep-running cultural pessimism as well. This attitude was aptly expressed in Spengler's widely read *Decline of the West* (1918–22). In 1925 a German exhibition, "Die Neue Sachlichkeit," revealed that German artists took a matter-of-fact, or objective (*sachlich*), approach to their subject matter, in contrast to a romantic or an expressionist one. In so doing, their style took on the character of engineering drawings. Artists favoring technological forms and industrial and urban scenes predominated in the exhibition. Like Sheeler's, their paintings mostly lacked workers and metropolitan crowds. In some ways

LOUIS LOZOWICK,
***CORNER OF A STEEL PLANT*, 1929**

LOZOWICK, *HIGH VOLTAGE—COS COB*, 1929

LOZOWICK, *TANKS #1*, 1929

their work recalled that of the painters of landscapes devoid of the marks of human activity. However, the Neue Sachlichkeit artists were not seeking to portray the works of God-the-creator, but of man-the-maker, the inventor, engineer, system builder, and architect. Franz Radziwill and Carl Grossberg, two of the most influential Neue Sachlichkeit members, painstakingly delineated the interacting components of machines and the structural members of bridges and buildings. Their paintings of factories resembled architectural renderings. No grime, oil, rust, or workers intruded their messy complexities into a world of pristine technology. Some of the Radziwill and Grossberg compositions strongly recalled toy-train layouts or engineers' models of industrial complexes.[115] The Neue Sachlichkeit style, like that of the mechanistic Americans whom we have considered, emphasized a hard, metallic sharpness of highlighted details, incorporated schematization to omit confusing complexity, lacked traces of brush strokes and the painterly process, and, unlike the Futurist paintings, was rigidly structured and static. Neue Sachlichkeit compositions, however, were not simply representational. These artists reorganized machines and structures. They highlighted and selected their elements to make distinct impressions that captured more effectively than a realistic engineering or architectural rendition the essence of the technological artifact.

Grossberg, who had also studied architecture, introduced discordancy into his work by occasional suggestions of surrealistic phenomena. In *Boiler in a Refinery*, he ostensibly portrayed the apparatus of industrial chemistry, but he did not indicate its overall purpose or how the technology worked. The looming form of the boiler and the chest beneath it suggest a great icon hung above an altar. This impression recalls Bertolt Brecht's poem "700 Intellectuals Pray to an Oil Tank." Grossberg's painting is reminiscent also of Lozowick's tank lithographs. One of Grossberg's most memorable paintings, *The Diver* (1931), shows a suspended deep-sea diving suit, presumably containing a person completely sealed off from the environment. Instead of being in an ocean, the diver stands in a steel-and-glass, cagelike, technological setting. With arms pathetically limp, the figure hangs helpless and puppetlike from cables, a prisoner of his own creations. Here Grossberg acknowledged his ambivalence toward technological progress.[116]

Radziwill also studied architecture.[117] In his oil paintings, he was preoccupied with the juxtaposition of technological artifacts, such as airplanes and ships, with unnatural cosmic events. On occasion he interjected people, but they are lifeless and without discernible relationship to the technological environment. For contrasting effect, he often placed modern technology in a rural setting. In the 1928 *Dorfeingang* (*Village Entrance*), he showed in the middle ground a locomobile, or self-propelled steam engine, standing functionless. Several rustic, steep-roofed, Germanic cottages frame a sterile, modern building with its obligatory flat roof. The confusion heightens with his juxtaposition of a modern

CARL GROSSBERG,
***BOILER IN A REFINERY*, 1933**

GROSSBERG, *DIVER*, 1931

FRANZ RADZIWILL, *VILLAGE ENTRANCE*, 1928

RADZIWILL, *THE HARBOR*, 1930

airplane high above old sailing ships. Water, canalized, cannot flow freely. A threatening sky contains incomprehensible cosmic formations. Radziwill also painted great steel steamships but, instead of placing them on a trackless ocean where they would be reduced to scale, he docked them where they dwarf stoic human figures and adjacent facilities. In *Hafen mit zwei grossen Dampfern* (*Harbor with Two Great Steamers*, 1930), one ship, bow up, threatens to bear down on hapless observers in a tiny boat, while the other ship walls them in. Small sailing ships are also caught helpless by this monstrous technology.

Grossberg and Radziwill, and Sheeler and Lozowick, portray technological forms and industrial landscapes in similar fashion, though the German and the American painters reveal subtle differences in their attitudes toward technology. Radziwill and Grossberg had difficulty casting aside a deep cultural conditioning fraught with pre-modern symbols and values. The handsome, imposing forms and implied values of the medieval, baroque, and neoclassical German architecture and art juxtaposed with the forms of the modern technological era caused a tension—often creative in effect—among some Germans of aesthetic sensibility. On the other hand, the Americans had less problem in abandoning forms and values of a shorter, less deeply rooted, and less aesthetically rich American past. The paintings of Sheeler and Demuth reflect a less ambiguous acceptance of technological culture than those of Grossberg and Radziwill, who, like many other Germans and Europeans, could not completely mask feelings of foreboding and alienation in the face of the technological transformation. As we have seen, the National Socialists exploited this antimodern and antitechnology attitude.

8 TENNESSEE VALLEY AND MANHATTAN ENGINEER DISTRICT

The prowess of the independent inventors, the well-publicized achievements of the industrial research laboratories, and the organization and management of large systems of production spread the belief that America could invent and produce its future by design. Taylorism coupled control of the work process with planning; system builders like Henry Ford and Samuel Insull set production goals and designed controls suitable to fulfilling them. During World War I the federal government, if somewhat tardily and sometimes ineffectually, planned industrial production. After the war engineers and industrialists such as Morris Llewellyn Cooke wanted to carry government planning over into peacetime, as in the case of Pennsylvania's plan for Giant Power, but, after industry recovered from a postwar slump and business as usual became the keynote, the movement for government involvement in planning lost momentum. Before the onset of the Great Depression and the inauguration of Franklin Delano Roosevelt, the technological transformation of the nation was seen as largely and rightly a private enterprise. The rhetoric of the politicians and the public-relations ideology of the great corporations persuaded most Americans that the spreading systems of production were the fruit of free-enterprise capitalism. They ignored the impressive development of new industrial technology in Germany, where many utilities were government-owned, and where government participation in industrial development was commonplace. A few liberal-minded engineers and industrialists, such as Gerard Swope, president of the General Electric Company, went so far as to join with Herbert Hoover, first when he was secretary of commerce and then as president in 1929, to encourage, in the spirit of scientific management and conservation, other engineers and industrialists voluntarily to plan and coordinate in order to utilize efficiently human and natural resources. Hoover, as a trained engineer who had experience in mining engineering, found the challenge congenial.

Liberal social scientists who struggled to increase government's role in deploying production technology on a large scale remained on the defensive until the coming of the Great Depression. Then they spoke more boldly of the value of planning and, specifically, of government planning for the development of entire regions. Planning with modern technology in a democracy could fulfill, they believed, the age-old vision of men and women creating an ideal environment—the New Jerusalem that would compensate for the fall from grace and the exodus from Eden.[1] Their enthusiasm for planning often found expression in plans for regional transformation through electrification. Electric power was to be the technological agent for the transformation of regions. Progressive politicians would prepare the legislative ground; and social scientists—some spoke of "human engineers"—would preside over planned development. Such goals, as we have seen, had been defined by historian and social critic Lewis Mumford and others of like mind as early as 1925.[2] Advocates of regional planning established associations. Mumford, Benton MacKaye, Clarence Stein, and a handful of architects and planners organized the Regional Planning Association of America in 1923. Stein became chairman of the New York State Commission on Housing and Regional Planning and chairman of the Committee on Community Planning of the American Institute of Architects. MacKaye planned and promoted the Appalachian Trail. In 1925 the International Town, City, Regional Planning and Garden Cities Congress met in New York. The Russell Sage Foundation financed a *Regional Plan for New York and Its Environs*, published in 1929. Urban and regional sociologists established centers and research programs at Columbia University, the University of Chicago, and the University of North Carolina. Howard Odum's Institute for Research in the Social Sciences at Chapel Hill became well known for its studies of Southern regionalism.[3]

REGIONAL PLANNERS

Mumford, an ardent regionalist, spread his ideas and those of liberal social scientists, architects, and planners among a wider segment of the public. Patrick Geddes, the Scottish biologist and sociologist, deeply influenced Mumford and other regional planners with his views of a region as an ecological system. All of them believed in the beneficial potential of modern technology if it were kept under the control of planning and reforming social scientists supported by enlightened civic leaders rather than under the aegis of profit-motivated capitalists and industrialists. They argued that the damage done to the physical environment during the first Industrial Revolution by the misuse of the steam engine, the railroad, and the factory in the black regions of Pennsylvania, New Jersey, and Ohio could be corrected by an informed and benign use of technology, especially electrical. They blamed the growth of these congested, grim, and grimy regions on the lack of vision and effective leadership among social reformers during the first Industrial Revolution. Those social reformers had not compensated for the lack of social sensitivity on the part of mechanical engineers, most of whom had been in the service of captains of industry whose single-minded goal had been the increase of material wealth. They deemed the need for social or "human" engineers to have been as necessary then as it would be during the new industrial revolution.

The reforming social scientists, educators, social workers, and enlightened civic leaders, such as those who wrote for the social-science periodical *The Survey*, were the new "human engineers." They were gradually establishing standards, they believed, to measure human well-being as shown by individual fulfillment, fruitful and harmonious social relationships, and social creativity. The new human engineers would use these standards, not simply the accumulation of material wealth, to determine the ways in which technical possibilities could be used.[4] The new human engineers, they argued, should develop large plans for regional development based on electric-power grids or networks. They saw Governor Gifford Pinchot's Giant Power plan for the Commonwealth of Pennsylvania as such a forward-looking concept.

Those persuaded that regional planning and electric power, along with the automobile, the telephone, and related technologies, would usher in a new era made their essential program manifest in two issues of *The Survey*, the first in 1924 dedicated to "Giant Power" and the second in 1925 to "Regional Planning."[5] Those contributing articles advocating Giant Power or some variation of regional electrification through power grids included Governor Pinchot; Joseph Hart, an associate editor of *The Survey*; Cooke, the ally of Governor Pinchot and later head of the Rural Electrification Administration under Roosevelt; Sir Adam Beck, chairman of the Hydro-Electric Power Commission of Ontario, Canada, a government-owned and -operated regional power system; Governor Alfred E. Smith of New York; and Samuel Gompers, president of the American Federation of Labor. The issue also included interviews with Ford and Herbert Hoover, then secretary of commerce. Notable contributors to the "Regional Planning" issue a year later included Mumford; Stein; Stuart Chase; Governor Alfred Smith; Joseph Hart; and Robert Bruère, the industrial editor of *The Survey*. *The Annals* of the American Academy of Political and Social Science, another influential social-science journal, also supported Giant Power with a special issue subtitled "Large Scale Electrical Development as a Social Factor."[6]

The overarching logic combining electric power and regional development conveyed in these issues was straightforward: The nature of power use has shaped various eras of modern history. During the era of coal and steam, power transmitted over long distances by rail and distributed for short distances by leather belt resulted in concentrations of industry and population at grimy mines, near grim factories, and at the nexus of rail lines. In the new era, power from electric generating plants at coal mines and at dam sites would be transmitted over long distances by high-voltage electric networks, or grids, and over short ones by lower-voltage systems. As we have noted, since power transmission and distribution by wire cost less than by rail and belt, population and industry could be dispersed to more life- and culture-supporting environments. Mumford named the move back to the uncongested countryside from megalopolis "the fourth migration."[7] The first migration settled the continent; the second settled the industrial districts; and the third gathered population in New York and other financial centers. Mumford did not want the fourth to go unplanned, as had the other three. Growth without planning, he predicted, would "pile up embarrassing conditions. . . . Why should we not plan so as to reap advantages from the Fourth Migration?"[8]

Balanced regional economic development could be achieved by making electricity available to farmers and village workshops. The ends sought by Mumford and his allies as they promoted rural electrification extended beyond mere economic rationality. Observing the migration of population from the countryside into the city, they began to idealize the rural values and society disappearing before their eyes. In an unsigned preface to the special issue on regional planning of *The Survey*, Mumford described the social scientists, politicians, and academics who contributed articles. He found their views tinged by nostalgia. One of them lived in a Brooklyn flat and wrote "wistfully" of Connecticut villages; another was "a prophet crying in the wilderness of the city, his heart in the crumbling country towns"; and another "clings" to a tiny communal patch of green in the midst of city tenements.[9] These and the other enthusiasts for the new revolution wanted to piece together their fragmented spirits by creating green work and cultural communities. Like many other Americans, they were experiencing loss and change as man-made America displaced the frontier, leveled the wilderness, and even invaded the Great Plains.[10]

Believing that electrical-power transmission, the internal-combustion engine, and alloys, in company with other twentieth-century inventions, had ushered in the neotechnic era, Mumford looked forward to the next era, the biotechnic. This was one in which biological and social sciences would bring in a new society and culture framed in a regional setting. He believed that the geographic region should replace the industrial metropolis as the framework for social and cultural activity. In defining regionalism, he recalled that, before the coming of the railroad in the United States, industrial activity and population had been dispersed within geographical regions. In the Northeast, waterwheels driving mills of various kinds allowed small industrial communities to flourish. Canals permitted sufficient exchange of raw materials and finished goods for a limited regional economy to develop. In New England before the Civil War, a balanced and rich regional culture had flourished in a "golden day."[11] In the biotechnic era, there would be a dispersal once again of industry and population and, at the same time, a web of interactions and transactions bringing a highly developed but loosely knit regional economy and a cultural efflorescence.[12] Not only would power transmission and motorized vehicles permit the spread of a network of communication and transportation, but they would also allow it to climb into the upland regions that the railroads, their grades usually limited to two percent, did not reach. Writing in the 1930s, Mumford pointed out that, along the Tennessee Valley, the hilly uplands with their salubrious climate were beginning to show the energizing and civilizing effects of the new technology as it was employed by the Tennessee Valley Authority.[13] He and other regional planners of his day hoped that the TVA would prove the efficacy of regional-planning techniques in the Tennessee Valley.

Regionalism appealed so strongly to Mumford in the 1930s because he saw it as a way first to transcend and then to displace the mechanistic ordering of the individual and the society that had taken place during the paleotechnic era of iron, steam, and factory concentration: a world, he wrote, "whose cold mechanical perfection was described by physics and astronomy."[14] Mumford thought of an organism, in contrast to a mechanism, as growing rather than expanding, as taking on an evolving, complex pattern rather than as having a fixed geometric form, as interacting with the environment rather than as existing in functional relationships with other machines, and as involving the biological phenomena of inheritance and memory. Conceiving of a human society as an organism, he saw it absorbing, sustaining, and adding to a complicated social heritage. The region, he stressed, was the geographical area in which lived a human society-as-organism.

For Mumford, the region, however, was not a simple geographic unit with unchanged and unchanging characteristics. Not only had soil, climate, flora, and fauna changed over time, but humans using technology had altered, suppressed, and enhanced them. Through social organization and culture, humans had also given regions originally defined by geography alone a social and cultural character. For instance, regions had distinguishing dialects, dress, and political practices. The region, then, for Mumford was a highly complex concept. It clearly was not an area defined by unchanging physical frontiers, and certainly not a political or technological unit. He lamented equally the unnatural areas of industrial concentration and the imposition by nation states of artificial political boundaries that destroyed organic regions.

Mumford provided examples of regions in the United States. New York State, when rounded out by the inclusion of northern New Jersey and part of Connecticut, possessed the diversity of resources and habitats, the balance of agricultural and industrial activity, and the requisite individuality and coherence to constitute a region. Whereas the New York region was defined primarily by geography, natural and man-made, and economics, New England found its regional identity primarily in the cultural heritage of its golden day. Nathaniel Hawthorne, Ralph Waldo Emerson, and Henry David Thoreau felt deeply their regional roots, thinking of themselves first as New Englanders, then as Americans.

Mumford's plans for balanced regional development would make use of the new technology not, he insisted, chiefly for industrial development and private gain, but for the conservation and enhancement of life. "One uses the word life in no vague sense," he wrote; "one means the birth and nurture of children, the preservation of human health and well being, the culture of the human personality, and the perfection of the natural and civic environment as the theater of all these activities."[15] In the future biotechnic era that was evolving within the neotechnic, Mumford foresaw the artist, scientist, architect, teacher, physician, singer, musician, and actor using a greater proportion of resources than the industrial engineer. He also envisaged—and desired—the decline of the market economy, for it brought the production of goods and services that were profitable, not those that were socially desirable but unprofitable. Quality housing for lower-income persons was a major objective of the new regional development. Modern building technology, dispersal of population, social planning, and new values could effect this change. Having learned to deal with the mechanical and electrical in a fruitful and systematic way during the neotechnic era, "we have still to invent," he wrote, "that wider system of order which will assist in the transformation of our social relations: one of its symbols is the regional plan itself."[16]

THE TENNESSEE VALLEY

Although Ford had more down-to-earth regional-planning objectives than Mumford, he nonetheless envisioned a scheme for the development of the Tennessee Valley. The Tennessee Valley attracted the interests of persons as diverse as Mumford and Ford, and of a large segment of the American public, because of the valley's poverty coupled with its potential for economic growth. In an era when great dams and hydroelectric-power plants constituted high technology, and when networks of high-voltage power-transmission lines made possible the spreading of hydroelectric power over an entire region, the Tennessee River and its tributaries offered the forward-looking the opportunity to combine modern technology and modern planning in the cause of regional development. In addition to power, dams along the rivers could provide flood control and navigation.

The possibility of energy development and regional planning in the Tennessee Valley appealed to Ford's system-building instincts. In 1921–22 he offered to lease the Wilson Dam and its associated power plant for a hundred years from the federal government and to complete the largest concrete dam in the world. In addition, he offered to purchase from the U.S. government the two nearby government-owned nitrogen-fixation plants. These facilities had been under construction during World War I at Muscle Shoals, Alabama, on the Tennessee River, to provide the power needed to manufacture nitrogen used in explosives and fertilizers. Ford said not only that he intended to construct a large waterpower-based industrial complex for the manufacture of fertilizer, aluminum, steel, and automobile parts in the vicinity of Muscle Shoals, but that he also planned to use power from various sites along the Tennessee River to promote the economic development of the Tennessee Valley region. He promised to produce in the nitrogen-fixation plants low-cost fertilizer for the region's farmers, and to have the larger of the two plants ready for quick conversion to explosives production in case of war. One newspaper reported that he planned to build a city seventy-five miles long—one presumes along the Tennessee River—and that this creation would be one of the nation's great industrial centers. Later Ford made clear that the seventy-five-mile-long "city" would be a chain of industrial towns fulfilling his goal of having workers and their families living in small communities and enjoying the benefits of a semirural life. He promised to make every effort to protect the health and social welfare of the workers and their families. He also promised that eventually he would turn over the completed project to the people of the area or to the federal government.

Regional-planning organizations, the National Grange and the Farm Bureau Federation, Southern newspapers, community leaders representing agricultural, industrial, and realty interests, the American Federation of Labor, and a majority of members of Congress from the region supported Ford's proposal. Ford enthusiasts throughout the nation enthusiastically applauded and looked forward to another Ford industrial miracle. Ford insisted that his motives were public-spirited. "If Muscle Shoals is developed along unselfish lines," he said, "it will work so splendidly and so simply that in no time hundreds of other waterpower developments will spring up all over the country. . . ." "In a sense the destiny of the American people for years to come lies here on the Tennessee River," he added.[17]

Opposition to the Ford offer stemmed from various quarters, among them those who wanted the Tennessee Valley to be developed by the government, not private capital. Governor Pinchot severely criticized the Ford plan because it did not conform to conservationist ideals and called for a hundred-year lease at a time when the law provided for a maximum fifty-year term for private hydroelectric projects on public waters. Senator George Norris of Nebraska, a staunch champion of public-power development, blocked passage of congressional legislation that would have facilitated the transaction. The National Fertilizer Association and various privately owned utility spokesmen argued that Ford's monetary offer and the terms of the lease exploited the government and the people. The government rejected Ford's offer, but not before he had held up his grand scheme and a special mode of financing it as a counter and contrast to exploitative financial projects advanced by those he identified as Wall Street financiers and "international Jewry." He blamed them for blocking his efforts to fulfill a private vision and a public need. Ford was engaged at the time in a program to "educate" Americans about the evils of "international Jewry." Responding to Ford's anti-Semitism, Samuel Untermeyer, a noted Wall Street lawyer, characterized him as "ignorant on every subject except automobiles."[18]

Senator George Norris shared with Ford the belief that the American people might find their future along the Tennessee River. His vision, not Ford's, prevailed in 1933 with the coming of the Roosevelt New Deal and the establishment of the Tennessee Valley Authority. As chairman of the Senate Committee on Agriculture, Norris had responsibility in the 1920s for legislation pertaining to the Muscle Shoals dam and nitrogen-fixation plants for fertilizer manufacture. At first uninterested, Norris soon took up the gauntlet for government ownership and operation of the Muscle Shoals facilities. He went even further and proposed development of the Tennessee Valley through hydroelectric power, flood control, and soil conservation. He studied successful precedents for public power in Los Angeles and the province of Ontario, Canada.[19] In the face of various schemes for private ownership, this Nebraskan in his sixties tirelessly and resourcefully maneuvered for twelve years through three hostile presidential administrations until the like-minded Roosevelt came to office. The TVA victory for the Republican senator resulted from his patience, his commitment to his goals, and his parliamentary opportunism, including *ad hoc* alliances with the Democrats. Initially a Republican, in time he became an independent, rallying support for his projects from congressmen of various persuasions.

Norris passionately believed in the promise of the electrical age. Having as a boy experienced the heavy work, the sweaty summers, and the freezing winters of a simple Ohio farm, he welcomed the "fascinating" possibilities of electricity "lightening the drudgery of farms and urban homes, while revolutionizing the factories."[20] Persuaded, as were other progressive politicians, that the privately owned utilities, especially the large holding companies or trusts, were frustrating the dream of social

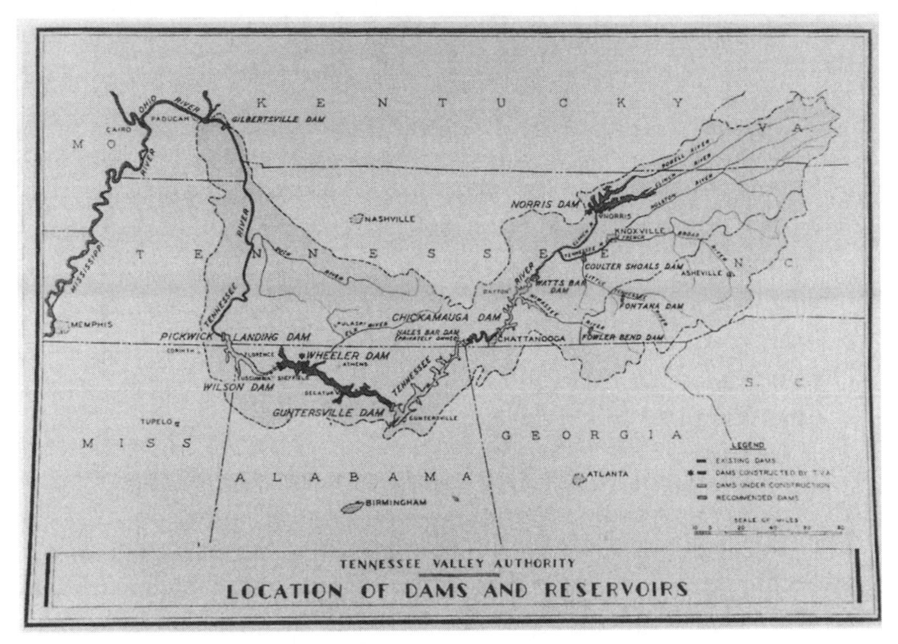

TVA SYSTEM, C. 1936. MUSCLE SHOALS (WILSON DAM) IS TO WEST OF WHEELER DAM.

transformation through electrification, he directed his resentment and cutting rhetoric against the "power trust." Like Pinchot, Norris wrote in 1926 that, unless countermeasures were taken, "we will find ourselves in the grip of privately owned, privately managed monopoly and it will be extremely difficult to shake off the shackles that will then be fastened upon all of us."[21] Insull's insistence that he was bringing the benefits of the electrical age to the nation obviously fell on deaf ears. Norris could only have nodded vigorously in assent when Roosevelt, campaigning for president, made reference to Insull as a lone wolf, an unethical competitor, and a reckless promoter.[22]

Not only determined to wrest control of the electrical revolution from private hands, Norris, like Pinchot and other conservationists and regional planners, also looked forward to a government-coordinated, systematic development of several of the nation's river valleys as systems. The electric power generated could, he argued, be used to pay for regional development. Greatly pleased when Roosevelt, before his inauguration, invited the senator to accompany him through the Tennessee Valley region, Norris took heart when he heard him say at Montgomery, Alabama:

> Muscle Shoals gives us the opportunity to accomplish a great purpose for the people of many States and, indeed, for the whole Union. Because there we have an opportunity of setting an example of planning, not just for ourselves but for the generations to come, tying in industry and agriculture and forestry and flood prevention, tying them all into a unified whole over a distance of a thousand miles so that we can afford better opportunities and better places for living for millions of yet unborn. . . .[23]

For a nation plunged into the Great Depression, these were stirring words projecting a powerful vision that private utilities, with their concentration on the profitable generation of power, could not match.

With the coming of Roosevelt and the New Deal in 1933, progressives like Norris advocating public power and those advocating regional planning took hope. As governor, Roosevelt had supported the regional planners in his own state of New York. Among the "Brain Trusters" who wrote his early speeches and advised him on policy were those who wanted regional planning extended on a national level to the regulation and coordination of industrial production. The administration intended to plan land use, population distribution, energy development, fiscal and monetary policies, and prices and production through various new agencies such as the National Recovery Administration, the Soil Conservation Service, the Resettlement Administration, the National Planning Board, the Rural Electrification Administration, and the TVA.

The new administration's TVA legislation included several sections authorizing planning, for which Arthur E. Morgan, whom Roosevelt had already named chairman of the governing commission of the TVA, was responsible. Morgan was an engineer imbued with the high-minded problem-solving approach to technical and social problems characteristic of the most imaginative and public-spirited members of his profession. During the interwar decades the engineering profession enjoyed a prestige

PRESIDENT FRANKLIN D. ROOSEVELT, ELEANOR ROOSEVELT, AND ARTHUR MORGAN AT NORRIS DAM, 16 NOVEMBER 1934

unmatched earlier or since. A self-taught water-control engineer, he and the firm he established accumulated a record of notably successful projects, including a huge flood-control project in 1913 for the Miami River Valley in Ohio. Like so many others conditioned by engineering experience and values, he believed that an optimum solution to technical, economic, and even social problems could be found. As TVA head and commissioner, he was also distrustful of those who, with what he viewed as value-laden considerations, confounded what he had decided was a reasonable and objective solution to a problem. He scorned those who tailored policy to suit needs for political alignments and support. Those who disagreed with his carefully thought-through and worked-out positions he tended to consider irrational. As a result, he would righteously defend his solution to a problem and dismiss the counterproposals of others. Like so many other leading engineers and scientists of his day, Morgan disdained "politics," which he saw as the intrusion of the extraneous into the world of the builders. Politics, in his opinion, often meant favoritism to individuals and pandering to economic, regional, and other interest groups.[24] He was a far wiser student of nature than of human nature.

An ascetic, almost mystical utopian, Morgan, seeing an opportunity to spread his philosophy and methodology, in 1920 accepted the presidency of small Antioch College in Yellow Springs, Ohio. Founded by Horace Mann in 1853, Antioch had not yet become, as its founder intended, the Harvard of the Midwest. With the support of imaginative and public-spirited engineers and industrialists like Charles Kettering, Morgan established a program that gave students an opportunity to fashion a community. This program combined learning with doing, by means of a work-study program and the founding of campus industries. Morgan wanted to instill the entrepreneurial spirit. Antioch, with Morgan, soon attracted nationwide attention. He wanted the college to educate philosopher-engineers who were technically competent, dedicated to truth, and generalists with creative drive. He approved of work, self-reliance, integrity, efficiency, balance, symmetry, and a holistic approach.[25] In addition to his engineering style, he brought the fervor of a reforming educator and a determination to imbue the young with his own characteristics.

Morgan the holist admired the generalist. He discounted the specialist, as had Edison and Ford. Morgan systematically approached problems so that each component in the system could be considered in relation to the "current of events and in relation to all other components."[26] Drawing on his own experiences as a hydraulic engineer and a generalist, he praised those who, in building a great dam, coordinated and integrated the work of the geologists, lawyers, civil engineers, and mechanical engineers, as well as the work of social scientists who deal with social problems of population resettlement and labor relations. Speaking to the American Sociological Society in 1936, he said that one explanation for the breakdown of human culture is too many specialists and too few generalists. "A human culture," he added, "does not long remain stable on a plane much higher than that of the generalized culture of the governing class."[27]

Morgan believed in planning. From his water-control projects he had learned that myriad regional factors must be systematically taken into account as one drained, diked, dammed, confiscated, flooded, channeled, and rehabilitated. As a holistic thinker, he believed not only that rivers and their valleys must be treated as systems, but that the political process had to conform to nature's imperatives.[28] As TVA commissioner, he argued that the TVA should carry out its agricultural program not in isolation but as one element in a designed social and economic order.[29] He also spoke of using the Tennessee Valley as a laboratory in which models for planning throughout the nation could be tested.[30] He intimated that those who misused the land should be deprived of it, and that a campaign should be launched against wasteful local governments in the region. He spoke of cooperatives and the use of a kind of local money. Possibly because Rexford Tugwell, an advanced New Deal planner, and others counseled him not to disclose prematurely his intended experiments and thereby arouse suspicion and opposition, Morgan did not reveal more of what he specifically envisioned in a "laboratory" for planning.[31] He continued to speak often but loosely of "social and economic planning."[32]

As constituted in 1933, the TVA was, as Morgan argued, not intended to be primarily a power-generating organization. He did not want the TVA's broad mission to become hostage to the explosive public-power issue.[33] Roosevelt at the beginning seemed to be of this persuasion. In his April 1933 message to Congress, he said his plans for the Tennessee Valley transcended power development and involved flood control, prevention of soil erosion, afforestation, elimination from agricultural use of marginal lands, and distribution and diversification of industry. Such goals along the Tennessee, he argued, would lead logically "to national planning for a complete river watershed involving many States and the future lives and welfare of millions." He added, "this in a true sense is a return to the spirit and vision of the pioneer. If we are successful here we can march on, step by step, in a like development of other great natural territorial units within our borders."[34] The sponsors of the TVA did not define "planning" precisely but, like so many other terms employed then, it was a magic word and a captivating concept for them.[35] Roosevelt described the TVA to Congress as transcending

> . . . mere power development . . . a corporation clothed with the power of government but possessed of the flexibility and initiative of private enterprise . . . charged with the broadest duty of planning . . . for the general social and economic welfare of the Nation. . . . It is time to extend planning to a wider field. . . .[36]

Secretary of Interior Harold Ickes called the TVA a magnificent regional-planning experiment. One 1934 study interpreted it as the greatest experiment in regional planning outside of Soviet Russia.[37] In 1937 Norris introduced a bill with a proposal, patterned on the TVA, to establish seven regional-planning and power authorities to develop other river valleys, but the proposal died in the Senate.[38]

Morgan saw his charge as fulfilling the so-called planning sections of the TVA act. These authorized surveys and general plans would guide the expenditure of funds to bring orderly physical, economic, and social development. As an action-oriented engineer, Morgan insisted that those who executed also plan. Taking again the holistic approach of the system builder, he wrote in 1937 that "in any important undertaking sociological components are as real and important and as deserving of analysis and treatment as the engineering or financial or legal components, all of which should be treated so as to result in an integrated program."[39] With the passage of time, however, the difficult, messy complexity and conservatism encountered in the valley forced Morgan and his supporters to modify plans to bring about a social transformation of the region. In so doing they disappointed more progressive regional planners throughout the nation, who had enthusiastically called the TVA "a prevision of Utopia" in which a new civilization would arise.[40] Instead of fulfilling the grand vision, Morgan and his associates had to limit themselves to planning regional education and health facilities, especially in cases of population resettlement necessitated by dam construction, the flooding of land, and other construction projects. If schools and hospitals were inundated by new lakes, then the TVA cooperated with local government in funding and establishing new facilities to incorporate physical improvements and new approaches. A TVA relocation service took into account the financial resources, family size, religion, and preferred type of farming of the people to be moved. If water control stimulated health hazards, as from mosquitoes, then the TVA introduced a program that could be taken over permanently by local authorities. Instead of simply constructing temporary housing for thousands of workers on the Norris Dam, Morgan had constructed a permanent town named Norris, designed and administered by the TVA. It served as a model for expanding communities.[41] These measures, however, were a far cry from what many social scientists and social critics, like Mumford, had in mind by regional planning.

In general, the conservative inertia of existing government departments and agencies, conservative congressmen, and the stubbornness of established interests and patterns on the local level blunted the thrust of planning programs. The result was a complex and often contradictory mix of partial and local successes that fell far short of the desired coordination and systematization so dear to planners.[42] The TVA, however, during the tenure of Morgan as chief commissioner, kept alive the hopes of regional planners. Even though TVA planning failed to fulfill the broad concepts implied in the president's early statements, the agency struck out boldly by taking over functions traditionally assigned to federal agencies and departments. Within the Tennessee Valley the TVA administered agricultural services usually provided by the U.S. Department of Agriculture; it built dams and assisted in navigation, work usually left to the U.S. Reclamation Service and the Army Corps of Engineers; and it conserved forests, a function generally performed by the U.S. Forest Service. Prudently, however, the TVA sometimes arranged joint activities in the valley with other government agencies.[43]

THE POWER FIGHT

Because Arthur Morgan was an engineer, the president named Harcourt Morgan, a university president, and David Lilienthal, a lawyer, to serve as the other two commissioners. Harcourt Morgan was an expert on Southern agriculture, a champion of balanced resource development, and president of the University of Tennessee. Lilienthal, a thirty-three-year-old lawyer, had established a reputation as an aggressive public-

ABOVE: WASHDAY PRIOR TO THE TVA

LEFT: DRUGSTORE, NORRIS, TENNESSEE, 1937

BELOW: NORRIS, TENNESSEE, 1937

HARCOURT MORGAN, ARTHUR MORGAN, AND DAVID LILIENTHAL

power advocate when he was a member of the Wisconsin Railroad Commission, which regulated electric and other utilities.

During the five years in which Arthur Morgan was head commissioner, Lilienthal showed little enthusiasm for the broad-scale planning approach initially proposed by Morgan. Lilienthal saw the prime goal as public power and straightforward economic development. Keenly sensitive to political interests, he realized that the support or acceptance of public power by political and economic leaders in the region and by congressmen could be jeopardized if the TVA imposed radical economic and social reform in the name of planning. Later, Senator Kenneth McKellar of Tennessee boasted that his efforts kept the TVA focused on dam building rather than half-cocked schemes "to uplift the ignorant and benighted people of the Tennessee Valley."[44]

While Arthur Morgan focused on construction and planning, Lilienthal expended his energy on the public-power issue. His uncompromising ways and his insistence that a public-utility enterprise was a public business exacerbated the tension between public-power advocates and the private utilities in the Tennessee Valley. In Wisconsin, earlier, as a member of the Wisconsin Railroad Commission, he caused friction and animosity between the commission and the private utilities, but he successfully broadened the commission's powers. Still earlier, in Chicago as special counsel to the city, he had won a $20-million refund for telephone-company customers. Roosevelt and Norris expected him to deal with the thorny issue of public-power development by the TVA. Unlike Arthur Morgan, who favored negotiations among reasonable men, Lilienthal was ready for a rough-and-tumble fight with the "Power Trust." He also had a well-developed sense of political power, the art of achieving the possible. One friend said of Lilienthal that he was very ambitious and likely "to steal the show." He and Roosevelt found each other temperamentally and politically sympathetic. The president showed considerable affection and substantial support for the younger man. After World War II, Lilienthal became the first chairman of the Atomic Energy Commission (AEC). Historian Tom McCraw observed that Lilienthal invested "whatever he was doing with cosmic importance. Sometimes, as in the Atomic Energy Commission, this attitude was justified."[45]

Within months after the establishment of the TVA, the public-power issue began to overshadow other TVA initiatives. Public and congressional attention began to focus on the power struggle between the TVA and the Commonwealth and Southern Corporation, an electric-utility holding company with major interests in the Tennessee Valley. A legal confrontation between public and private power lasted from 1933 until 1939. Wendell L. Willkie, an experienced trial-and-utilities lawyer and newly named head of Commonwealth and Southern, spearheaded the resistance of private utilities throughout the nation to public power. Willkie, a Wilsonian Democrat and a man of great personal charm, ably mounted a strong case against public power and negotiated resourcefully with the TVA to limit the spread and use of TVA power. He later ran against Roosevelt for president and campaigned against the expansion of government ownership in general. Progressives and liberals saw Lilienthal as a champion in the fight against Willkie and the Power Trust spoken of by Pinchot, Norris, Roosevelt, and others.

At the start of the power fight, the TVA sought to acquire a market for electricity from the Wilson Dam. For this, the TVA needed to establish municipal and rural cooperative electrical utilities as customers in territories then served by Commonwealth and Southern's power-and-light companies. The TVA wanted these small utilities to take power in bulk from the transmission lines of the TVA and distribute it to residential, commercial, and other customers. The TVA also needed a rate policy. With Lilienthal making policy, the TVA announced that it would build a transmission line from the Wilson Dam in Alabama to the projected Norris Dam in Tennessee and sell electricity in the territory along the transmission line, preferably to municipally owned plants and rural cooperatives. In this territory, the TVA would serve as a yardstick to measure the rectitude of private-utility rate schedules in the region, a major assignment of the TVA. It took a bold step in 1933 when it announced its 3-2-1 formula for residential rates (3¢ per kilowatt hour for the first 50 kilowatt hours; 2¢ for the next 150; and 1¢ for the next 200; and for all over 400, 4 mils per kilowatt hour). The national average cost of electricity at the time was 5.5¢ per kilowatt hour. Lilienthal also inaugurated a policy of publicizing the use of electrical appliances and making consumer credit available. In this way he hoped to raise the load factor and consumption of electricity so that rates could be lowered. In so doing he followed, without acknowledgment, the precedent of Insull and other innovative private-utility heads.

Willkie countered the TVA's plan to establish municipal and cooperative utilities as customers with a proposal that Commonwealth and Southern buy all of the TVA power and save the TVA from having to find customers and depending for funds on an unpredictable Congress.[46] After rejecting this proposal, Willkie then offered to sell to the TVA the Tennessee Electric Power Company, a Commonwealth and Southern subsidiary owning the distribution network in the territory along the proposed high-voltage line. But Lilienthal found the price too high. Alarmed that Lilienthal would carry out the threat to go into the Commonwealth and Southern territory, duplicate facilities, and take away municipal and rural customers, Willkie agreed in 1934 to sell territories and facilities to the TVA. As part of the agreement, the TVA conceded that for five years, or until completion of the Norris Dam, it would not spread its power sales into other Commonwealth and Southern territory. When power became available from Norris Dam, then the TVA would need to find new markets for its power. Subsequently, numerous disagreements over the details of the contract delayed the turnover of properties to the TVA, thus heightening animosity between the public and private power camps.

In 1936 more conciliatory parties suggested that the TVA and the private utilities in the Tennessee Valley region pool their power in a great transmission system, from which power would be sold by an organization administering the pool to various private and municipal utilities and to rural cooperatives for distribution. The plan resembled Pinchot's Giant Power plan and the Pennsylvania, New Jersey Interconnection, or pool, that private utilities had established.[47] Advocates argued that the pool would bring power to the region at the lowest possible costs, for the facilities of private- and public-power generators would be harmoniously used, not competitively duplicated. Arthur Morgan wanted the pool. He approached problems in a way he believed to be entirely pragmatic, "not based upon any abstract theory of government," but one favoring private or public power, depending on which was more efficient and economical.[48] Lilienthal was open to negotiations. Willkie favored the pool.

Roosevelt called a conference in September 1936 to encourage negotiations. The backgrounds of those invited reveal the involvement of high-level political and financial interest groups in an issue that might on the surface seem to be simply a technical and economic one. Alexander Sachs, an economist with Lehman Brothers and a marketer of securities, originated the plan. Later, during World War II, he played a critical role in bringing the attention of his friend Roosevelt to the threat of a German atomic bomb. Others attending the conference included Frederic Delano of the National Resources Committee, the New Deal's informal planning agency; Cooke, in addition to being head of the Rural Electrification Administration, also an originator of the Pennsylvania Giant Power plan; Owen D. Young, head of the General Electric Company; Thomas Lamont of the investment firm of J. Pierpont Morgan and Company; Frank McNinch, chairman of the Federal Power Commission; Samuel Ferguson of the Hartford Electric Light Company, an expert on pooling; Wendell Willkie, Lilienthal, and Arthur Morgan. Harcourt Morgan was ill. Roosevelt presided.

Negotiations broke down after nineteen Southern power companies entered a suit in May 1936 challenging the constitutionality of the TVA. Willkie insisted that his Commonwealth and Southern Corporation was not responsible. Lilienthal and Norris's conviction that the private interests wanted to destroy the TVA seemed confirmed, however. Lilienthal and Norris now opposed the plan on the grounds that it would blunt the effectiveness of the TVA as the symbol and substance of bold public-power projects. They thought it would force the TVA into an alliance with private-utility forces, which, they believed, they had even more reason to distrust. Roosevelt acquiesced and ended the talks. Tension between Lilienthal and Arthur Morgan increased during negotiations over the pool because Morgan, in drawing up a technical memorandum, used as an adviser a man who had been closely associated with Insull. Lilienthal took this as one more indication of Morgan's insensitivity to the public-power issue; Morgan saw Lilienthal's reaction as his inability to place technical matters on an objective plane.

After the pooling plan was dropped, the U.S. Supreme Court, by challenging the constitutionality of the TVA, decided against utilities that had helped disrupt the pooling plan. By 1939 the utilities had exhausted every chance for a judicial elimination of the TVA. Willkie then sold to the TVA the Tennessee Electric Power Company, thus giving the TVA a large and integrated market for electricity. The "power fight" of the 1930s was over; the future of the TVA seemed secure. Lilienthal summed up the long campaign between public and private power as one in which he had put "some vague tremors into the most brazen crowd in the country, the utility industry. That has been exciting, and it does have importance and permanence." Stimulated by the fight itself, Lilienthal also found his strong commitment to social progress fulfilled. Cheap power and electrical cooperatives improved living and working conditions among the poor farmers in the "underdog" South, where conditions were "so much worse than anyone expects." During the wearing and emotionally draining controversies, Lilienthal had enjoyed, and found some relief from the tensions during the power fight, making speeches before country crowds in which he demonstrated small-farm electrical machinery and gadgets like grinders and brooders. He was "like an Indian root doctor," he recalled.[49] However, those benefited were white farmers, for the TVA acquiesced in white supremacy and, in general, left blacks out of the programs. The model town of Norris was entirely white.[50]

THE MORGAN-LILIENTHAL FEUD

During the first three years of the TVA, a feud developed among the commissioners. Arthur Morgan confronted Lilienthal directly. Harcourt Morgan usually supported Lilienthal. An *ad hoc* arrangement did not resolve the conflict. It delegated engineering and construction responsibilities, economic and social planning, forestry, and overall "integration of the parts of the program into a unified whole" to Arthur Morgan; electric-power legal issues to Lilienthal; and agricultural policy, including fertilizer production and rural-life planning, to Harcourt Morgan. The arrangement actually violated the intended spirit of coordination and systematic interaction.[51] Arthur Morgan believed that Lilienthal, with Roosevelt's backing, was causing the power-generation tail to wag the TVA dog. Lilienthal's struggle with Arthur Morgan was even more emotionally draining and wasteful of the energy of the commissioners than the power fight. Their differences climaxed in 1937, when Morgan published an article in *The Atlantic Monthly* stating that he differed with the other two commissioners on power policy and attacking "public officials" who use arbitrary coercion and false propaganda to achieve their ends. Later he spoke of a conspiracy of the other two commissioners against him that involved "log-rolling," through which Lilienthal received support for his power program in return for his support of Harcourt Morgan's agricultural policies.[52] Morgan labeled as propaganda Lilienthal's claims that TVA electric rates were an effective yardstick against which to measure private-utility rates. Morgan pointed out that TVA rates were subsidized. He summed up his attitude toward Lilienthal in a letter to Representative Maury Maverick of Texas in which he referred to evasion, intrigue, and sharp strategy that "makes Machiavelli seem open. . . ." At the height of the confrontation, when Morgan, who claimed his ethics were those of the New Testament, made a small and personal gesture of reconciliation, Lilienthal, the embattled tactician, interpreted it as a sign of weakness. Morgan declared that an attitude of boyish candor could mask hard-boiled, selfish intrigue.

Other differences among the commissioners included the willingness of Lilienthal and Harcourt Morgan to negotiate with a local, politically influential figure who was making claims of doubtful validity against the TVA for property damage. Arthur Morgan also challenged the others for their advocacy of a fertilizer program that he found technically and economically unsound.[53] The straightforward engineer was disturbed by the political considerations warping the decisions of his fellow commissioners. In conversation with Lilienthal, Roosevelt said of Arthur Morgan that he would be grand in a planning section of the Public Works Department. "You know," Roosevelt added, "he doesn't know *anything* about power. . . . He is a human engineer—he likes the idea of getting people out of the coves and onto a better way of living. You aren't," Roosevelt said laughingly to Lilienthal, "a human engineer."[54] The remark was not lost on Lilienthal, with his keen political instincts.

Lilienthal and Harcourt Morgan interpreted the *Atlantic Monthly* article as impugning their integrity. Roosevelt asked Arthur Morgan to substantiate his serious charges, which he refused to do, insisting that he would deal only with a congressional committee. Affronted, his authority challenged, Roosevelt, after giving Morgan several opportunities to substantiate his charges, asked for his resignation. When it was not forthcoming, Roosevelt forced it in March 1938. He said that the public owed a debt to Morgan, but that he had proved himself temperamentally unfit to exercise divided authority.[55] Norris privately reacted more strongly. He found that Morgan had treated the president insultingly and disrespectfully. Morgan's charges were comprehensible to Norris only "on the ground that the man has lost his reason in his insane jealousy against his fellow members. . . ."[56] In 1938 a joint House-Senate committee rejected Arthur Morgan's charges. After Morgan's ouster, Harcourt Morgan became chairman of the commission, and James Pope, a former U.S. senator from Idaho, became the third member. Lilienthal took over the chairmanship in 1941 and served until 1946, when he became head of the Atomic Energy Commission.

In 1940, after Morgan's departure and the establishment of public power in the region, Lilienthal publicly assumed Morgan's role as a planner and system builder:

> For the first time in the history of the nation, the resources of a river were not only to be "envisioned in their entirety"; they were to be developed *in that unity with which nature herself regards her resources*—the waters, the land, and the forests together, a "seamless web"—just as Maitland saw "the unity of all history," of which one strand cannot be touched without affecting every other strand for good or ill.[57]

Lilienthal began to say that he and the TVA had pioneered in grass-roots planning. He celebrated this policy in his highly enthusiastic book, *TVA: Democracy on the March* (1944). Arthur Morgan, however, should be associated with the early planning, and Harcourt Morgan more directly with the grass-roots approach. Harcourt Morgan insisted that, whenever it was possible, the responsibilities of the TVA be delegated to, or executed in cooperation with, local authorities, such as land-grant colleges and their extension services. Again, however, it was Arthur Morgan who encouraged self-help cooperatives, subsistence homesteads, and rural zoning.[58] Despite the national publicity that centered on Lilienthal, Arthur Morgan, during his five-year chairmanship, "was responsible for some of TVA's most noteworthy achievements—the splendid engineering work, the enlightened labor policies, and almost all the social experiments."[59] The construction program under Arthur Morgan accounted alone for more than three-fourths of the TVA budget.

It is ironic, considering his conflict with Arthur Morgan the engineer, that Lilienthal so enthusiastically sang the praises of engineers in *TVA: Democracy on the March*. (Morgan, however, does not find place in the index.) Lilienthal wrote, "there is almost nothing, however fantastic, that (given competent organization) a team of engineers, scientists, and administrators cannot do today. . . . Today it is builders and technicians that we turn to: men armed not with the ax, rifle, and bowie knife, but with the Diesel engine, the bulldozer, the giant electric shovel, the retort—and most of all, with an emerging kind of skill, a modern knack of organization and execution." He warned that technology without moral purpose "may indeed be evil," but he hastened to add that moral purpose made effective by respect for the unity of nature and the participation of the people in development would result in immense benefits for society from the work of "inventors, engineers, and chemists."[60] He felt that these principles were respected in the TVA, the largest construction-and-engineering work carried out by any single organization in history.

Many TVA engineers and architects, including Arthur Morgan, were remarkably accomplished. The chief electrical engineer on the project, Llewellyn Evans, from a public-power background in the Pacific Northwest, was described by a colleague as a prophet and visionary who not only had a firm grasp on the technicalities of power generation and transmission, but also developed the radically new electricity rate structures of the TVA to stimulate and control economic and social development in the valley.[61] Roland A. Wank, Hungarian-born chief architect for the TVA, designed dams whose form and symbolic power fired the confidence of a people mired in the depression in their country's technological prowess and vision. Inscribed on the dams were the moving words "Built for the People of the United States."

World War II brought dramatic changes in the TVA. What to many had seemed a bold social experiment now settled into the ways of a well-established electric light-and-power utility. During the war, the TVA and the press made much of its contribution. Aluminum for airplanes, explosives for bombs and artillery, and countless other energy-intensive products flowed from the valley to the front. Only after the war could it be made known, however, that the Manhattan Project for the manufacture of atom bombs had made prodigious use of TVA power at Oak Ridge, where there were plants for separating fissionable isotopes from uranium. The demands of the defense industry and an increasing response to the policy of lowering electricity prices solved the TVA's problem of finding a market for its low-cost electricity. Ironically, at the same time this caused a serious new environmental problem. After exhausting the principal waterpower resources through the construction of numerous dams, the TVA had to resort to coal-fired steam plants to meet the demand for electricity. The only alternative would have been to lower demand, which would have contradicted the long-standing policy of stimulating

INTERIOR OF POWERHOUSE AT FONTANA DAM. CONSTRUCTION BEGAN JANUARY 1942. INSCRIPTION READS: "BUILT FOR THE PEOPLE OF THE UNITED STATES OF AMERICA."

energy consumption in the valley. The TVA had overcome its political enemies and tamed the Tennessee River, but in the postwar world it had to face irate conservationists.[62] During the Cold War the Atomic Energy Commission, headed by Lilienthal from its inception in 1946 until 1950, continued to expand its facilities in the valley, and the TVA began building huge coal-burning plants. The Kingston Steam Plant, completed in 1956 near Oak Ridge, was the largest in the world. In the 1950s more than half the TVA's enormous power output went to two AEC plants, at Oak Ridge, Tennessee, and Paducah, Kentucky.[63] When the TVA resorted to using strip-mined coal, it became a prime target for the conservationists and environmentalists. As a conservation agency, the TVA lost its "flawless reputation." Many of its one-time enthusiasts had their ardor dampened by headlines like "TVA Ravages the Land" and "Power Corrupts: The TVA and Coal Operations Ruin the Kentucky Hillsides." Subsequently, its massive shift to nuclear power complicated and deepened the controversy. One might give a twist to Lord Acton's phrase and say of the TVA, "Power tends to corrupt and absolute power corrupts absolutely."[64] Supporters argued that strip mining and nuclear power were in accord with TVA's policy of low-cost power; however that may be, the Authority was caught in an energy-cost-ecology dilemma, for it felt bound to continue to serve its market with low-cost power. By the 1960s critics were charging that the TVA was not an example of creeping socialism, as conservatives had feared, but of creeping conservatism. The TVA had acquired a bureaucratic and technological momentum; its sense of mission had "faded into an appreciation of good bookkeeping."[65] From path-breaking to bookkeeping within a few decades may be the pattern of large technological systems, but the Tennessee Valley during World War II would witness the birth of yet another massive project that fed on the energy generated by the aging TVA.

MANHATTAN PROJECT

The path-breaking involvement of government in planning and developing a large technological system in the Tennessee Valley was eclipsed during World War II by the government's role in the Manhattan Project,

KENTUCKY DAM. CONSTRUCTION BEGAN JULY 1938. ROLAND WANK WAS CHIEF DESIGNER AND ARCHITECT.

or "Manhattan Engineer District," its formal code name. The project had major facilities in Oak Ridge, Tennessee, and elsewhere throughout the country. It used energy developed by the TVA. The history of this project, which produced atomic bombs during World War II, has often been told. Usually such histories focus on the atomic physicists and chemists, the fabrication of the bombs at Los Alamos, New Mexico, and the dropping of one of the bombs on Hiroshima. This is understandable, for the *dramatis personae* at Los Alamos were articulate and imaginative scientists led by the highly intellectual, cultured, and charismatic J. Robert Oppenheimer. After the war he became a national, even tragic, figure, because of the security-clearance hearings to which he was subjected. The atomic scientists have also attracted attention because a number of them played prominent roles after the war as advocates of nuclear-arms control or nuclear disarmament.

On the other hand, we can see the Manhattan Project as a continuation of the growth of large production systems, a development that we have followed since the late nineteenth century. Thousands of workers, engineers, and managers as well as scientists labored at the heart of the Manhattan Project. But it was more than a scientific endeavor. The Manhattan Project was an industrial development-and-production undertaking dependent on scientific laboratories and scientists for essential technical data and theoretical understanding of various processes. Despite frequent assertions that the Manhattan Project was without precedent, the relationships among scientists, engineers, and managers in inventing and developing the bomb were in many ways analogous to the relationships we have encountered among them at innovative production companies such as General Electric, American Telegraph and Telephone, and Du Pont. After invention and development, the Manhattan Project became a centrally controlled and coordinated production system. In this it was similar to Ford's automobile production system and to Insull's giant electric light-and-power system—with the major exceptions that the military played the role of system builder and that the government funded the project. Yet there was precedent even for this in the construction projects planned and supervised by the U.S. Army Corps of Engineers and in the government funding of the massive TVA. Considering the ever-increasing scale of technological-production systems, the successful mobilization of invention and science before and during World War I, and government success with the TVA, it seems, in retrospect, almost inevitable that the government and the military, two of the century's most rapidly expanding and powerful bureaucracies, should become involved in the twentieth century's most characteristic activity—technological-system building.

We have also observed that independent inventors, like Nikola Tesla and Lee de Forest, turned over their inventions to industrial corporations for development. Similarly, industrial firms under contract to the government contributed substantially to the development of the discoveries and inventions of atomic scientists. An ingenious experiment and an

OTTO HAHN AND FRITZ STRASSMANN FISSION EXPERIMENT ON WORK TABLE

imaginative interpretation of it leading to the development of atomic bombs occurred in Berlin in December 1938, when German scientists Otto Hahn and Fritz Strassmann designed laboratory tabletop apparatus that split—they were convinced—the uranium atom. As with breakthrough inventions made by independent inventors, there had been a long string of prior discoveries, inventions, and theoretical explanations in atomic physics leading up to Hahn and Strassmann's achievement. Among those making these contributions were Ernest Rutherford and James Chadwick in Britain, Irène Curie and Frédéric Joliot in France, Enrico Fermi in Italy, Niels Bohr in Denmark, Lise Meitner in Germany, and Ernest O. Lawrence in the United States. After 1938, scientists working independently in universities continued to make critical inventions and discoveries that culminated in the bomb. In their freedom from large organizational constraints, they were similar to the independent inventors of the late nineteenth century. After the establishment of the Manhattan Project in 1942, however, most of the leading atomic physicists and chemists in the United States, including those who had emigrated from fascist Europe, were drawn into Manhattan Project laboratories at the University of Chicago, University of California (Berkeley), and Columbia University. There they experienced the constraints often felt by industrial research scientists.

Although it is part of the continuing history of large technological systems, the Manhattan Project was unprecedented in its concentrated expenditure of human resources for the manufacture of a single product—atomic bombs. The Ford River Rouge plant, some of the Soviet regional production complexes, and prewar German and U.S. projects for the development of new gasoline-production processes bear comparison as complex technological systems extending over large areas and involving a host of workers, engineers, and managers. The Manhattan Project set a precedent in the extensive employment and the influence of highly trained physicists and chemists, who interacted creatively with industrially experienced engineers, metallurgists, and skilled machinists. The project also differed from earlier ones in that no system builders comparable to Ford or Insull emerged. Brigadier General Leslie Groves seemed a likely candidate for the role, but he never provided the inspired technical leadership given by Ford, nor did he elicit a collective creativity during the Manhattan Project similar to that which Ford had stimulated at the Highland Park plant as the assembly-line system evolved. Many scientists on the atomic-bomb project considered Groves contentious, ambitious, heavy-handed, sharp-tongued, egotistical, and bureaucratic, but those closer to him respected his judgment and ability to get to the heart of complex matters.[66] Among army officers, Groves's subordinates feared him, only a few liked him, but he obtained the respect he demanded, for he was known as an intelligent, strict taskmaster completely dedicated to the tasks assigned, among which had been supervision of the building of the Pentagon building outside Washington, D.C.[67]

In the case of the Manhattan Project, the problems were too complex and the knowledge and skill needed to solve them too specialized for any individual to assume the role of system builder. Insull, with his years of experience working at Edison's side in his laboratory, supervising manufacturing at Edison General Electric, and mastering the details of utility management during the early years in Chicago, could integrate in his

VANNEVAR BUSH, JAMES CONANT, LESLIE GROVES, AND UNIDENTIFIED OFFICER (LEFT TO RIGHT)

mind the various functions and decisions. Ford could speak the languages used throughout his enterprises and make decisions from an intimate knowledge of the processes under way. As we have seen, he kept out persons with highly specialized knowledge. Groves was never able to penetrate the highly specialized world of atomic physics and chemistry. Scientists doing science remained strangers to him. The chemical-engineering expertise that prevailed in the project lay outside his engineering experience. If, like Ford, he had disdained and dismissed specialists, there would have been no Manhattan Project. Groves presided over the formal and informal committees of scientists, engineers, and managers. He drew on his military training and experience, depended on his insights into the competence and leadership abilities of others, and on his familiarity with organizational tactics and strategy. System building by committee, a development that a Ford could not imagine, was the rule during the Manhattan Project. As a result, the Manhattan Project, like a committee, tended to probe its way to decisions, trying first one solution, then another, or even several simultaneously, and reaching its goal by gathering momentum through the enrollment of human and material resources. Taylor, Ford, and Insull would have been appalled by the inefficiency.

In September 1942 the then–Colonel Groves of the U.S. Army Corps of Engineers was named military head of the secret atomic-bomb project. He asked that his appointment as brigadier general go through before he assumed his responsibilities. He felt he needed that increased prestige and authority in dealing with the highly accomplished civilians with whom he would have to interact. Groves's small supervisory and science-advisory committee included Vannevar Bush, electrical engineer and former dean at the Massachusetts Institute of Technology, and James Conant, chemist and president of Harvard University. Bush, experienced in engineering and industry, headed the wartime Office of Scientific Research and Development, and Conant, familiar with industrial chemistry, chaired the National Defense Research Committee. Through these two agencies they presided over the mobilization of U.S. science for defense and war. Bush and Conant also participated in an OSRD advisory committee, the S-1, on which served scientists heading atomic research at the universities of California (Berkeley) and Chicago, and at Columbia University. In the following months the laboratory work that had been pursued on a limited scale under contract to the OSRD was transferred to the Manhattan Project and came under the jurisdiction of Groves and his supervisory committee. Bush considered the relationship of Groves to the committee as comparable to that of a vice-president in charge of operations in relation to a board of directors.[68]

Groves faced serious organizational and managerial problems. He had to persuade wartime government agencies to respect the priority of his secret project and allocate scarce resources to it. He had to make contracts with engineering and industrial corporations to build plants and install processing equipment. Modes of cooperation between highly independent academic scientists and highly organized industrial managers and engineers of different mind-sets and styles had to be established. Tensions among scientists and engineers over decision making increased, especially among émigré atomic physicists at Chicago and the Du Pont industrial

ERNEST O. LAWRENCE, ARTHUR COMPTON, BUSH, CONANT, KARL COMPTON, AND ALFRED LOOMIS (LEFT TO RIGHT)

managers and engineers. Groves had to persuade unions of skilled workers to forgo some hard-won autonomy in the name of a project about which they could not be fully informed. He and the military gave secrecy and security a high priority, but the scientists on whom he had to depend had a long and strong tradition of open exchange. Perhaps most worrisome to him, Groves also found that the design and construction of production plants, and the commitments to enormous expenditures, had to go ahead without sufficient technical data and with the probability of numerous failures. Otherwise there would be no possibility of meeting the objective of having bombs within three years. Groves only half-jokingly considered purchasing a house on Capitol Hill in Washington, D.C., near the Congress, in order to be only a short walk away from the interminable hearings that would follow if, after the war, Congress first learned of the Manhattan Project as a colossal failure.

ARTHUR HOLLY COMPTON
(1892–1962)

After being promoted to general in September 1942, Groves was ready to approach the impressive Nobel laureates who headed the laboratories and the highly intelligent, proud, even arrogant, independent scientists, émigré and U.S.-born, who worked with them. In Chicago he found Arthur Compton, Nobel Prize–winner for physics in 1927, who had organized the Metallurgical Laboratory (a code name) early in 1942 to do theoretical studies and experiments needed to design the bomb. Later the lab took on the responsibility of developing theory and technical data needed to produce plutonium. At the height of its activity, the staff of scientists, technicians, engineers, and administrative personnel in Chicago numbered more than two thousand. Compton attracted to Chicago, among other leading atomic physicists, chemists, and engineers, three eminent émigré physicists, Nobel laureate Fermi, Eugene Wigner, and Leo Szilard. Later, Nobel laureate James Franck, another who had left Europe in the face of fascist oppression and racism, joined them. Groves may have felt discomfited in the presence of the nuclear physicists and Nobel Prize–winners at Chicago, but he tried to counter their doubts about his competence. In an early Chicago meeting, he acknowledged that he had no Ph.D., but pointed out that he had ten years of formal education, college and postgraduate. "That would be the equivalent of about two Ph.D.s," he ventured.[69]

The émigré physicists feared that the atomic scientists whom they had known in Germany would quickly follow up on the work of Hahn, Strassmann, and Meitner and build a bomb. Fermi and Wigner had carried on research, some of it at Columbia University, to investigate the nature of uranium fission. They believed that a chain reaction could be achieved in a properly designed lattice, or pile, of uranium and graphite blocks, but they lacked the uranium and graphite to test their hypothesis on a sufficiently large scale. In 1939 Wigner and Szilard had prodded Einstein to sign a letter to Roosevelt warning him of the danger and urging him to large-scale support of atomic science and bomb production. They elicited the support of Alexander Sachs of Lehman Brothers, whom we last encountered proposing power pooling to the TVA, to deliver the letter to his friend Roosevelt. Sachs, given to parables, opened with a story about Napoleon's rejecting as preposterous the proposal of Robert Fulton to build warships without sails. Sachs compared what he now had to propose with the steamboat inventor's project.[70] After the Einstein letter, the U.S. government did appropriate modest funds for atomic research. Much importance has subsequently been given to the intervention of Szilard, Wigner, and Einstein. It was not this letter, however, but a technical report from a British committee known as MAUD in the fall of 1941 that persuaded Bush and Conant, Roosevelt's principal science advisers, that a major push for the bomb was needed. The MAUD report concluded that a uranium bomb could be made by either side and was likely to lead to a decisive result in the war. Realizing that the United States, not Britain, had the resources for a large project, Roosevelt accepted Bush and Conant's recommendations and found funding to accelerate research.

Being from a European tradition, in which scientists often headed industrial enterprises, Wigner, Szilard, and many of their associates at the Chicago lab felt uncomfortable with the military—and later the industrial—management of the Manhattan Project. In the spring of 1941 the U.S. Army had named as principal contractor for the bomb project the firm of Stone & Webster, which had a long history both of engineering, management, and financial consulting, and of large-scale construction in the electrical-utilities field. When the Chicago scientists, Wigner and Szilard among them, learned that they would be working with or for an industrial contractor, they resisted. Szilard drew up a memorandum in which he lamented the prospect of an authoritarian industrial contractor stifling science that needed to be done properly by the best scientists interacting along "democratic" lines. After Stone & Webster sent one of its best engineers to brief the Chicago scientists on plans for bomb production, the scientists at the Met Lab became highly agitated. They found the engineer ignorant of atomic physics. Szilard now felt that not only had an authoritarian organization moved in to take over from the democratic scientists, but an ill-informed one as well.[71]

Groves, who depended on industrial contractors, became critical of the scientists. He was disturbed by the Chicago lab's lack of technical data needed for the design of production processes, and by what he took to be the indecision of the scientists in gathering these data. In the fall of 1942 he learned that the scientists had no experimental proof, only predictions based on theory, that plutonium, the explosive they proposed for bombs, could be produced in an atomic reactor, or pile. Nevertheless, Groves, flying in the face of conservative engineering practice, decided to go ahead "at full speed" with the construction of full-scale atomic piles, or production plants.[72] Needing to draw up production schedules and set plant capacity, he asked the Chicago scientists the amount of plutonium needed for each bomb. At that time only infinitesimal amounts had been made in laboratories. Chicago gave him a figure that he assumed to be "within twenty-five or fifty per cent," and he "was horrified when they quite blandly replied that they believed it was correct within a factor of ten." Groves compared his position to that "of a caterer who is told he must be prepared to serve anywhere between ten and a thousand guests,"[73] a macabre metaphor from our post-Hiroshima perspective.

Discussions among the scientists and engineers at Chicago about the critical problem of cooling the atomic piles revealed the uncertainties, complexities, contradictions, and tensions among those from differing cultural and professional backgrounds. Cooling with gas or with liquid offered the principal choice. Gases removed heat less efficiently than liquid but were expected to interfere less than liquids with the nuclear reaction producing plutonium in the pile. Gases needed high pressures and complex containers, compressors, and pumps; liquids required pipes that would complicate pile design. The availability of various liquids and gases of promising characteristics further complicated the decision. The engineering contingent at Chicago opted for gaseous-helium cooling in the summer of 1942. Compton, not enthusiastic about this recommendation, delayed making a decision. Szilard then spoke of aimless drift at Chicago, attributing it to Compton's desire to avoid controversy and to army security restrictions that prevented the free exchange of information among atomic scientists in the U.S.[74] Szilard, imaginative and bold, proposed the exotic liquid bismuth as a coolant. Wigner explored the possibility of a water-cooled pile. Dismayed by the discord, and tending to judge scientific doubts as indecision, Groves launched into the speech that was to become his trademark. To him, a wrong decision was better than none at all. If there were several likely solutions, he urged the scientists to try them simultaneously, for achieving a solution was more important than saving money. Even though they might have only limited data from which to proceed, he felt an urgent need to move ahead with the designing and construction of full-scale plutonium piles, or production plants.[75]

Made impatient by the scientists, Groves turned to industry, with which he had established good working relationships in doing construction work for the army. He insisted on bringing the Du Pont Company into the project, and he persisted until that reluctant corporation agreed to take the risky plunge into the unknown waters of the world of nuclear science. In the fall of 1942 he first approached the powerful corporation he so greatly admired, which had worked with the Army Corps of Engineers on earlier projects. Groves knew that he "was asking Du Pont to embark upon a hazardous, difficult and perhaps impossible undertaking, at a time when it was already straining under the terrific war burden it was carrying." Du Pont's inexperience in nuclear physics, the lack of technical data that could be acquired only over a long period of laboratory research

WALTER CARPENTER, PRESIDENT OF DU PONT (FAR RIGHT)

and semiworks, or small-scale testing, operation, and the physical operating hazards all discouraged its participation. Du Pont also recalled with distaste the investigations of the Senate Munitions Investigation Committee of 1934, when extremists called the Du Pont family "merchants of death" because of profits they had made from explosives during World War I. When Groves, however, told the management and board of directors that the president of the United States and the secretary of war judged the project to be of the highest military importance, Du Pont acquiesced.[76] The well-tailored, impressive president of Du Pont, Walter Carpenter, remained courtly, courteous, and firm throughout the negotiations. After stressing that his was an American company with a long history of contributing to national defense and of not shirking patriotic duty, Carpenter, on behalf of the board of directors, accepted the project. In December 1942 the Du Pont Company formally undertook to design, construct, and operate plutonium-production piles and separation plants. A new division of the explosives department was assigned the project. Carpenter laid down the condition that Du Pont make no profit. The contract stipulated a fixed fee of $1.00, with the government paying all costs of the project. On hearing the news, leading Du Pont engineers who would have the responsibility for the project "all went to the [Hotel Du Pont's] Brandywine Room to have a drink of condolence and commiseration."[77] Groves, looking back, found the action of Du Pont's directors "a true display of real patriotism."[78] Du Pont took the position that its part in the Manhattan Project stemmed from patriotic duty and that the company "would . . . have absolutely no pecuniary or vested interest in the Manhattan Project."[79]

After Groves brought the Du Pont Company into the plutonium project, its engineers, along with the Chicago scientists, faced the troublesome problem of choosing a coolant for the piles. Du Pont engineers argued that water would corrode uranium, thus blocking the flow of the coolant, and causing a violent explosion of the high-temperature water. Before making a final decision on cooling, the Chicago scientists constructed an experimental atomic pile. This was to provide data for estimating the probability that the nuclear reaction in the pile would either slow down or stop because of the neutron-absorbing characteristics of the cooling system. By December 1942 Fermi had the pile in operation underneath Stagg Field, the University of Chicago football stadium. From data derived from the pile, the Chicago scientists decided that water was the superior coolant. Du Pont, backed by Groves, did not immediately take the scientists' advice. Du Pont considered using air for a relatively simple semiworks in Tennessee that would provide data for the full-scale production piles and helium as coolant for these. Du Pont later adopted the water cooling proposed by Wigner for the production piles. This was one more demonstration for the Chicago scientists, especially the senior émigrés and the younger Americans, that scientists, with their theoretical understanding and imagination, could direct development better than conservative engineers bound by traditional practice.[80]

It soon appeared that the Chicago lab might become a subsidiary facility providing advice and data to Du Pont. Realizing that their knowledge of nuclear physics was far greater than Du Pont's, many scientists at Chicago continued to believe that the Metallurgical Laboratory should preside over the design and construction of the production facilities. Several of them told Groves that they could design and construct the plutonium plant if he would provide them with fifty to a hundred junior engineers and draftsmen. He found the suggestion absurd.[81] The confidence of the scientists arose from their assumption that, from a limited supply of fissionable material, only one or two bombs would be built, not a production line of bombs, as Groves intended. The American military argued that, until victory was assured, it had to have the capacity to deliver bombs repeatedly. Some scientists countered that the Germans might limit themselves to a small-scale project, make one or two bombs before the Americans had gotten into production, exploit the psychological effect of the incredible destructiveness of these, and demand immediate surrender.[82]

Szilard, the Hungarian, was the scientist most voluble in his opposition to the military-industrial take over, but his anxiety was shared in various measure by Wigner, Fermi, and most of the other major scientists.[83] Their doubts about the capacity of industrial engineers not familiar with nuclear physics were compounded by the widespread belief among European intellectuals that large industrial corporations and the military had contributed greatly to the rise of fascism in Europe. Wigner had been taught in Europe that such large industrial organizations as Du Pont were the enemies of democracy.[84] Younger American physicists rallied to support the émigrés, because, as Compton observed, of their distrust of the authority embodied in any bureaucratic organization. Compton nearly precipitated a rebellion when he first told his scientists that he favored an industrial corporation's taking charge of construction and production.[85] At Chicago some of the competitiveness that had developed during World War I between the Naval Consulting Board of inventors, engineers, and industrialists and the National Research Council with its physicists became discernible again. Groves exacerbated these feelings in Chicago when he contended that scientists were without discipline. "You don't know how to take orders and give orders," he complained to Compton, who rejoined that the discipline of scientists was self-discipline in the pursuit of the facts.[86]

By early 1943 scientists at the Metallurgical Laboratory believed it had become a contributing but clearly subordinate affiliate of the Du Pont

ALBERT EINSTEIN AND LEO SZILARD

Company.[87] The new role discouraged the experimental physicists working under Fermi and the theoretical physicists under Wigner. Wigner had watched as Du Pont first rejected and then, after delaying, accepted his proposal for water-cooling the production piles. This meant, he noted, that the air-cooled semiworks Du Pont chose and had under construction would not serve as a good model for the liquid-cooled full-scale works. He was deeply discouraged when Du Pont invited neither him nor any of his associates to join the company's design group. "By February, 1943, Wigner had lost all hope. Du Pont seemed to be floundering, but Greenewalt, as Du Pont liaison, steadfastly refused all offers of help."[88] Wigner, believing that he might be an obstacle to cooperation, offered his resignation, but Compton persuaded him to take a month's leave, then stay on.

Szilard also became deeply discouraged by and resisted the military-industrial leadership. Groves later observed off-the-record that Szilard was the type whom any employer would have fired as a troublemaker. Richard Rhodes, in his history of the Manhattan Project, ventures that "Groves seems to have attributed Szilard's brashness to the fact that he was a Jew."[89] Groves and Szilard engaged in a long-term confrontation. Szilard sharply criticized Groves's compartmentalization of scientists and the restraints, on grounds of security, that he placed on the exchange of information. Szilard insisted that imagination and inventiveness in science required freedom of exchange and exploration. He might have cited the independent inventors of the late nineteenth century to support his argument. When Szilard bent the security rules and also badgered the Du Pont engineers who came in to take over pile design, Groves drafted a letter calling Szilard an enemy alien and proposing internment for the duration of the war. Arthur Compton was able to forestall Groves's sending the letter and taking action.[90] Late in 1942 Szilard believed he had found leverage to press his points when he announced that he intended to file a patent application, probably with Fermi, on the chain-reaction inventions that the two had made before they secured the financial support of the government. Szilard was no stranger to invention and patents, having applied in Germany, between 1924 and 1934, for twenty-nine patents, alone or with Albert Einstein (most of the joint patents concerned home refrigeration).[91] The matter moved up to Bush, who argued that Szilard did not have a case for his patent applications, because he had not disclosed his prior inventions to the University of Chicago at the time of his employment. A government legal adviser informed Szilard in 1943 that in his opinion Szilard did not have a patentable invention. Just in case Szilard did obtain a patent, Groves informed him, the government would negotiate for patent rights only after he agreed to turn them over to the government. Groves coupled this with a statement that the government would also negotiate Szilard's contract with the University of Chicago, which was up for renewal. Szilard probably interpreted this correctly as a demand that he trade his patent rights, if any, for the opportunity to continue working on the bomb. Before the confrontation, Szilard had spoken of turning over any payment for the patent rights to a government corporation under the direction of scientists who would promote the development of atomic energy. Ten years after the war Szilard and Fermi were issued a patent for their invention of the nuclear reactor.[92]

Greenewalt had the unenviable task of serving as liaison between Du Pont and the Chicago lab. He was head of the research unit of the Du Pont department responsible for the plutonium project; after the war he became president of the company. Greenewalt tried to allay the Chicago physicists' fears that their lab would become a field station of the Du Pont explosives department. He told them that the Du Pont research team would ask Chicago for information derived from experiments and would bring Du Pont's designs, based in part on this information, to Chicago for review. Prudently, Greenewalt declined Groves's request that he become the "boss" of the physicists, for Greenewalt realized that men such as Szilard and Wigner would respond to the leadership only of a distinguished physicist. Greenewalt assured the scientists that Du Pont was in the project "to do a job" rather than to "pick their brains" and "make a barrel of money out of atomic energy."[93] He reminded them of his company's success in applying basic research in the development of nylon. When Groves and Roger Williams, the head of the Du Pont plutonium project, decided that a small-scale plutonium reactor was, for reasons of safety, not to be built at Argonne, near Chicago, but in thinly populated Tennessee, Greenewalt had to respond to "a nasty situation badly handled," because Compton and the other scientists had not been consulted.[94] Greenewalt had some success in persuading Compton to reorganize the management of research at Chicago. He found that the scientists were reluctant to freeze engineering designs, preferring to introduce changes as new information suggested them. Ultimately he persuaded some of the scientists that the time-consuming complexities of engineering construction dictated hard-and-fast decisions so that the construction process could be begun. Despite his tact and patience, Greenewalt did not succeed in working well with Wigner, who caused him "much grief."[95]

HANFORD PRODUCTION REACTORS

The unrest at Chicago did not prevent Groves and Du Pont from forging ahead. In February 1943 Groves acquired about a half-million acres in Hanford, Washington, for the plutonium-production piles, plutonium-separation plants, and worker housing. Such vast acreage provided the isolation needed to distance population in the region and to maintain secrecy. The lonely site provided cooling water from the Columbia River and electric power from a high-voltage transmission line connecting the Grand Coulee and Bonneville hydroelectric dams. Construction then proceeded concurrently with the acquisition of experimental data by Chicago scientists and the design of the facilities by Du Pont, a most unusual and nerve-racking experience for the engineers.

On 13 September 1944 one of the production piles was ready for loading with thousands of slugs of uranium. Construction had employed more than forty-two thousand persons during the peak of activity the previous June. Massive equipment had been assembled, with the tolerances usually reserved for fine watchmaking. Welders of superior skill and special training had provided leakproof containers for highly radioactive products. Skilled carpenters, millwrights, pipefitters, and electricians also solved unprecedented problems. Low productivity from pipefitters was the only persistent labor problem at Hanford. Groves personally tried to persuade the head of the International Union, M. P. Durkin, later secretary of labor, to intervene, but Groves recalled that "he did not seem to be at all interested in our pleas."[96]

Initially at Hanford there were three giant production piles, or reactors, for producing plutonium, and four plants for separating the plutonium from other products of the chain reaction taking place in the production piles. Glenn Seaborg, the young chemist from Berkeley who had discovered plutonium in February 1941, developed the separation process. Later he and others discovered that "plutonium is so unusual as to approach the unbelievable." Under some conditions plutonium can be hard and brittle, and under others, as soft and plastic as lead. "It is fiendishly toxic, even in small amounts."[97]

HANFORD ENGINEERING WORKS

Arthur Compton, Fermi, Roger Williams (engineer in charge of the project for Du Pont), and Greenewalt were on hand in September 1944 to witness the loading of the first completed pile with slugs of uranium, a culmination of the thought and labor of many thousands of persons. Two weeks later the pile contained sufficient uranium to produce a plutonium-producing chain reaction. For several hours the power level and the chain reaction increased as predicted, but then the power began to decline and continued to do so until, by six-thirty in the morning, the pile had shut itself down completely. Deeply disappointed, and alarmed that Hanford might have become a monstrous fiasco, scientists and engineers explored numerous explanations for the seemingly catastrophic failure. The next day, however, the pile became active again, and the nuclear reaction reached the previous level—only to shut down slowly again. Physicist John Wheeler then offered a diagnosis. He had been investigating the possibility that the reaction was being "poisoned" by one of the by-products of plutonium production. He conjectured that the poisoning substance was absorbing some of the neutrons released by uranium fission, leaving an insufficient number of neutrons to maintain the chain reaction. Wheeler knew the characteristics of the pile well, for he, unlike most of the Chicago scientists, had worked closely and agreeably with the Du Pont engineers. He was their "favorite scientist."[98] The time that it took for the pile to reactivate after shutting down led him to believe that xenon-135 was the poisoning agent. Earlier, Greenewalt had offered the same hypothesis. The chain reaction started up again, Wheeler reasoned, when the xenon isotope, with a half-life of a few hours, died out. Fermi readily agreed with this diagnosis, but, in order to cross-check, inquiries were made about the performance of an experimental pile that had been built by the Chicago scientists at Argonne. When the pile was run at full power there, the scientists recorded similar effects. This confirmed Wheeler's identification of xenon. Groves, however, seethed with anger when he learned of the episode, for he had earlier ordered the Chicago scientists to operate the experimental pile at full power, only to have his orders ignored. If his orders had been followed, xenon poisoning would have been identified earlier. Compton apologized but added, to mollify Groves, that the Chicago scientists, instead of running the pile at full power, had used it to make an important new discovery about nuclear reactions.

The Chicago scientists often denigrated the overly conservative Du Pont designs, especially in the matters of pile coolants. They accused Du Pont of cautiously lowering the risk of failure to protect its reputation at government expense and at the cost of delaying the project. Now, in Hanford, conservative design saved the day. Alerted by Wheeler's studies about pile poisoning, the Du Pont designers had provided in the Hanford piles capacity for uranium in excess of that required by theory, experiment, and calculation. (The Metallurgical Laboratory scientists cited this as an outstanding example of "extravagant conservatism.") But, with this excess capacity, uranium in the pile could be increased and override the poison effect. Other piles at Hanford were accordingly modified, and by the end of December the production process was under way in two piles; a third was ready about six weeks later.[99]

OAK RIDGE AND ELECTROMAGNETIC SEPARATION

Besides building reactors at Hanford to produce plutonium, the government contracted for the construction of plants at Oak Ridge, Tennessee, to produce another explosive, uranium-235. Because a number of production processes, companies, universities, and prominent scientists and engineers were involved at Oak Ridge, we need an overview of activities there before we follow developments in detail. In the fall of 1942 the army chose the site near Knoxville because of the availability of electric power from the TVA, adequate water supply, railway connections, and sparse settlement. Following experimental work at the University of California, Berkeley, at Columbia University, and elsewhere, three major processes for the production of the uranium isotope 235 were installed at Oak Ridge: electromagnetic separation, gaseous diffusion, and thermal diffusion. As in the case of plutonium production at Hanford, university scientists and industrial corporations cooperated in U-235 production. University of California physicist Lawrence and his Berkeley laboratory designed the electromagnetic process. At the same time, Stone & Webster was constructing the small air-cooled plutonium facility, or semiworks, intended to provide technical data for the design of full-scale facilities at Hanford. When the mammoth scale of the Oak Ridge complex became clear to Groves, he hastened to bring in other industrial corporations as contractors. Not looking for "long beards," since "he already had so many Ph.D.s that he couldn't keep track of them,"[100] Groves persuaded Tennessee Eastman Corporation, an operating—not research-and-development—subsidiary of Eastman Kodak, to become operator of the electromagnetic, or Y-12, plant. Subsequently, Westinghouse Electric and Manufacturing Company, General Electric, and the Allis-Chalmers Manufacturing Company accepted contracts for the manufacture of equipment for the electromagnetic-separation facility.

Scientists at Columbia University under the leadership of Harold Urey, a U.S. chemist who had won the Nobel Prize in 1934 for the isolation of heavy hydrogen, carried out the preliminary research on the gaseous diffusion process for uranium-235 production. Early in 1943 Grove asked the M. W. Kellogg Company, a large and experienced engineering-and-construction organization, to design and build the gaseous-diffusion plant at Oak Ridge. Kellogg established a subsidiary, the Kellex Corporation, to do the job and bear the stigma of possible failure. Then Groves named a Union Carbide and Carbon Company subsidiary, Carbide and Carbon Chemicals Corporation, to assist in design and construction and then to operate the plant. The Chrysler Corporation and Allis-Chalmers Manufacturing Company were among the other subcontractors. Philip Abelson, a physicist, worked out the design of the thermal-diffusion process for separating U-235, the third Oak Ridge production process. Abelson, funded by the U.S. Navy, had initially proceeded quite independently of the Manhattan Project. In June of 1944 the H. K. Ferguson Company of Cleveland, Ohio, accepted a contract to construct the thermal-diffusion plant after Groves decided to incorporate Abelson and his process into the Manhattan Project. Mrs. H. K. Ferguson, young widow of the company president, headed the Cleveland company and negotiated with Groves.

The interactions between the scientists and the industrialists at Oak Ridge present some variations on the Chicago Met Lab–Du Pont themes. Lawrence, born in small-town prairie South Dakota, educated at the universities of South Dakota, Minnesota, Chicago, and Yale, professor at the University of California (Berkeley), and Nobel Prize–winner for physics in 1939, became the Edisonian figure of the Manhattan Project. His personality was quite different from those of Wigner and Szilard. In the world of Sinclair Lewis's *Main Street* he would have been known as a "booster"—in his case, of science. Like Edison, he had almost an aversion to mathematical reasoning. As was not the case with many of the other scientists of the Manhattan Project, he and Groves reacted sympathetically to each other from the start. Inventor and builder of complicated machines, master of his own laboratory, enthusiastic promoter and money raiser for his projects, and strong-willed and driving head of a band of young laboratory associates, Lawrence greatly influenced the character of the Manhattan Project and the development of American science. Like Edison, he generated a legend with his concentrated and sustained bursts of energy and his ability to rest by taking quick naps while he was sitting in the laboratory. The graduate students working with him referred to him as Maestro. Having studied engineering as an undergraduate at the University of South Dakota, Lawrence later, as an experimental physicist, took the experimental, problem-solving, machine-building approach of an engineer. He and his laboratory assistants, many of them doctoral candidates or postdoctoral assistants at the University of California (Berkeley), anticipated the future course of physics as they collectively experimented with big machines of their own design and construction. In a sense, they were polarized around their machines. Later, during the making of explosives for the atom bomb, Lawrence's focus on machines led him to neglect difficult chemical problems.

Shortly after leaving a position at Yale for an associate professorship in the ambitious and rising Berkeley physics department, Lawrence had the inventive idea for a machine that helped solve a problem on which European and American physicists were working. Their goal was to generate high-energy particles capable of penetrating the nucleus of an atom to reveal its inner structure. In 1929, while he was browsing through a German journal, the *Archiv für Electrotechnik*, his eye caught the diagrams of a Norwegian engineer. Rolf Wideröe had designed a device that repeatedly used low voltages to boost the energy of ions successively, instead of using enormous amounts of energy to achieve the effect in one stage. The precisely tuned boosts to the particles were analogous to pushing a child in a swing higher and higher by making appropriately timed and placed pushes.[101] The Lawrence invention was like an electric motor with a revolving stream of atomic particles instead of a revolving armature. When the particles gained sufficient acceleration, they were drawn off as a beam of high-energy particles.[102] Lawrence's inventive insight was seeing the possibility of increasing the energy boost by using a circular track for the path of the accelerating particles rather than by using the straight tubes of the Norwegian. Over the next few days he exuberantly told friends, "I'm going to bombard and break up atoms!" and "I'm going to be famous!"[103]

Energetically raising money from the National Research Council and other sources, and spending increasing time in the laboratory surrounded by graduate students, Lawrence experimented with various designs of his circular, magnetically resonating accelerator. He and M. Stanley Livingston, a graduate student and later assistant, by 1932 had in operation an accelerator, later called a cyclotron, that generated one-million-volt hydrogen ions, or protons. When Livingston brought the machine to the point where he could write on the blackboard "1,000,000 volts," Lawrence danced around the room.[104] Controversy later arose over the respective contributions of Livingston and Lawrence to the invention of the electromagnetic beam-focusing that had solved critical problems on the way to achieving the successful machine. In the patent application of 26 January 1932 for the cyclotron, Lawrence filed as a solitary inventor, behavior reminiscent of Edison the laboratory head. Livingston later became bitter about the omission. Lawrence, however, did not exploit the patent financially.[105]

Greatly encouraged by results with the cyclotron, Lawrence acquired an old building that the Engineering School was abandoning and named it the Radiation Laboratory. In order to build a larger cyclotron, he finagled as a gift an abandoned eighty-four-ton magnet once intended for use in trans-Pacific radio transmission by the Federal Telegraph Company. Postdoctoral fellows, learning of the pioneering work, sought places in the Rad Lab. Before the end of 1932 the powerful new machine with the giant magnet generated more than four million volts. Earlier in the year, by using a machine operating at only 125,000 volts, John Cockroft and Ernest Thomas S. Walton at the Cavendish Laboratory at Cambridge University, had gotten a jump on Lawrence by disintegrating the nucleus of the lithium atom with high-energy protons. Lawrence congratulated Cockroft and Walton, then proceeded to use his far more powerful machine to bombard and transmute many more elements. For the penetration and exploration of atomic structure, scientists were now no longer dependent on the limited amount of radium in the world as a source of

SZILARD (LEFT FOREGROUND) AND ERNEST O. LAWRENCE, 1935

alpha rays, or particles. Rad Lab cyclotrons soon produced new isotopes of chemical substances and furthered experimental cancer treatment. Lawrence became internationally known for his imaginative, large-scale engineering approach to solving scientific problems and for his impressively successful fund-raising. The Radiation Lab became a mecca for nuclear physicists and chemists. Physicist Luis Alvarez, who had been a graduate student in the highly individualized atmosphere of the physics department at the University of Chicago, observed, "In Berkeley no one had a room, since the Radiation Laboratory was one large wooden building with no doors between various research areas. The cyclotron, on which everyone worked . . . didn't belong to any one person."[106]

As a pivotal figure in U.S. nuclear physics, Lawrence, not surprisingly, saw in 1941 the possibility of a fission bomb. In September, when he

SIXTY-INCH CYCLOTRON AT LAWRENCE RADIATION LABORATORY

and Conant were in Chicago to receive honorary degrees, Lawrence urgently warned Arthur Compton and Conant that Nazi Germany might be the first to make an atomic bomb. He told them of the possibility of both plutonium and uranium-235 bombs.[107] After listening to Lawrence's information and persuasive argument, Conant, then head of the National Defense Research Committee, asked, "Ernest, you say you are convinced of the importance of these fission bombs. Are you ready to devote the next several years of your life to getting them made?" Lawrence was startled. His agenda, as usual, was already filled with work, much of it defense, into which he had thrown himself with customary enthusiasm. He hesitated only momentarily before replying, "If you tell me this is my job, I'll do it."[108]

Lawrence then turned to the adaptation of his cyclotron for experimentation in the production of fissionable materials. In 1940 he had obtained a grant of over $1 million from the Rockefeller Foundation for building a mammoth 4,900-ton, 184-inch cyclotron. Lawrence and his associates began redesigning it as a mass spectrograph for separating uranium-235 for the atom bomb. Groves, interested in Lawrence's project, visited Berkeley. Dressed in sports coat and gray flannel slacks, and full of youthful enthusiasm, Lawrence met him at the station. Driving atrociously and at reckless speed, he conveyed the general up to see the cyclotrons and mass spectrograph. Impressed by the purposeful bustle of the lab, Groves was nevertheless disappointed by the small output of uranium. Undaunted, Lawrence then led him to see the new 184-inch machine located on Radiation Hill, with its lovely view of the San Francisco Bay and its bridges. The general, builder of the largest building in the world, learned that the 184-inch machine had the largest magnet in the world. Lawrence explained to Groves how the magnetism would separate the lighter U-235 isotope from the heavier U-238. When Groves anxiously asked how much of the fissionable U-235 had been produced, Lawrence replied that there was nothing appreciable as yet. "This is still all experimental, you see. . . ."[109] Later, critics pointed out that Lawrence always promised to deliver with the next machine what the existing one could not.

Groves, however, liked the experimental, engineering, and entrepreneurial style of the enthusiastic and accessible young Nobel Prize–winner. He committed the resources needed to the electromagnetic process, despite doubts expressed by scientists elsewhere about the feasibility of large-scale production of U-235. Engineers of Stone & Webster, the company so sharply criticized by the Chicago scientists, were given the task of helping Lawrence develop his plans for mass production of U-235. Unlike the Chicago–Du Pont relationship, with its science-engineering tensions, the electromagnetic project was dominated by Lawrence and his mostly American-born young scientists. As increased war work drained the supply of physicists and technicians, Lawrence ingeniously and relentlessly recruited staff for the lab. He persuaded a friend, a philosophy professor, to become manager of laboratory facilities, and an astronomy professor to become director of scientific personnel. He asked his staff to help him recruit anyone with the promise of doing anything well. Because many of those recruited were inexperienced in their new jobs, he cautioned them about being discouraged about mistakes, for he, as an explorer of the unknown, where there are no experts, had made many. Like Elmer Sperry and other independent inventors in the past, he told his young physicists to start out beyond theory and invent around the difficulties.[110] Sperry had recommended jumping into deep water to learn how to swim.

Backed by Groves and respected by the Stone & Webster engineers, Lawrence persuaded them late in 1942 that construction should begin on the electromagnetic-production facilities (Y-12) in Oak Ridge, Tennessee. Groves designated the plant "Y-12." The machines, designed in California and built in Tennessee, were dubbed "calutrons." When all were installed, their combined magnets were a hundred times heavier than, and dwarfed, the one-time giant magnet of the 184-inch cyclotron. The large machines needed unprecedentedly high vacuums. The Manhattan Project borrowed $400 million worth of silver from the U.S. Treasury to use in the magnet coils. The process was designed to produce about a hundred grams of U-235 each day.[111]

When it was first tried out on a large scale, in September 1943, the process failed to approach expectations. Countless electrical shorts, leaks in vacuum tanks, chemical-equipment breakdowns, lack of spare parts, inexperienced operators, and long shut-downs for extensive repairs caused despair among some of the project's supporters—but not Lawrence. Some felt he behaved like the captain of a sinking ship who, to instill confidence among his crew, insisted that everything was going according to plan. Realizing that Radiation Laboratory scientists, like inventors, had the knowledge and tinkering ability to make the things that they had helped design work, he dragooned more than a hundred physicists, engineers, and technicians away from the intellectual excitement and exhilarating vistas of the Berkeley hills to the mud and poor quarters of the newly industrialized Tennessee countryside. There they found utter confusion and extremely poor morale among the Tennessee Eastman operators, who had to contend with the trouble-plagued machines. Even Groves was pessimistic, but seemed "under Ernest's thumb."[112] When Groves tactlessly reminded Lawrence that his reputation was at stake, he cheerily rejoined that he already had a secure one, so the general had best look to *his*. Even Lawrence and his people could not solve all the problems and infuse everyone with the "Berkeley spirit." Forty-eight giant magnets had to be sent back to Allis-Chalmers, the Milwaukee manufacturer, for cleaning and rebuilding. They had been designed with their heavy current-carrying bands so close that electrical shorts were inevitable. With great ingenuity and *ad hoc* tactics, Lawrence and a host of engineers and scientists had enough equipment back on line by February 1944 to produce appreciable quantities of U-235. By November 1944 a substantial part of the plant was operating at about eighty-percent efficiency. In April 1945, partially enriched uranium from the thermal-diffusion and gaseous-diffusion processes was used as feed to increase the output from the electromagnetic process. Groves, the engineers, and the scientists had come late upon the idea of feeding output from one process into another to enrich the product, or increase the percentage of the U-235 isotope in the uranium. Natural uranium contained only 0.7 percent of the U-235 isotope. A product enriched to eighty percent was considered sufficient for making a bomb.[113] Armed couriers then regularly carried disguised suitcases packed with U-235-enriched uranium to Los Alamos, New Mexico, where scientists and engineers were designing and assembling bombs.

Throughout, Lawrence maintained his public show of optimism and sustained his relentless activity, but at a cost to himself. Colds, bronchial disorders, and nagging backaches plagued him. In late 1943 a close friend found him hospitalized and more completely "beaten" than ever before. He was depressed by failures and delays and talk of his project as a monstrous boondoggle.[114] Even after the electromagnetic plant produced U-235 for the bomb, Lawrence's project could be spoken of as a failure when measured against expectations.[115] He had promised that his process had a great advantage over others because it would completely separate U-235 in one stage; ultimately, it became necessary to adopt a two-stage process. He had also said that his process would become the principal producer of U-235, which it did not. Lawrence's electromagnetic process may have hastened the production of sufficient explosives for the first uranium bomb, by only a day or so. In retrospect, we can also see that his tendency to discount immediate failures and even neglect to correct them as he turned to new designs and to predicting future successes flawed his style. The official historian of the Manhattan Project has observed that "late in 1945, long after the electromagnetic plant had failed to meet its expectations . . . Lawrence was still proposing bigger and better equipment."[116] Yet his optimism in the face of setbacks encouraged others to persist.

GASEOUS DIFFUSION AND THERMAL DIFFUSION

Groves, Bush, and Conant also pushed the gaseous-diffusion process as another means to produce U-235. In principle, gaseous diffusion involved the familiar process of using a porous membrane to separate the lighter isotopes in an element from heavier ones. To separate on an industrial scale the fissionable U-235 isotope from the heavier U-238 in uranium, an unprecedentedly immense cascade of filters, or barriers, had to be constructed. Diffusion by barrier, however, was a process known to chemical engineers. In the case of the Oak Ridge process, a gaseous compound of uranium was repeatedly pumped through the cascade until an enriched uranium with a higher percentage of U-235 was obtained. Drawing on prior work done in Britain, a theoretical group at Columbia University under Urey investigated U-235 separation before the Manhattan Project was organized. Then the research at Columbia was expanded greatly and consolidated under Urey's direction as the "SAM Laboratories." Contracts for construction and operation of a gaseous-diffusion plant at Oak Ridge were let, as we have noted, to Kellogg's new entity, Kellex. Carbide and Carbon Chemicals was to assist in design and construction, and be the operator.

Percival C. Keith, vice-president of Kellogg, a voluble chemical engineer from Texas, recruited engineers for Kellex. Like many other leading U.S. chemical engineers, Keith had studied under Warren K. Lewis at MIT. Known as the father of chemical engineering, Lewis had long served as a consultant to leading petroleum-refining companies and had chaired several major Manhattan Project advisory committees at critical junctures in the project's history. Keith enticed engineers into working for Kellex by comparing the work that they would do—which he could not define because of security—to that of medieval artisans who devoted a lifetime to the patient creation of stained-glass windows at Chartres, or to that of Lorenzo Ghiberti, who had given decades to creating the bronze portals with their sculptured reliefs for the baptistry at the Florence Cathedral. Keith had little patience with the scientists at Columbia University who, "like most academicians, seemed to Keith unpredictable and ineffective, prone to wander off the straight paved road of practical progress into interesting, but irrelevant theoretical byways."[117] Groves surely found these notions congenial.

Choosing the material for the barrier became a thorny problem. The uranium gas would pass for hundreds of miles through a cascade of barriers with their pumps and piping. Some materials might react with the gas to corrode the barrier and restrict flow. The barrier material had to remain strong and durable under extreme operating conditions. In addition, it had to be suitable for fabrication on a mass scale. Experimental testing of various materials became imperative. Late in 1942 the favored candidate was a nickel barrier. By 1943 the Columbia team had completed a pilot plant for the production of a nickel barrier at about the same time that construction of a full-scale plant for production of the barrier was under way, another instance of a chronology that frustrated orderly engineering development. After the pilot plant had produced a sufficient quantity of the barrier for thorough testing, problems appeared. The material proved too brittle and structurally weak; it was subject to corrosion and blockage; and the quality was not uniform.[118]

In the summer of 1943 Urey became discouraged by the lack of progress despite the great effort. Relations became strained between him and young John Dunning, a Columbia scientist who was an enthusiastic advocate of gaseous diffusion. Characteristic of his tendency to exaggerate failure as well as success, Urey's pessimism led Groves, Bush, and Conant to enlarge on Lawrence's electromagnetic process. They cut back on the gaseous-diffusion plans, believing that partially enriched uranium from this process could be used as a feed for the Lawrence process in which the uranium, made up of several isotopes, could be "topped off," or finally enriched with sufficient U-235 isotope, to be used as an explosive in a bomb.

Keith, too, became discouraged by the test with the nickel barrier. Then he learned of an alternative nickel-powder barrier being tested by Clarence Johnson, a young Kellex engineer, with help from the Columbia team. The nickel powder would be sintered and formed into a strong tube of metal. Keith wanted to drop prior work and move ahead on the production of the new barrier. Urey, however, believed that the new material would also in time prove troublesome. He further feared that a shift would destroy the morale of the Columbia team, who had invested so much in the old barrier, which, in his view, might still be improved. So Urey refused to divert his laboratory to the new sintered nickel-powder approach championed by Keith. Groves, urgently driving to obtain production of fissionable materials, decided that both should be developed. Angered, Urey fired off an impassioned letter to Groves arguing that no more of their scarce scientists and resources should be given over to the project and, if a full-scale gaseous-diffusion plant were built, that it should follow British design.[119] Unable to ignore the disruptive strains in the gaseous-diffusion project, Groves relieved Urey from any real responsibility by naming Lauchlin Currie, a Union Carbide engineer, as associate director of the SAM lab. With Groves's backing, the soft-spoken, diplomatic Southerner effectively took control. Diplomatically he insisted that he was merely an engineer, present to assist the world-famous chemist.[120] Urey accepted the conditions, for he had not asked for the job of lab director; he hated administration; he had long suspected that his name had been used to attract scientists to the project. He also believed that he would be used as a scapegoat in case of failure.[121]

The momentum of the gaseous-diffusion project was too great to abandon. More than ten thousand workers were constructing buildings for the plant at Oak Ridge. There were more than nine hundred Kellex and seven hundred Columbia people involved. Manufacturers, such as the Chrysler Corporation, already had subcontracts. Keith, deeply committed to the new barrier, was determined to have production of U-235 under way by early 1945. He did not see as a way out of the morass of problems the detached and unhurried approach of the British, who came in as consultants. Early in 1944 Groves made the difficult decision to push production of the new barrier and only to continue research on the older. He set a year as the target date for commencing operation of the plant at Oak Ridge. British consultants considered this a reckless decision. Groves then had the problem of scrapping work that had already been done on the old barrier and changing the direction of the gaseous-diffusion project, which had already taken on mammoth proportions. In the spring of 1945 the gaseous-diffusion plant with the sintered nickel-powder barrier was operating sufficiently well for him to resort to the ingenious plan of using outputs as feeds, or inputs. Uranium, first enriched by a thermal-diffusion plant, was fed into the gaseous-diffusion plant for further enrichment, then finally into the electromagnetic-separation plant for the enrichment needed for use in a bomb. On the other hand, the gaseous-diffusion plant soon reached such a level of efficiency that its output could be used without topping off.

The delays and disappointments experienced in 1944 with magnetic separation and with gaseous diffusion plagued by barrier problems heightened Groves's interest in the thermal-diffusion process for enriching uranium or increasing the proportion of U-235. This process had been long neglected by him and his advisers Bush and Conant. Physicist Philip Abelson headed the small-scale development that originated outside the Manhattan Project. In June 1944, spurred on by Oppenheimer, who had learned of the recent progress in thermal diffusion, Groves appointed a committee that included Warren K. Lewis to make an evaluation. It found that the process was enriching uranium only slightly, but the committee saw the possibility of using the output as a feed.

Abelson and thermal diffusion did not fit into the Manhattan Project pattern at all. Abelson, codiscoverer of neptunium, the first transuranium element, was the first scientist in the United States to produce several

LEE DUBRIDGE AND
PHILIP ABELSON, 1940

hundred pounds of uranium hexafluoride when it was critically needed for experimentation. He was, however, cut off from the Manhattan Project because he worked for the U.S. Navy at the Naval Research Laboratory. There, on a small scale, he developed thermal diffusion as a means of separating uranium isotopes. His apparatus consisted of columns of hot pipes inside cold pipes, with uranium hexafluoride flowing in the space between the cold and hot pipes. The lighter U-235 isotope tended to collect along the surface of the hot pipes and rise to the top of the system, where the enriched product could be removed. Simple as it was in operation, it required prodigious amounts of steam. The navy saw the apparatus as a means of obtaining uranium fuel for a nuclear-powered submarine.

Groves had first learned of the project shortly after taking charge of the Manhattan Project, but he also knew that Roosevelt did not want the navy involved in the atomic-bomb project. In addition, the reports he obtained about thermal diffusion persuaded him that it was an exorbitant consumer of energy, or steam, and could not produce for three or four years uranium sufficiently enriched to be used in bombs. The situation was further complicated because of poor relations between Dr. Ross Gunn, in charge at the Naval Research Laboratory project on the one hand, and, on the other, Bush and Conant. Bush and Conant had left him out of early committee deliberations dealing with uranium. Some critics later suggested that the episode also reflected the ancient army-navy rivalry, but Groves rejoined, not altogether persuasively, that this was limited to football games. As a result of these and other considerations, Abelson proceeded independently.

When Oppenheimer renewed Groves's interest in 1944, Abelson had a hundred-column plant under construction at the Philadelphia Navy Yard, where steam was available. Armed with a positive report from the Lewis committee, and having found a contractor in Mrs. Ferguson's company, Groves ordered that a full-scale plant be constructed at Oak Ridge. In order to save time, Abelson's semiworks design was to be replicated exactly. Immense quantities of steam were available at Oak Ridge from a plant built to supply the gaseous-diffusion plant not yet ready for operation. Working at a furious pace punctuated by inevitable errors and an accident costing the lives of two men, Abelson and the contractors achieved the remarkable feat of having the plant partially in operation by September 1944.[122]

LOS ALAMOS

With the arrival, early in 1945, of small quantities of plutonium from Hanford and uranium-235 from Oak Ridge, the frustrating effort to design bombs without sufficient test materials was over. In the fall of 1942, advised by Arthur Compton and others, Groves had named Oppenheimer to lead the search for the design of the bomb and to preside over its assembly. Groves, in company with Oppenheimer, then chose a remote site of haunting grandeur in the Jemez Mountains, thirty miles from Santa Fe, for the location of the laboratory. Oppenheimer attracted major theoretical and experimental physicists, chemists, metallurgists, and explosives experts and their families to the distant site, which came to be thought of as the birthplace of the atomic bomb.

Waiting for the fissionable material, Los Alamos scientists worked out a highly complex theory as the basis for the design of both a plutonium and a U-235 bomb. After simulating the behavior of the fissionable materials, physicists could gather needed data about explosive nuclear reactions, such as information about the design that would most effectively harness the explosive potential of the fissionable materials. Physicists also had to calculate the amount of fissionable material needed for the bombs. Metallurgists explored the exotic properties of laboratory quantities of plutonium and U-235; and explosives experts searched for means of setting off, or igniting, the bomb. Some of the work was extremely dangerous. A fearless thirty-three-year-old Canadian, Louis Slotin, performed experiments characterized as "tickling the dragon's tail." Shortly after the war he died of exposure to gamma rays sustained when he disconnected with his bare hands a lethal device that, because of his having let a screwdriver slip, threatened his colleagues in the room.[123]

A young physicist, Seth Neddermeyer, invented a method for initiating the nuclear explosion of the plutonium. At first rejected by his colleagues, Neddermeyer's invention ultimately saved the plutonium-bomb project when conventional means completely failed. Naval Captain William S. Parsons, the head of the ordnance division at Los Alamos, with an outstanding reputation for artillery and ballistics, had earlier rejected the Neddermeyer "implosion" method. He found it improbable, because of its complexity and disappointing test results. Neddermeyer's persistence caused friction between him and the captain. But Oppenheimer, a gifted leader in dealing with his proud and independent staff, was won over to Neddermeyer's implosion method when the brilliant Hungarian mathematician John von Neumann calculated that it would work. Oppenheimer then brought in George Kistiakowsky, a distinguished scientist and ex-

J. ROBERT OPPENHEIMER, ENRICO FERMI, AND LAWRENCE

plosives expert, to act as buffer between Parsons and Neddermeyer and to pursue the implosion method.[124]

As was often the case with scientists elsewhere in the Manhattan Project, the physicists, chemists, and metallurgists at Los Alamos proved themselves especially inventive. A thirty-three-year-old physicist, Charles Critchfield, conceived of an "initiator," called Urchin, first received skeptically by Oppenheimer, that ultimately supplied the neutrons needed to ignite the plutonium during implosion. Cyril Stanley Smith, a British-born American metallurgist, invented a solution to the problem of blistering on the matching surfaces of plutonium hemispheres, which was so severe that it threatened to delay testing of the bomb. He filed down the hemispheres and then inserted gold foil between the facing halves of the bomb's plutonium core to achieve the necessary perfectly smooth surfaces and fit. Robert F. Christy of the "theoretical division" invented a "gadget" that took advantage of an unexpected characteristic of plutonium to make possible the use of extra compression rather than extra size to achieve a critical, or explosive, mass. Chemists, under the direction of young Joseph Kennedy, developed ingenious methods for preparing plutonium compounds of almost unbelievable purity, and the metallurgists under Smith invented means of converting these to metal, and of fabricating the metal into desired shapes despite unprecedented changes in metal density when it was heated.

After the bombs were assembled with their charges of U-235 and plutonium, Groves and his advisers decided to ship "Little Boy," the U-235 bomb, directly to the Pacific theater of operations. There was not enough material for a test bomb, and its performance was not considered problematic. A gun method for exploding the uranium bomb was considered reliable. On the other hand, they decided to test "Fat Man," the plutonium bomb named for Winston Churchill, at Alamogordo, a lonely desert site and bombing range 210 miles south of Los Alamos. The probability of the effective performance of the plutonium bomb, with its implosion and other innovations, was not as high, and there was on hand sufficient plutonium for another bomb to be used against Japan. Called "Project Trinity," the horrendous test explosion of the plutonium bomb occurred on 16 July 1945. So often described in print and film, the episode has become a legend. A B-29 bomber, *Enola Gay*, dropped Little Boy, a uranium bomb, on Hiroshima and it exploded at 8:16 A.M. on 6 August. Contemporary estimates placed the number of dead at at least 100,000: official statistics counted 70,000 fatalities up to 1 September, with 130,000 wounded, of whom 43,500 were known to be severely wounded. Subsequent estimates ranged up to 140,000 dead by the end of 1945; perhaps 200,000 after five years. The bomb destroyed more than half of the more than 70,000 buildings in Hiroshima. On 9 August another B-29, *Bock's Car*, dropped the Fat Man, with its plutonium explosive, on Nagasaki, an alternate target. Haze and smoke obscured the primary target, Kokura Arsenal on the north coast of Kyūshū. The bomb exploded over Nagasaki at 11:02 A.M., leaving 70,000 to die by the end of 1945 and 140,000 altogether over the next five years.[125] The Japanese surrender offer reached Washington on 10 August.

ASSEMBLING THE BOMB AT THE TRINITY TEST SITE

Even before the bombs were dropped on Hiroshima and Nagasaki, some leading physicists tried to prevent their use. Once the threat had passed that the National Socialists would have the bomb first, a group of physicists participating in the Manhattan Project tried to persuade the military and the politicians not to use it in combat. Franck, Szilard, and several other leading scientists working on the Manhattan Project prepared a report in June 1945, before Hiroshima, recommending that the bombs, if exploded, be used only in an uninhabited area in a demonstration designed to persuade the Japanese to surrender. They argued that an effort should be begun to bring international control to prevent an arms race and a holocaust. Immediately after the war, hundreds of young scientists, especially physicists and others in the Manhattan Project, became involved in a movement to inform Congress and the people about atomic energy and to ensure its civilian and international control. Their support helped bring defeat of the military-oriented May-Johnson Bill and the passage of the McMahon Bill, which established the Atomic Energy Commission. The more active among them organized the Federation of American Scientists, and a group in Chicago published the influential *Bulletin of the Atomic Scientists.*[126] The scientists protested the use of the bomb and the likelihood of military control of atomic energy. They warned the world of the dangers of nuclear proliferation, but they did not unite in protesting the participation of physicists in the massive technological system that had produced the bomb. With a few notable exceptions, they did not refuse to participate further in the growth of the nuclear-production system, military and nonmilitary. The Atomic Energy Commission would preside over the nuclear-production system.

ATOMIC ENERGY COMMISSION

After the efforts of the Great Powers to establish international control of atomic energy failed, the United States sought means to continue the production of atomic weapons and to explore other possibilities for the use of atomic energy. It should be recalled that several of the Chicago scientists had wanted only to produce a few bombs with small-scale facilities rather than to establish production lines and the massive facilities now at the disposal of the government. The bill authored by Senator Brien McMahon was originally intended to exclude the military from substantial influence over these facilities and the development of atomic energy, but it was subsequently amended by conservative senators to increase the role of the military. Almost a year after the Hiroshima bombing, the president signed the bill and established the Atomic Energy Commission (AEC).[127]

At the end of 1946 the production facilities were transferred from the Manhattan Project to the AEC, which was comparable in size to the nation's biggest business enterprises. Its properties and organizations included production plants at Hanford and Oak Ridge; the Clinton laboratory at Oak Ridge and the Argonne laboratory near Chicago, both operated by the University of Chicago; the laboratory and bomb-assembly facility at Los Alamos; and laboratory installations at, and operated under contract by, California (Berkeley), Chicago, and Columbia universities. Also transferred were about two thousand army personnel, four thousand government workers, and almost forty thousand industrial-contractor employees. The wartime investment amounted to more than $2.2 billion, and expenditures were continuing at the rate of hundreds of millions annually. The authors of the Atomic Energy Act establishing the AEC called it radical, observing that

> The Act creates a government monopoly of the sources of atomic energy and buttresses this position with a variety of broad governmental powers and prohibitions on private activity. The field of atomic energy is made an island of socialism in the midst of a free enterprise economy.[128]

Put in the context of the history of modern technology, the action taken by the American government was even more radical than contemporaries realized. Atomic scientists had already pointed to the possibility of using uranium and plutonium as low-cost fuels to produce heat and steam to drive turbines in electric-power stations, ships, submarines, and other vehicles. Enthusiasts argued that atomic energy could bring technical and social changes comparable to those that had occurred during the British Industrial Revolution, when coal as a fuel displaced wood, water, and wind as the prevailing energy sources. The U.S. government, then, in a remarkable action, was taking responsibility for presiding over an anticipated industrial revolution that many expected would be of even greater magnitude than the British Industrial Revolution, one of the epochal events of history.

Who was to head the Atomic Energy Commission? Not surprisingly, President Truman named Lilienthal, who then headed the TVA, which had set the precedent for government involvement in funding and operating a large technological system. Of the four others the president named to the AEC, only one, Robert F. Bacher, was a scientist. A physicist, he had been an important figure at Los Alamos. Sumner Pike was a former member of the Securities and Exchange Commission, Lewis Strauss a partner in the investment firm of Kuhn, Loeb & Company, and William Waymack the editor of a Des Moines, Iowa, newspaper and a public director of the Federal Reserve Bank of Chicago. The last three brought to the commission considerable experience and influence in the world of finance.

During the year and a half between the end of the war and the transfer of authority to the AEC, Groves had not been inactive. He hoped that the possibility of nonmilitary nuclear power would induce General Electric to take over operation of the Hanford plutonium works from Du Pont when it decided to withdraw from its wartime responsibilities there. Even before the war ended, Du Pont's chief executives had carefully considered the commercial advantages that might come from the company's exploitation of its wartime experience with nuclear energy. But its difficulty in attracting leading physicists to do industrial research and its reluctance to be once again branded "merchants of death" helped persuade Du Pont in the fall of 1945 to forgo substantial investment in the new field.[129] GE agreed to take over Hanford, but also persuaded the government to fund another General Electric research laboratory, later named the Knolls Atomic Power Laboratory, near the GE plant and research lab at Schenectady, New York.[130] At Oak Ridge, the inefficient thermal-diffusion plant and some of the electromagnetic-separation facilities were shut down, and the gaseous-diffusion production facilities were expanded. Monsanto Chemical Company took over from the University of Chicago the operation of Clinton laboratory at Oak Ridge. Groves decided that the combination of bomb production and research functions at Los Alamos would be divided, with bomb assembly moving to Sandia Laboratory, near Albuquerque, New Mexico, and coming under army control. The army would contract with industrial corporations for the last stages of bomb manufacture, formerly done at Los Alamos. Los Alamos was to do more fundamental weapons research, including investigation of the feasibility of a hydrogen bomb. The general also named a committee of scientists and engineers, including Arthur Compton and Lewis, to advise him on research and development. Groves accepted the committee's recommendation that national laboratories should be established and funded by the government, initially at Argonne and at Brookhaven, Long Island, to carry on primarily fundamental, nonclassified research with equipment too expensive for universities or industry to afford. Consortia of universities were to engage in research by using the large experimental machines, such as particle accelerators, and atomic piles at the national laboratories. Groves also authorized research funds for the University of Washington, the University of Rochester, Iowa State University, Columbia University, the Massachusetts Institute of Technology, and the Battelle Memorial Institute.[131] The mammoth government, industry, and university complex that would soon preside over atomic-energy developments was taking shape. Construction and operation of production and research facilities would be subcontracted to industrial corporations and universities, as had been the case in the Manhattan Project. It and the AEC set a pattern for other major postwar research, development, and production projects such as those of the Defense Department and the National Aeronautics and Space Administration. By contrast, both TVA engineers and workers, who were government employees, had constructed and operated the dams, power plants, and other facilities.

In 1947 the Atomic Energy Commission faced a complex array of issues and interests, many of them conflicting. Governed by a commission of presidential appointees, the AEC was destined to feel political crosscurrents arising in Congress and various government departments, such as War, State, and Commerce. Large corporations would also exert influence because of their political power, presidential and congressional. The original, common, ultimate objective of obtaining bombs before the war's end usually had dampened conflicting interests, but repressed differences began surfacing after 1945. The AEC and the military, especially Groves, interpreted differently the provisions in the Atomic Energy Act about the control of the last stages of the assembly and the custody of the bombs. Engineers challenged the dominant influence of scientists in making research and technical decisions within the AEC laboratories, and especially in the administration of the AEC laboratory at Argonne. University-operated laboratory facilities sometimes competed with one another and with those administered by industrial contractors for priority in reactor projects. Two AEC committees, an influential General Advisory Committee, headed by Oppenheimer, which represented the scientific establishment made powerful by its Manhattan Project successes, and a Military Liaison Committee, which included Groves and was concerned about military needs, kept their often-conflicting interests before the AEC commissioners. A major issue involved the priority to be given to bomb production as compared with the development of atomic reactors for power. According to one source, Lilienthal, new head of the AEC, had to report to President Truman in April 1947 that there were no operable bombs in the atomic arsenal.[132] After 1947, as the Cold War brought pressure from the Congress and the administration to strengthen the nuclear deterrent, bomb manufacture took priority, but work on the reactors continued. Opinions within the AEC were divided about emphasis to be given to power reactors for nonmilitary and military uses. Leading AEC scientists differed about which type of reactor to give research priority. Some advocated breeder reactors, which would produce plutonium as well as power, some believed in one kind of coolant and moderator, while others had different choices. Industrial contractors, such as General Electric, had to decide whether military or nonmilitary development of atomic energy furthered their long-range commercial interests. There were differences among scientists and within universities as to how deeply they should become involved in richly funded classified military projects. AEC tensions and conflicts heightened as its expanded program revealed shortages of experienced scientists and engineers and of materials such as uranium. Differences of opinion also arose about sources of uranium, from the possibility of increased domestic prospecting and mining to reliance on external suppliers, including South Africa. An alternative was increased expenditure for piles, and separation processes that would use uranium more efficiently. Lilienthal and others were learning about the complexity of the loosely federated government-industrial-military complex now dominating the course of late-twentieth-century technology. The day had passed when American system builders like Ford and Insull could virtually isolate their empires from government, military, and university influences.

The role of General Electric during the early AEC years reveals the labyrinthine character of a postwar technological system with the AEC as its nucleus. After the company took over the operation of Hanford from Du Pont, problems with the plutonium piles and the separation-process plants increased. As the three wartime piles aged, the graphite used as a moderator of fast neutrons tended to swell and block passages for the loading and removal of uranium and irradiated slugs. The wartime chemical-separating process removed plutonium for bombs from the slugs, but left the uranium to be disposed of as waste, with other highly radioactive fission products, in giant underground tanks at Hanford. The AEC had to decide whether General Electric should replace the aging piles with ones of similar design, or construct piles of improved design. General Electric also had the overall responsibility for research and development of a new process called Redox that separated and saved uranium as well as plutonium from the irradiated uranium slugs. Plants for this process were expected to be the single largest government construction process in history. In 1947 and 1948 various AEC committees and administrators found the General Electric management at Hanford moving indecisively and slowly on the pile and the separation problems. Conferences with General Electric executives also revealed their ambiguous position on the tasks of its AEC-funded Knolls Atomic Power Laboratory. It was unclear whether GE was interested only in designing and developing a power-generating breeder reactor for nonmilitary purposes, or in one that might serve both military and nonmilitary goals.[133]

As relations between the Soviet Union and the United States deteriorated, the AEC, under continuing pressure from the administration and the military, pushed weapon production at the expense of reactor development. In June 1948 Oppenheimer summed up the General Advisory Committee's attitude with the remark "We despair of progress in the reactor program."[134] Before an audience of influential scientists in April, Admiral Earle W. Mills, chief of the navy's Bureau of Ships, said that perhaps less than one percent of the designing needed for a submarine nuclear reactor had been completed, and he blamed the AEC. He called for more engineering and industrial involvement. Lewis Strauss, an AEC

SZILARD READS THAT THE SOVIET UNION HAS DETONATED AN ATOMIC BOMB, 1948.

commissioner at the meeting, was heard to say to Mills, "I never thought an old friend would do that to me."[135]

The AEC moved some reactor research from the Clinton to the Argonne laboratory, but opinion about the best course to follow remained divided, and research efforts remained dispersed among several projects at Argonne, including a fast breeder reactor and a high-flux materials-testing reactor. Work continued on an intermediate power reactor at GE's Knolls laboratory and on planning and design for a water-cooled thermal reactor for submarine propulsion at Clinton. The submarine reactor was a conservative design, depending insofar as possible on components available from industry. Into this ambiguous complex of AEC reactor design and development, Admiral Mills and Captain Hyman G. Rickover, later known as the creator of the nuclear navy, moved to provide a focus. A key to their success was a proposal appealing both to those desiring more emphasis on reactors and to those favoring military needs. Of key importance also were the drive and determination of Rickover.

RICKOVER THE SYSTEM BUILDER

The U.S. Navy had been left out of the Manhattan Project. During the war the navy had proceeded on a modest scale toward the distant objective of nuclear propulsion through its support of Philip Abelson's thermal-diffusion process. This was the process that Groves belatedly took over in the spring of 1944, when the gaseous-diffusion and electromagnetic processes were not producing according to expectations. The navy, nevertheless, was determined not to be left out when the AEC took responsibility for atomic-energy developments. Before the AEC assumed responsibility in January 1947, the Naval Research Laboratory in 1946 had circulated a report by Abelson and two assistants that called for a nuclear-powered submarine to be in operation within two years. The navy also dispatched a small contingent of officers and civilian naval engineers to the Clinton laboratory at Oak Ridge, Tennessee, to participate in the early planning for an experimental reactor. Among them was Rickover, a naval-engineering officer with a master's degree in electrical engineering from Columbia University but no prior experience with nuclear energy. He had an impressive record of directing the design and procurement of electrical apparatus for the navy during the war. Rickover possessed a reputation for single-minded dedication to technical excellence. He did not leave the details of design, development, and inspection of new equipment to industrial contractors, but became deeply involved with his hard-working and closely supervised naval staff in the details of the entire process of development, procurement, and installation. Other naval officers usually preferred to exercise overall supervision of contract letting and coordination, and then to assume the role of customers not involved in the process of manufacture. At the same time that Rickover won the reputation for technical proficiency and achievement, he also became known as a blunt, severely critical, devastatingly frank leader who created hard feelings with his hard driving. Although he was a U.S. Naval Academy graduate, he did not conceal his contempt for the spit and polish and the social traditions of naval service. He was also known for relentlessly pursuing his particular technical projects without consid-

REAR ADMIRAL RICKOVER ABOARD THE *NAUTILUS* DURING SEA TRIALS, JANUARY 1955

eration of the broader objectives of tradition-minded line officers. Rickover, though senior officer among the naval contingent at Oak Ridge, was not placed in charge. Some superiors were concerned that he might take over and define the project without taking into consideration contrary points of view and interests.[136]

In characteristic fashion, however, Rickover soon took over the preparation of the regular fitness reports of the other officers, evaluations that greatly affected promotions and placement in the navy. This gave him the authority he sought to direct the activities of the small naval contingent at Oak Ridge. Soon the others were following his lead in conscientiously attending Oak Ridge seminars and in painstakingly reading and summarizing reports on reactor science and technology. Fired with enthusiasm for nuclear propulsion, Rickover shaped the contingent into a nucleus for the development of atomic power in the navy. He established within the group a discipline and an esprit de corps that became a legend at Oak Ridge.[137]

Rickover quickly established himself as a pivotal figure in the navy's search for its role in the development and use of atomic energy. He sought support, both in the navy and in the Atomic Energy Commission, for nuclear submarine propulsion. Improved submarine-detection techniques at the end of the war had revealed the need for a deep-running submarine not dependent on air-breathing engines. Therefore, influential submarine officers supported Rickover. By contrast, Captain William S. Parsons, who had played a leading role as a munitions expert at Los Alamos and served on the AEC Military Liaison Committee from 1946 to 1949, opposed emphasis on nuclear propulsion, concerned that it would divert the navy from the goal of developing and delivering nuclear weapons. Outside the navy, among physicists, Rickover found enthusiastic support, especially from Lawrence. Characteristically, he told Rickover that the project must be conceived of on a scale large enough to attract attention and gain support. Edward Teller, who had been a physicist at Los Alamos and soon become known as the father of the hydrogen bomb, also warmly endorsed nuclear propulsion for the navy. Teller usually reacted enthusiastically to imaginative new ideas, evaluating them intuitively. He cautioned Rickover, however, that few engineers could adapt to radical ideas such as nuclear power, and that many scientists were likely to wander from the technical goal. On the other hand, physicists on the AEC General Advisory Committee gave higher priority to navy projects other than nuclear propulsion.[138]

In this sea of currents and countercurrents, Rickover resorted to the bureaucratic tactic of drafting letters for superiors committed to nuclear propulsion to sign. Admiral Chester W. Nimitz, chief of naval operations, and John L. Sullivan, secretary of the navy, obliged. In obtaining Nimitz's support, Rickover enlisted influential submarine officers as allies. In December 1947 the letter sent from Nimitz to Sullivan stated the need for an atomic submarine capable of launching a missile with a nuclear warhead. Memoranda from Sullivan to Defense Secretary James V. Forrestal and to Bush, who continued to advise the government on science-and-technology policy, recommended that the Atomic Energy Commission and the navy's Bureau of Ships work out a mutually acceptable procedure for designing, developing, and constructing the submarine.[139] In time, Admiral Mills, who initially had doubts about Rickover as a leader because of his abrasive style, appointed him liaison officer to the Atomic Energy Commission. Mills, who wanted an atomic submarine, believed that the task needed a hardheaded, ruthless system builder and that Rickover filled that bill. The commission, encouraged by Rickover, formed a modest Naval Reactors Branch at the Argonne laboratory, where general reactor research was centered. Rickover soon took charge of the efforts of both the Atomic Energy Commission and the navy to design and develop a nuclear submarine. He wore two hats: under the navy's Bureau of Ships he headed the Nuclear Power Branch, named Code 390; for the Atomic Energy Commission, he headed the Naval Reactors Branch. Unlike persons less determined and experienced in dealing with bureaucracy, he did not fall neglected between two stools. Instead, he used one to reinforce the other.

He then created a complex organization interrelating the AEC, its General Advisory Committee of prominent scientists, its Argonne laboratory, the navy, and industrial contractors. Believing, with other naval officers, that the Argonne scientists had too much influence on reactor design, Rickover was particularly concerned to increase the role of industry in the submarine project.[140] Realizing that AEC physicists had other priorities and were not accustomed to the driving, narrowly focused engineering and industrial style, Rickover gradually took into his own hands increased responsibility for the submarine project. Following the precedent of Groves, he believed that he could exercise more effective control over industrial managers and engineers than over physicists doing research in an academic tradition. So Rickover, as Groves before him, turned to major industrial contractors. Using his liaison position and waxing influence, he persuaded the Atomic Energy Commission to offer a contract to Westinghouse to construct a submarine nuclear-propulsion plant designated the Mark I. It was to be a pressurized light-water reactor. Rickover persuaded Westinghouse's chief executive officer that nuclear energy would have many industrial applications and that the company should expend substantial resources to establish itself in the field.[141] The company signed in December 1948. Research and design would continue to be an AEC Argonne and Rickover responsibility. Rickover also asked

General Electric to consider designing and developing for submarine propulsion a version of the intermediate power reactor on which it had been working. In time, the AEC contracted with GE to build a sodium-cooled reactor that could be used on submarines.

Rickover's practice of exercising close supervision of contractors, his knowledge of nuclear engineering, and the dual authority he possessed because of his navy and AEC responsibilities resulted in closer ties between the newly formed Westinghouse atomic power division and its laboratory with Rickover's naval reactors unit than with its own parent Westinghouse organization. The AEC historians Richard Hewlett and Francis Duncan characterized the company's atomic-power division as "theoretically a part of the contractor's organization but in many respects an integral part of Rickover's project. . . . [This] offered new and unexplored possibilities for managing engineering enterprises."[142] Rickover also succeeded in having the General Electric Knolls laboratory concentrate primarily on the nuclear-submarine project. In his dual roles as Atomic Energy Commission official and head of the navy's nuclear-propulsion program, Rickover was building an autonomous development enterprise dependent on outside support only for funding.[143] The results provide us a prime example of creative system building.

Critical problems had to be solved as the design and construction of the light-water submarine reactor progressed. Shielding a submarine crew from radiation, securing materials to meet unprecedented reliability and durability standards, and acquiring heat exchangers and controls for the propulsion plant proved especially demanding problems. In his own Naval Reactors Branch, he developed the scientific and engineering competence to initiate and supervise research and development on rare and unfamiliar materials in various industrial and university laboratory facilities. One of the materials, zirconium, Rickover had decided to use in the core of the reactor. Initially the Atomic Energy Commission took responsibility for developing sources and manufacture of zirconium of the quantity and quality needed, but by 1950 Rickover, discouraged by the slow progress, persuaded the commission to turn over the zirconium problem to him. His naval group then moved quickly on the basis of its detailed technical information and contracted for the construction of a manufacturing plant. This was a typical Rickover episode.

As development of the nuclear-propulsion unit proceeded, Rickover persuaded the Electric Boat Company—soon to be a subsidiary of the General Dynamics Company—to work with Westinghouse in the construction of the submarine hull and the water-cooled nuclear-propulsion unit. General Electric and Electric Boat also contracted to build a sodium-cooled reactor for propulsion and a hull, but this project fell several years behind the Westinghouse–Electric Boat pressurized-water design. In private, Rickover was sharply critical of General Electric as an industrial contractor, in contrast with his high regard for Westinghouse.[144] Rickover wanted national attention for the first nuclear-submarine project, since he assumed that this would facilitate additional public and congressional support. He arranged for President Truman to attend the keel-laying of the Westinghouse–Electric Boat submarine on 14 June 1952 at the Groton, Connecticut, shipyard of the Electric Boat Company. In order to expedite the project, Rickover had characteristically intervened in company management during construction to bring about a reorganization of the company. AEC historians Hewlett and Jack Holl have observed that "Rickover and his staff were as unyielding and unforgiving as was the technology they were attempting to master."[145]

Named *Nautilus*, like two previous navy submarines and Jules Verne's famous fictional craft, the submarine was ready for sea trials in January 1955. S.S.N. *Nautilus* performed remarkably. Commander Eugene P. Wilkinson, its first captain, wrote: "The results of the tests so far conducted definitely indicate that a complete re-evaluation of submarine and antisubmarine strategy will be required."[146] In April, *Nautilus* steamed thirteen hundred miles submerged to San Juan, Puerto Rico, a distance ten times greater than any other submerged submarine had previously traveled. Her speed was unprecedented. The performance of the submarine during operations of the Atlantic Fleet proved spectacular. *Nautilus* could overtake surface ships, avoid detection, and in some cases evade torpedo attacks.

This signal achievement brought Rickover national prominence and

ATOMIC SUBMARINES S.S.N. *SEA WOLF* AND S.S.N. *NAUTILUS*

enthusiastic support from the members of the U.S. Congress concerned about national defense and prestige. When members of the Congressional Joint Committee on Atomic Energy, who were greatly impressed by his achievements, learned that he had been passed over for promotion once more and that thirty-nine other captains had been selected, they decided to hold up all the promotions until the navy's selection process could be investigated. There was talk of bringing civilians into the selection process. The promotion came in spite of the opposition of high-ranking, influential naval officers who found Rickover lacking in those qualities of well-rounded leadership and social ease traditional among seagoing line officers. His single-minded pursuit of his projects, his focus on technology, and his harsh criticisms had not won him a reputation as a team player. Some supporters believed that as a Jew he was discriminated against. Without question, he displayed those outsider characteristics common among the great independent inventors of the past. Captain Rickover, after years of being passed over by promotion boards, became

Admiral Rickover. The nation's press took satisfaction in the victory of the achievement of the outsider over the establishment.[147]

Subsequently, Rickover also assumed a leading role in the introduction of a nuclear navy with submarines and surface ships, including large aircraft carriers. When they were equipped with long-range missiles and nuclear warheads, the nuclear submarines became a major component of the nuclear-deterrent policy of the United States. A navy Special Projects Office under Admiral William Raborn presided over the design and construction of the solid-fuel Polaris missiles placed on nuclear submarines in the 1960s. Unwritten orders from the chief of naval operations, Admiral Arleigh Burke, kept Rickover out of this project until it was well under way. His participation, it was feared, would bring his domination and threaten the close cooperation of various naval bureaus that the more diplomatic Raborn had cultivated.[148] Because Rickover was so well known, however, the press often mistakenly identified him as the director of the outstandingly successful Polaris program.[149]

SHIPPINGPORT

Rickover also presided over the design and construction of America's first public-utility nuclear-power plant. In essence, it was a modified naval reactor. In so doing, he lastingly imprinted such characteristics as reliability and durability onto the country's nuclear-power industry. These characteristics had been introduced primarily to respond to military concerns. Though it was publicly heralded as the power plant that inaugurated the era of public-utility electricity from the atom, the plant proved to be more a prototype for future development than a commercial enterprise. Economy of construction and operation were not the highest priority. In 1953 President Eisenhower's administration called for a nonmilitary power reactor that could be used by utilities throughout the world. Because electricity cost more in Europe, some nuclear proponents expected initially a larger market there for nuclear-power plants than in the United States. The American government and industry wanted Europeans to use reactors designed and manufactured in the States. Eisenhower's close advisers on the National Security Council feared that the

Soviet Union might enter the world market first with a power-generating reactor and establish with its design a foothold which the United States could not dislodge. From long experience, industry knew that an early sale brought continuing orders for replacement and improvement parts and the training of personnel who would favor the reactor design with which they had had their initial experiences. This possibility, if fulfilled, would, in the opinion of the president's advisers, be a major triumph for the Soviets in the Cold War.[150] So, despite the country's being caught up in the Korean War, the Soviets having exploded a thermonuclear device in August 1953, and a shortage of plutonium for atomic weapons, the Eisenhower administration gave a high priority to the nonmilitary reactor. Eisenhower also insisted that he wanted to lead the world in an Atoms for Peace program, which would develop nuclear power for peaceful purposes.[151]

A number of interest groups, including those wanting to exploit nuclear power commercially, immediately became involved. Private- versus public-ownership advocates wanted private manufacturers and utilities to take over, insofar as national security allowed, the development of nuclear power for nonmilitary purposes. A few campaigned against private ownership, preferring that the government control the new power source. Scientists advising and employed by the Atomic Energy Commission saw the need for a thorough research program to ensure the most advanced design. Rickover did not want the Atomic Energy Commission to be diverted from naval reactors to the designing of public-utility plants. The president and his administration wanted early results and a tight budget. The military and the Atomic Energy Commission envisaged civilian reactors breeding plutonium for military uses. A compromise emerged with the decision made by the Atomic Energy Commission to have Rickover and his Naval Reactors Branch modify the design being developed for an aircraft-carrier reactor and take charge of the civilian-reactor project. The nuclear-carrier reactor was given up for the time being. Rickover was to use Westinghouse for additional design and other subcontractors for construction in a pattern not unlike that being used with the nuclear-submarine project. Critics of AEC and Eisenhower administration policy argued that the rush to develop a nonmilitary reactor based on a military one resulted in an inferior design.

Private utilities were asked to submit proposals to design and construct the electrical-power components and to operate the nuclear plant.[152] The Atomic Energy Commission accepted the offer of the Duquesne Light Company of Pittsburgh to provide the site, build the turbogenerator plant, assume $5 million of the cost of developing and building the reactor, and operate and maintain the entire facility. The AEC would own the reactor and sell the steam to the company. The commission chose Stone & Webster to serve as consultant and architect-engineer, and the Dravo Company of Pittsburgh to provide various construction services. As they had with the nuclear-submarine project, Rickover's group of naval officers and civilian engineers were closely involved in all phases of decision making and construction. For them, safety of operation was the overriding consideration. As a result, instruments for monitoring and controlling the behavior, especially the temperature, of the fuel in the core of the reactor and coolant temperatures and flow became critical design problems. The naval and Westinghouse engineers and various subcontractors also gave particular attention to the strength and durability of the large vessels for containing the reactor and the coolant system. Design proceeded expeditiously, because many of the components were scaled up from naval reactor designs, especially the prior work on the submarine plant.

In the Shippingport project, Rickover's style of leadership proved on occasions more a problem than a solution. Accustomed to working with engineers, managers, and skilled technicians on an individual basis or in small groups, he did not adapt well to working with masses of construction workers building a large power station. Rickover admitted that he did not understand the traditional ways of the construction industry and the labor unions. Groves had also been perplexed by labor problems during the Manhattan Project. Rickover's naval associates "held their breath during his visits to the site, for fear that a blunt question to a foreman or a sharp reprimand to a worker would cause a walkout."[153] The project also tried Rickover because various equipment and labor problems threatened to delay completion beyond the 1957 deadline set by Eisenhower's administration and the Atomic Energy Commission. Adding to the woes of Rickover, who took inordinate pride in fulfilling his commitments, was an exceeded budget, caused in part by the need to use extensive overtime labor in order to meet construction schedules. Yet the reactor began sustained operation (went critical) on 2 December 1957. On 23 December, the target year, the reactor reached its full power rating after the Duquesne company personnel had taken over operation.

During the next two years, seminars provided hundreds of engineers from government and industry information about the design and operation of the exhaustively documented reactor. Although the pressurized, light-water reactor had higher capital costs and cost more to operate than coal- or oil-fired power plants, the technical performance of the reactor, as shown by a battery of tests and data, was judged to be excellent. Ten of the first twelve utility reactors operating in the United States within a decade followed the basic design of Shippingport in light water as a moderator-coolant. Most of these reactors also, like Shippingport, used slightly enriched uranium.

LEGACY

In 1954, with the passage of an Atomic Energy Act, the U.S. Congress and the Eisenhower administration greatly accelerated the growth of the nuclear industry. The legislation represented a compromise among those in the administration, Congress, and industry who preferred that private enterprise develop atomic energy, others who wanted a cooperative arrangement between government and private companies, and some who wished the industry to be nationalized. It allowed private corporations to build and own nuclear-power plants, but the government continued to own and control the fuel. The AEC was also instructed to share more of the results of nuclear research with private enterprise. Forestalling a new TVA, the act forbade the AEC from constructing its own plants to generate and supply electricity. The TVA, on the other hand, would build nuclear plants. It was a major supplier of power to AEC installations. In 1955 the AEC invited U.S. utilities and manufacturers to present proposals for the construction of atomic-power plants designed to demonstrate technical and economic feasibility. Government funds would be available for research and development. Among those responding were the Nuclear Power Group, which included some of the largest utility companies in the nation, and the Bechtel Corporation of San Francisco, a large consulting-and-construction firm from whose managerial ranks Caspar Weinberger and George Shultz later moved to Washington to become Cabinet members.

General Electric, Westinghouse, Babcock and Wilcox, and Combustion Engineering established reactor-manufacturing facilities. They were supported and encouraged by government subsidies for domestic research and development, as well as the Eisenhower administration's offer of subsidies to foreign governments who would buy U.S. experimental reactors. Low-interest loans from the Export-Import Bank to foreign governments buying reactors also stimulated the market. A GE executive promised a young man entering the company that within ten or twenty years the company's nuclear-power business would be larger than the entire company in the 1950s.[154] Before the end of the Eisenhower administration, the Yankee Atomic Electric Company, a consortium of thirteen New England utilities, the Consolidated Edison Company of New York City, and the Pennsylvania Power & Light Company were constructing pressurized water reactors; and the Dresden Nuclear Power Station in Morris, Illinois, was under construction with a boiling-water reactor. The Rural Co-op Power Association of Elk River, Minnesota, was also building a boiling-water reactor. Consumers Public Power Company of Columbus, Nebraska, was constructing a sodium-cooled fast reactor.[155] In the 1960s General Electric stimulated the domestic market by offering to design and construct nuclear plants ready for service (turnkey plants), to be turned over to electric light-and-power utilities. Initially the company's price was artificially low, in order to prime the market and build up experience and manufacturing standards and routine that were expected to lower manufacturing costs. Estimates reveal that GE and Westinghouse lost about $1 billion on thirteen turnkey plants constructed in the mid-1960s. After breaking through the utilities resistance, the manufacturers saw domestic orders jump to twenty plants in 1966. Then the manufacturers raised prices and returned to the customary practice of passing on cost overruns to the utilities. Sales continued because of the assumption that costs would decrease over time. This was usually the case when small-scale manufacture rose to large-scale. Leading consulting-engineering and construction firms, such as the Bechtel Corporation and Stone & Webster, were also deeply involved in the inauguration of what some persons called the "atomic age." As Philip Sporn, a leading utility engineer and manager, observed, these were the bandwagon years.[156] Yet there were harbingers of problems to come because of the accelerated development. McCone, head of the AEC, observed in his personal notes:

> One receives the impression in travelling that so many companies have launched forward blindly into this field making huge investments in engineering organizations and plants and equipment that they now are rather desperately advancing exotic and extreme and sometimes unsound developments in the hope of gaining contracts against which to advertise their facility investment and to employ their organization.[157]

The Manhattan Project and the postwar Atomic Energy Commission activity were the culmination of modern technological-system building. In the memories of the participants, in official histories, scholarly publications, and newspaper and television reports, the Manhattan Project became history's most impressive technical, scientific, industrial, and organizational achievement. The Manhattan Project—symbolized by the Hanford piles and the mushroom cloud over Trinity—cast its shadow over the great construction and production achievements of the past, such as the pyramids, medieval cathedrals, Renaissance Venice, Baroque palaces, Industrial Revolution canals and railroads, regional electric-power systems, Ford's River Rouge plant, the Soviet Magnitogorsk, and the TVA.

Mark Hertsgaard, author of *Nuclear Inc.*, refers to an "atomic brotherhood," which he defines as a conglomeration and an interconnection between corporations in the nuclear-energy business and their government sponsors. By the 1980s the "brotherhood," involving twenty-four large international corporations, had become a large and powerful complex, perhaps the largest and most powerful enterprise in history. The involvement of electric utilities, eight of the nation's largest banks, and seven of its largest insurance companies increased the influence of the government–nuclear-industry complex.[158] Military and diplomatic interests reinforced the complex. The Du Pont Company had re-entered the nuclear industry on a large scale when it agreed in 1950 to design, construct, and operate a new plant for the production of bomb materials in South Carolina, on the Savannah River. AEC Commissioner Thomas E. Murray told an electric-utility convention in Chicago in 1953 that nuclear power was as important for national security as nuclear weapons. "For years," Murray concluded, "the splitting atom, packaged in weapons, has been our main shield against the Barbarians—now, in addition, it is to become a God-given instrument to do the constructive work of mankind."[159] He reasoned that the U.S. atomic industry must bind a world alliance through the supply of reactors and fuel to actual and potential allies. America's creating, building, and systematizing genius reached its zenith in a nuclear enterprise, or technological system, that was a culmination of almost a century of ever-expanding invention, industrial research, and system building that extended from Pearl Street to Hanford, Oak Ridge, and Los Alamos.

It is ironic that President Eisenhower, in whose administration the nuclear-energy production system had so greatly expanded, warned the nation against the process—against the military-industrial complex, as he called it. Eisenhower knew firsthand the "military-industrial," or the government, industry, and university complex. During his presidency, 1953–61, his administration presided over the extensive testing and manufacture of hydrogen bombs, the building of a nuclear-powered navy, the development of intercontinental missiles, and the establishment of the National Aeronautics and Space Administration (NASA). In his farewell address in January 1961, Eisenhower, wartime supreme commander of the Allied Expeditionary Forces (1943–45), told the American people:

> This conjunction of an immense military establishment and a large arms industry is new in the American experience. The total influence—

> economic, political, even spiritual—is felt in every city, every State house, every office of the Federal government. We recognize the imperative need for this development. Yet we must not fail to comprehend its grave implications. Our toil, resources and livelihood are all involved; so is the very structure of our society.
>
> In the councils of government, we must guard against the acquisition of unwarranted influence, whether sought or unsought, by the military-industrial complex . . .
>
> Akin to, and largely responsible for the sweeping changes in our industrial-military posture, has been the technological revolution during recent decades.[160]

Eisenhower's phrase "the military-industrial complex" has proved memorable. He used it broadly to include not only the military establishment and the arms-making private corporations, but the scientific-technological elite of the nation's universities as well. A government contract, he sagely remarked, has become among academics a substitute for intellectual curiosity.

Despite the president's warning, military projects with names and acronyms like Trident, ABM, Minuteman, and Starwars proliferated. In the military sphere, a "baroque arsenal" emerged.[161] The Manhattan Project, with its systematic linking of military funding, management, and contract letting, industrial, university, and government research laboratories, and numerous manufacturers, became the model for these massive technological systems. Earlier models for system building, such as Ford's, Insull's, and the TVA's, paled in comparison.

COUNTERCULTURE AND MOMENTUM

From the late nineteenth century until the end of World War II, Americans commonly considered invention, industrial research, and systems of production the sources of goods for the good life and an arsenal of weapons for the great democracy. A reflective minority lamented the rise of the grim industrial city, poison gases, and bombing raids, the monotony of the assembly line, and the unemployment caused by the replacement of workers by machines. Since the early nineteenth century, those of a more philosophical and critical cast of mind found technology spreading a spirit of materialism.[1] Generally, however, the attitude of Americans toward technology was one of enthusiasm until after World War II, when the temper of the times began to change markedly. Engineers, managers, financiers, workers, the military, and others dependent on modern large-scale production technology for livelihood and status continued to take a positive attitude toward technological change, and even spoke of progress in ways reminiscent of the nineteenth century. Nonetheless, countless others began to ask what had been lost and what new dangers had appeared as humans increasingly invented, organized, and controlled their material world. In the early 1970s a perceptive observer could write, "Contemporary society is characterized by a growing distrust of technology."[2]

Some realized that the inventing, organizing, and controlling were not limited to the world of the production of goods, the so-called material world, but involved services and information as well. A few began to realize that technology was a more embracing concept and activity than industry or economics, and that modern technology was becoming increasingly organized into large systems involving technical and organizational components. In the preceding chapters, we have followed the evolution of a number of such systems, including electric light and power, the automobile, and the production system for atomic bombs. The increasing awareness of the destructiveness of atom bombs and the threat that their proliferation posed for the future of civilization greatly stimulated among the public a counterreaction to technology. John Hersey's book titled simply *Hiroshima* (1946), with its graphic description of the horror of the bombing, slowly seeped into public awareness. Four decades later Jonathan Schell, in a moving and profound work, brought a realization that, with the bomb, fallible mortals with hair-trigger controls decide *The Fate of the Earth.*[3] Rachel Carson's *Silent Spring* (1962) called attention to the loss of natural sounds, smells, and vistas as the man-made systems of production, with their toxic substances, displaced nature. Barry Commoner, in *Science and Survival* (1966) and *The Closing Circle* (1971), wrote of an environmental crisis caused by human beings driving to conquer nature in order to produce material wealth. A series of destructive, offshore oil spills and smog alerts in polluted cities served as graphic reminders of the vulnerability of the natural and urban environments. The use by the U.S. military, as John McDermott put it, of capital- and managerial-intensive technological systems to lay waste Vietnam only heightened the public's anger toward, and anxieties about, technology. McDermott wrote about a system of destruction consisting of aircraft, rockets, bombs, shells, technical specialists, pilots, bombardiers, radar operators, computer programmers, accountants, and engineers.[4] Thoughtful Americans could no longer glibly associate technology with incandescent lamps, Model T's, and "better things for better living."

Whereas Hersey, Carson, and Commoner focused on particular disorders in the technological world, several authors in widely read books attacked the foundations of the technological society. They believed that the rational values of the technological society posed a deadly threat to individual freedom and to emotional and spiritual life. Theodore Roszak's *The Making of a Counter Culture: Reflections on the Technocratic Society and Its Youthful Opposition* (1969) presented the thoughts of those like Jacques Ellul, Herbert Marcuse, Norman Brown, Allen Ginsberg, Alan Watts, Timothy Leary, and Paul Goodman, who, in the name of a higher culture, were undermining the foundations of a culture dominated by systems of production and destruction. Roszak wrote of "the technocracy," an enemy for him and the like-minded young, far more formidable than emergencies like the Vietnam War, or chronic problems of racial injustice and poverty. In his view, technocracy was a system that knit industrial society, that arose out of rationalizing, planning, and modernizing. Technocracy involved the large-scale coordination of people and resources and responded to an imperative quest for efficiency. Engineers, scientists, managers, and entrepreneurs—the experts—were orchestrating the total human and industrial context.[5] Roszak's technocracy resembles the large technological systems we have considered. *The Greening of America* (1970) was Charles Reich's contribution about a revolution of the young against the values of technology. He asked whether Americans could develop a new consciousness that placed humanistic values above the values of a technological culture. Reich argued that technology increasingly shaped society. "What we have," he wrote, "is technology, organization, and administration out of control, running for their own sake. . . . And we have turned over to this system the control and direction of everything—the natural environment, our minds, our lives."[6] Reacting against what he took to be the values of a technological society, he called for a new reality that would remove the pressures of time constraints, schedules, and rational connections.[7] He might have sworn eternal enmity to system builders like Frederick Taylor, Henry Ford, Samuel Insull, and Admiral Hyman Rickover. In *Small Is Beautiful* (1973), E. F. Schumacher found an expression that epitomized the reaction against the large-scale systems of production.

Herbert Marcuse, in *One-Dimensional Man* (1964), brought to an audience of Americans the insights of the Frankfurt school of German philosophers and sociologists, who had long analyzed a social and political world organized systematically for production. He argued that the systems of production in modern capitalist and socialist societies repress the spirits and constrain the freedom of individuals. Highly organized, hierarchical production systems deny workers influence in the work process. Craftspersons cannot express their creativity and skill in the things they make; the worker on the production line is as much alienated from his or her work as is a cog in a machine. Marcuse was seen as one of the shrewdest critics of subtle technocratic regimentation and domination prevailing in the industrial nations, both capitalistic and socialistic.[8] Jacques Ellul, in a similar vein, wrote of *The Technological Society* (1964) and *The Technological System* (1977), and Lewis Mumford probed the depths of technological society in *The Myth of the Machine* (1970). Those critics examining the foundations of technological society were, like the European artists earlier in the century, making another discovery of America. Instead of seeing and sentimentalizing the United States as essentially a nation of democratic politics and free-enterprise economics, the writers and philosophers of a counterculture probed the depth and extent of the mechanization and systematization of America. They tried to comprehend, for the individual and society, the deeper meaning of the spread of large technological systems. They asked what problems arose from the man-made characteristics of built America, from—as Frederick W. Taylor said—no longer putting man first, but putting the system first.

MUMFORD AND THE MEGAMACHINE

Mumford spread seminal ideas among those who sought to understand their technological society. As we have seen, he experienced the technological enthusiasm of the interwar years and believed in the coming of a neotechnic age of technology. Then, during the 1960s and 1970s, he shared with many others a widespread disillusion with technology. Russell Jacoby has aptly called him a "public intellectual," a thinker and writer who addressed a literate, general audience, a person free of university and other institutional ties who chose the problems about which he wrote on the basis of what he valued as socially significant rather than on the basis of academic acceptability.[9] Having earlier written about American material, literary, and artistic culture in the nineteenth century, Mumford turned in the 1920s and 1930s to studying and writing about technology and regional planning. As a result, through him we have a window on characteristic problems of this century and an arresting and idiosyncratic array of proposed solutions. He sought to find a usable past and, in that and in utopian visions, a critical stance to allow him to take the measure of modern times. The titles of his nearly thirty books reveal his persistent concern with the individual and the society existing in a human-made world. We find in these titles references to technological utopias, technics and civilization, the culture of cities, art and technics, the pentagon of power, and the myth of the machine.

After training in radio operation as an enlisted man in the navy, without a college degree, Mumford independently pursued his literary and cultural interests. Having abandoned a youthful desire to be an electrical engineer, he entrenched himself in literary and bohemian New York, which, as we know, was tinged with an enthusiasm for discovering and defining the technological culture. Though his first books were cultural and literary histories, he never lost his youthful fascination with technical things, and he acquired a conviction that things and ideas technological must be an integral part of modern culture. Influenced by both his cultural and his technical interests, he wanted technology to be organic rather than mechanistic. He shared philosopher Alfred North Whitehead's "romantic reaction" against the "mechanical philosophy."[10] Like Samuel Taylor Coleridge, he saw organic forms evolving under the influence of forces shaping them from within. External forces, in contrast, shaped mechanistic forms. Coleridge believed that mechanical forms often did not suit the innate characteristics of the things on which they were imposed.[11]

Besides *Technics and Civilization* (1934), Mumford wrote *The Pentagon of Power* (volume 2 of *The Myth of the Machine* [1970]), about the history of technology. In *Technics and Civilization,* he expressed his belief that the second industrial revolution could bring a more organic technology. The good society would arrive when persons motivated like pure scientists used technology to create a "Green Republic."[12] "We must turn society," he wrote early in the 1930s, "from its feverish preoccupation with money-making inventions, goods, profits, salesmanship, symbolic representations of wealth to the deliberate promotion of the more humane functions of life."[13] The reader of *Technics and Civilization* then could hardly have anticipated Mumford's profound pessimism forty years later in *The Pentagon of Power.* He no longer envisioned the control of technology passing to men free of avarice and anger. During the writing of *The Pentagon of Power,* he was "driven, by the wholesale miscarriages of megatechnics, to deal with the collective obsessions and compulsions that have misdirected our energies, and undermined our capacity to live full and spiritually satisfying lives."[14] By 1970 Mumford had witnessed the spread of large and more complex technological systems, especially the military-industrial production complex. The use of physics, once considered pure, to design the bombs that destroyed Hiroshima and Nagasaki appalled him. He lost hope that "pure" scientists and other persons with their disinterested values would take over control of technology to derail the megamachine. The immediate cause of his pessimism, the news of Hiroshima and Nagasaki in 1945, so stunned him that discourse became impossible for him for days afterward.[15] The mammoth organization needed to make the bombs was sufficient evidence for him of the involvement of the physicists in a horrendous megamachine. No longer did he expect them to humanize technology. He wrote, "to become the 'lords and possessors of nature' was the ambition that secretly united the conquistador, the merchant adventurer and banker, the industrialist, and the scientist, radically different though their vocations and their purposes might seem."[16] Science was becoming increasingly irrelevant to human intentions other than those of the corporate enterprise or the military establishment. Scientists no longer simply sought order in the universe and reported what they found; rather, they selected those aspects of nature that could be organized systematically into a mechanistic world view. They confused what they thought to be their higher order of reality with a sterile higher order of abstraction.[17] Galileo Galilei became the symbol for Mumford of the many scientists who had transformed a complex world—Baroque, for Galileo—into a quantitative, objective, sterile wasteland. He wrote of the crime of Galileo.

As he wrote the *The Pentagon of Power*, the concept of the "megamachine" became as much an obsession with him as organicism had earlier. He saw megamachines as shaping the course of human history, as deadening life insurgent, and as likely to bring civilization to a violent end in a nuclear holocaust. In *Technics and Civilization* the word "machine" had begun to take on new meaning for him, one that anticipated his later use of "megamachine." He began to distinguish between "machines," or specific objects like the printing press or power loom, and "machine," a shorthand reference to the entire technological complex, embracing tools, machines, knowledge, skills, and arts. The machine was becoming for him the physical embodiment of the mechanistic means to ends, or a technological system. When the idea had taken form as the megamachine, it had become a technological system of interchangeable parts, inanimate and animate, centrally organized and controlled, dependent on a priestly or scientific monopoly of knowledge, and ensuring the power, glory, and material well-being of an elite. The megamachine imposed purely mechanical forms on every manifestation of life, "thereby suppressing many of the most essential characteristics of organisms, personalities and human communities."[18] In reducing life to quantitative, mechanical, and chemical components, human beings denied life.

Mumford believed that the megamachine originated in Egypt, and that such machines had subsequently existed throughout history. The Egyptians, he concluded, had organized a gigantic machine, or system, its parts made up mostly of humans. The chief system builder was the deified ruler. Loyal scribes, messengers, stewards, and gang bosses managed his hordes of labor. Observations by priestly astronomers of the sun-centered heavenly regularity became the metaphoric justification for the absolute authority of the ruler.[19] Today, modern, bureaucratically administered, military-industrial projects, such as the Manhattan Project, are megamachines, but with a critical distinction from the older ones. In the early megamachines the center of authority lay with the absolute ruler; in the modern megamachine authority is centered in the system itself. Those who control megamachines and those who are regimented by them share the mechanistic, power-centered world view. The myth of the machine for Mumford embodied the widely held, mistaken belief that megamachines are " 'absolutely irresistible—and yet ultimately beneficial,' provided one does not oppose them. 'That magical spell still enthralls both the controllers and the mass victims of the megamachine today.' "[20]

Captivated by the idea of the megamachine, Mumford verged in his later writings on technological determinism, a philosophy he had previously studiously avoided. To describe contemporary society, he used the metaphor of an automobile, filled with passengers and without a steering wheel, rushing downhill toward an abyss. Yet his despair was not unalloyed. Usually he denied that technology invariably shaped society, arguing instead that technology was a part of the culture shaped by values. The machine, he believed, was usually the product of consciousness, not the creator of it.[21] Thus philosophically armed, he cold muster glimmers of hope for the future of the human race. He insisted that he was not a "prophet of doom." "On the contrary," he wrote, "the whole effort of my work is to diagnose, at an early stage, the conditions that may, if they are uncorrected, undermine our civilization."[22] Since, for instance, the forces of violence and destruction embodied in the military-industrial production complex manifested long-lived misanthropic values, these forces, despite their momentum, could be countered if there were a revolution in values during a neotechnic era.

ELLUL'S TECHNOLOGICAL SYSTEMS

Jacques Ellul, a French philosopher and cultural critic, also explored the implications of megamachines, or technological systems.[23] Ellul's *The Technological Society* (1964) also shaped attitudes among those reacting against "the system" and seeking to establish a counterculture. All-embracing technological systems had swallowed up the capitalistic and socialistic economies, and were, for Ellul, a far greater threat to our freedom of action than authoritarian politics. Denying the primacy of politics, Ellul argued that the state is no longer as influential a factor in shaping lives and history as technological systems. For him, political activity is unreality; technology is reality.[24] The order, method, neutrality, organization, and efficiency of technological systems have transformed humans, he believed, into a technicized component of technological systems, a component devoid of will, and analogous to one in a machine or megamachine.[25] Ellul explained that modern organizations integrate so well into technological systems because the drive for efficiency characteristic of modern technology also prevails in modern organizations. This drive has transformed them, he pointed out, into bureaucracies whose characteristics harmonize with the technical developments and production activities over which they preside. Efficiency has transformed organization into bureaucracy.[26]

According to Ellul, technology has changed the natural and cultural environments by fragmenting their realities and then recombining the discontinuous fragments into technological systems that are operational, instrumental, and organized for problem solving. The old complexity of culture and nature has been replaced by one in which technology has become the material environment and the shaper of culture. Neither nature nor culture now determines social structure. Technological systems have become the determiner.[27] These systems mediate between humans and nature and between men and women. They are brought in contact with one another through the connecting links, or networks, of the systems. Young people today, he insisted, are not liberally educated but trained to function in technological systems. Since they are not liberally educated, they have no basis from which to criticize the systems that embrace and direct them.[28] Ellul believed that technological systems not only make a new human environment, but also modify people's very essence. Humans must adapt themselves, as though the world were new, to a universe for which they were not created. Today we can no more dream of challenging the technological milieu than a twelfth-century man could have dreamed of objecting to trees, rain, a waterfall. Humans, he added, have "no intellectual, moral, or spiritual reference point for judging and criticizing technology."[29]

Ellul is a technological determinist influenced by Darwinian environmentalism. Charles Darwin thought that a natural environment selected characteristics and shaped natural organisms. Social Darwinians in the late nineteenth century, generalizing from Darwin, argued that the social environment selected the fittest humans to survive. Ellul, seeing the technological environment displacing the natural one and subordinating the political, economic, and social, thought that technological systems shape the character of human beings. He summarized his position by stating that the new technical milieu is artificial, autonomous, and self-determining; that its growth is driven by accumulated means rather than ends. The milieu is a system because all its parts are so mutually interactive that it is impossible to separate them. The technical component is inseparable from the others as well.[30]

Ellul believed that we are concerned today about problems arising in our technological society that superficially embody deeper, less easily perceived ones. These include, for example, air pollution and urban congestion, which he was confident would be solved as the technological systems evolved. He was also not concerned that technological systems would corrupt our morals. Instead, the systems would simply deny us the essential moral choices. The real problem for him was whether humans can master the technological systems and lift the growing burden of technological determinism. About this he was doubtful. Humans have already been "technicized." But politicians do not understand technological systems well enough to control them, and scientists and engineers are so specialized that their thinking cannot embrace the scope of technological systems, with their interacting technical, political, economic, and social components. He did not expect Marxists to master the technical milieu, because they embrace it uncritically.[31]

Ellul suggested that we, like Esau, are selling our birthright for a mess of pottage, that the price we pay for a cornucopia of goods and services is slavery. We fail to see that, as technology solves problems, it also creates them, as in the case of automobiles, congestion, and pollution. Furthermore, we cannot choose good technology and reject bad, for these qualities are inseparably intermixed within technological systems. He also lamented our enthusiasm for technology. Sputniks, he wrote, "scarcely merit enthusiastic delirium," and "it is a matter of no real importance whether man succeeds in reaching the moon, or curing disease with antibiotics, or upping steel production. The search for truth and freedom are the higher goals. . . ."[32] Ellul was not referring to political freedom but to freedom from the deterministic forces of technology, especially those stemming from the momentum of large technological systems of production, communication, and transportation.

APPROPRIATE TECHNOLOGY

In the late 1960s and early 1970s the books and articles of Ellul, Mumford, Marcuse, Roszak, and others critical of technological systems and concerned about their deterministic nature influenced the generation of activists who in the mid-1960s had focused on civil-rights issues and the destructiveness of the Vietnam War. Some of this generation of activists began to see modern technology, especially large technological systems, as the common cause of the cultural and social maladies about which they were protesting. Chemical corporations producing the napalm and destructive herbicides used in Vietnam also made, for example, the pesticides, herbicides, and pollutants fouling the environment at home.[33] Corporation heads presiding over large systems of production shuttled back and forth to the Department of Defense, where they managed systems of destruction. Talk of efficiency, order, centralization, and systematization was common to the military and to industry. As the activists became disillusioned by the unceasing organizing, campaigning, rallying, and frustrations of political action, or were intimidated and incapacitated by the countermeasures of the police, the army, and the courts, they turned increasingly to opposing "the system." They embraced decentralized and appropriate technology to counter large-scale technology.[34]

Indicative of enthusiasm for small-scale, appropriate technology was the popularity of *The Whole Earth Catalog* (1968), which identified the tools of a benign technology useful for shaping a new environment.[35] Assuming that hordes of people would move into small, self-sustaining communities, the editors of the catalogue described utensils and small machines that were high-quality, low-cost, and nonpolluting, that could be operated, maintained, and repaired by the lay person, and that were easily available by mail. Many of the items had been used by craftsmen and small farmers in nineteenth-century America. Langdon Winner has characterized *The Whole Earth Catalog* as expressing a vision of "a groovy spiritual and material culture in which one's state of being was to be expressed in higher states of consciousness and well-selected tools."[36] Many practitioners of the *Whole Earth Catalog*'s philosophy were trying, like the Amish, to reduce their dependence on large manufacturing and utility systems. By continuing to use draft animals, burning kerosene lamps, and utilizing watermills and windmills, the Amish and others who wished to live like them were cutting their ties with the networks of power and manufacturing.[37]

An article published in 1976 in the influential U.S. journal *Foreign Affairs* combined the aspirations of the appropriate-technology advocate and the closely argued, factual style of the scientist to make specific policy recommendations in the spirit of philosophical authors such as Mumford and Marcuse. Written by Amory B. Lovins, a British physicist and the British representative of Friends of the Earth, Inc., the article responded to the energy crisis of the decade and to long-term technical, economic, political, and environmental problems with a proposal that the industrial nations, especially the United States, take a "soft path" rather than a "hard path" in meeting future energy needs. Lovins defined the hard path by drawing on projections of government agencies, energy corporations, and electric-utility research institutes. The hard path envisaged the large utility systems' prevailing into a limitless future. Over the next half-century they would phase out oil and gas as their primary energy source for the generation of electricity and move toward a greatly increased—"massive"—supply of electricity generated from nuclear energy and coal. The soft path urged by Lovins called for "an alternate . . . future" that would phase out large electrical-supply systems, displacing them with small, decentralized sources of renewable energy by using soft technologies such as wind, sun, and vegetation. Soft technologies would be easily understood, usable without esoteric skills, and "accessible rather than arcane." Soft-energy technologies, Lovins stressed, are not "vague, mushy, speculative, or ephemeral, but rather flexible, resilient, sustainable and benign."[38] They flourish when people are thrifty, neighborly, humble, and dedicated to craftsmanship. Hard technologies are the creatures of large corporations, government agencies, and incremental projections of current practice. They are sustained by the likes of Samuel Insull.

Buttressing his argument with his easy familiarity with energy processes, Lovins pointed out that large power stations inefficiently convert three units of fuel into two units of waste heat and only one unit of electricity. Half the electricity bill is fixed electricity-transmission and distribution costs. Electric-power plants use massive amounts of energy to transform the primary energy of oil, gas, or coal into the higher form of electricity, which is then converted back into lower forms such as heat and mechanical force. Lovins asked, why not eliminate the costly transformation, transmission and distribution, and have consumers take energy directly from small, local sources such as windmills and water mills, solar generators, and low-temperature thermal-combustion processes? As an even more dramatic solution, he urged that energy consumption be reduced by a host of conservation measures and "technical fixes." Among the fixes were heat pumps and the more efficient burning of coal by the use of fluidized-bed combustion. He saw the possibility that all central generating stations could be eliminated but, instead of taking this radical approach, he recommended merely a reduction in their number and their increased resort to cogeneration, by which waste heat from industry would be used as an energy source by central generating stations.

Numerous technical, economic, political, and social benefits would come from following the soft path. Engineers and skilled labor would have interesting problems to solve in developing the new technologies. National resources saved could be used to fulfill other social objectives. The horrendous risks of nuclear energy need not be assumed. The proliferation of nuclear energy and the dangerous spread of nuclear weapons could be countered. The recommendations of the environmentalist would be addressed. Urban and industrial congestion would be reduced. Hierarchically organized, large-scale technology, which, according to Ellul, Mumford, and Marcuse, oppressed and dominated the individual, would be displaced. Social diversity and freedom of choice would increase.

Lovins is Samuel Insull stood on his head. Only about a half-century earlier, Insull, as head of Chicago's electrical utility and a vast electric light-and-power holding-company empire, had been seen as a sage exponent of technological growth and social betterment. He argued persuasively for centralized mass production of electricity. In the name of efficiency he and his engineers built larger and larger power stations. They expanded their networks for distributing and transmitting power over entire regions, because improved load factor would lower the costs of electricity. Insull waged a successful campaign to have industrial consumers give up their small power plants and instead connect to the spreading networks of power owned by him, his utilities, and his holding company. Few contemporaries then challenged the validity of the Insull argument that designated large-scale systems the appropriate technology.

Lovins insightfully perceived the conservative momentum of existing large-scale systems such as those presided over by Insull. Unlike technological determinists, who expected a new technology to bring dramatic social changes inexorably in its wake, he realized that, even though a new technology might be feasible, the attitudes and values that could bring about its use in the way reformers desired and prophesied might not prevail. He asked, quoting the popular comic-strip character Pogo, "so why do we stand here confronted by insurmountable opportunities?" The answer according to Lovins was, "apart from poor information and ideological antipathy and rigidity . . . a wide array of institutional barriers," including obsolete building codes, "an innovation-resistant building industry . . . opposition by strong unions to schemes that would transfer jobs from their members to larger numbers of less 'skilled' workers . . . fragmentation of government responsibility, etc."[39] "Any demanding high technology," he observed, "tends to develop influential and dedicated constituencies of those who link its commercial success with both the public welfare and their own." Those who have acquired the skill and professional techniques to manage and expand the systems are reluctant to acquire the new knowledge and skills that radically new technology demands. Money and talent invested in a large technological system give it "disproportionate influence in the counsels of government, often directly through staff-swapping between policy and mission-oriented agencies."[40]

The influence of the hard-path advocates notwithstanding, a number of organizations, some of them government-funded, began in the early 1970s to experiment with the use of soft and appropriate technology. Long-established and newly founded organizations dedicated to solving environmental, or ecological, problems applied themselves to ones involving both nature and large-scale technology. Commoner, in *The Closing Circle* (1971), supported this approach by arguing that advanced technology appeared to be the prime cause of environmental problems.[41] Unorthodox engineers and scientists contributed their expertise to inventing and demonstrating technology that was "soft" and characterized as ecologically benign, energy-saving, and small-scale. They intended that this technology displace capital-intensive, polluting, centralized, mass-production technology. Schumacher's *Small Is Beautiful* (1973) inspired many of the advocates of appropriate technology.[42] Other books appeared that heartened those who saw the dawn of a new age. Perhaps stimulated by the possibility that appropriate technology would counter the spread of oppressive technological systems, some of the writers who had been pessimistic about the future of the technological society wrote hopeful books. Five years after *One-Dimensional Man*, Marcuse published *An Essay on Liberation* (1969). Roszak's *Where the Wasteland Ends* (1972) proposed an alternative society to counter the social maladies described in *The Making of a Counter Culture*.

Alvin Toffler, in *The Third Wave* (1980), predicted a breakdown of the technological systems.[43] In a widely read and influential book, he foresaw the passing of the "second wave" of history, which corresponded roughly to the era of the British Industrial Revolution and the second industrial revolution in the United States and Germany. He welcomed the "third wave," which began about 1955, when white-collar workers in the United States began for the first time to outnumber blue-collar ones.[44] He defined waves by the prevailing technology, associating the third wave, for example, with the so-called space, electronic-communication, and computer revolutions. Each wave has its characteristic social arrangement and values. The second wave was one of standardization, specialization, synchronization, concentration, maximization, and centralization.[45] These are the same principles we have seen applied by system builders in the twentieth century. Toffler argued that these principles shaped both the society and individual behavior. Designating system builders "the integrators," he broadened the category to include the managers of large corporations. The third wave would, he believed, counter the specialization, synchronization, and other second-wave characteristics. It would substitute for them a holistic approach to problem solving, "flextime," the rise of small units of production, customized production, and a dispersion of managerial authority. He stressed the rise of the electronic cottage, where the white-collar person, working in a well-wired home, interconnected electronically with other workers, drew on central information storage, and became involved in teleconferences. Physical labor in industry would increasingly be done by robots. Toffler's book contributed greatly to the spread of these ideas about the future. These predictions are like those made in the 1920s by Ford, Mumford, and other enthusiastic supporters of the second industrial revolution. They believed, as we have seen, that electric power transmission and distribution would end urban congestion and distance the mine and the power plant from the small communities in which decentralized industry would take root. In the case of the third wave, the technology is at hand, as it was in the case of the second industrial revolution. Technological enthusiasts throughout history have tended to be utopian, assuming that available technology will be used to fulfill their particular visions of the future to establish a postmodern technology.

In architecture, too, a reaction against a modern, technological style took shape. Not essentially a rejoinder to the mass production of goods and the large technological systems that produce them, an architecture soon called postmodern provided a reaction, nonetheless, against values shaping the modern, or international, style of architecture, values shared with modern technology. Robert Venturi, an American architect, inspired many others of the profession to reconsider their commitment to the prevailing international style. In 1966 he published an influential manifesto the title of which, *Complexity and Contradiction in Architecture,* indicated his attitude. Architects, he insisted, should not be intimidated by the pure, clean, straightforward language of both modern architecture and modern technology. Unlike Behrens, Gropius, Le Corbusier, Mies, and other founders of the international style, his references were not to the values and principles of engineers, especially to Taylorism and Fordism. Instead, Venturi referred to the ultimate inconsistency in the contemporary world. He cited the mathematics of Gödel's proof, the T. S. Eliot analysis of difficult poetry, and the paradoxical quality of the painting of Joseph Albers. Unlike Mies, who argued that less is more, Venturi wanted an architecture that was rich and ambiguous, for he believed that less was a bore. He rejected the simplistic functionalism so popular among unimaginative disciples of the pioneers of modern international-style architecture. Venturi's distance from the industrial vocabulary, the grain elevators, and the factories of the early modern architects became even clearer when he, with Denise Scott Brown and Steven Izenour, published *Learning from Las Vegas* (1972), in which they asked for a consideration of signs and symbols other than those of the modern technological culture. They contemplated the values and aspirations of a popular, vernacular culture, feelings seeking expression as architecture and art. They believed that the architect should learn from the existing land- and urbanscape. The Italians of the Renaissance, for instance, had harmonized the vulgar and the Vitruvian. In the 1970s other architects, such as Michael Graves, James Stirling, Charles Moore, and Robert Stern, also helped a postmodern architecture to take root. Postmodern architecture, though springing from countercultural soil, was more clearly articulated than a postmodern technology or the postindustrialism championed by leaders of the counterculture.

MOMENTUM

A grievous flaw in the reasoning of enthusiasts for radically new technology, as contrasted with that of the advocates of postmodern architecture, lies in the former's failure to take into account how deeply organizations, principles, attitudes, and intentions, as well as technical components, are embedded within technological systems. Architecture and art do not seem so burdened by the past. Unlike Lovins, champions of a new style of technology fail to recognize the inertia, or conservative momentum, of technological systems. As we have seen, large-scale technology, such as electric light-and-power systems, incorporated not only technical and physical things, such as generators, transformers, and high-voltage transmission lines, but also utility companies, electrical manufacturers, and reinforcing institutions such as regulatory agencies and laws. Similarly, the inertia of the system producing explosives for nuclear weapons arises from the involvement of numerous military, industrial, university, and other organizations as well as from the commitment of hundreds of thousands of persons whose skills and employment are dependent on the system. Furthermore, cold war values reinforced the momentum of the system. Disarmament offered such formidable obstacles not simply because of the existence of tens of thousands of nuclear weapons, but because of the conservative momentum of the military-industrial-university complex. During the century of technological enthusiasm, we have watched Edison's Pearl Street supply system swell until it covered cities and then regions; we have seen Ford's system of production spread its network over a continent and overseas; and we have observed the scale of atomic energy escalate from the laboratory to a massive industrial complex. The number of persons involved has grown from hundreds to hundreds of thousands; investment of money from tens of thousands to billions of dollars. Massive systems, then, have a characteristic analogous to the inertia of motion in the physical world. Their mass of technical, organizational, and attitudinal components tends to maintain their steady growth and direction.[46]

A century ago, Karl Marx pointed out the ways in which vested interests, especially capital, shape the course of history. In fact, large technological systems represent powerful vested interests of another kind. Numerous persons develop specialized skills and acquire specialized knowledge appropriate for the system of which they are a part. A major change in the characteristics of the system or its demise would de-skill these people. The machines, devices, and processes in the system are the capital, but a special kind of hardware capital with characteristics that might be called "system specific." Changes in the system also make hardware capital obsolete. Faced with these eventualities, the people and the investors in technological systems construct a bulwark of organizational structures, ideological commitments, and political power to protect themselves and the systems. Rarely do we encounter a nascent system, the brainchild of a radical inventor, so reinforced; but rarely do we find a mature system presided over by business corporations and government agencies without the reinforcement. This is a major reason that mature systems suffocate nascent ones.

A number of historians, sociologists, and philosophers have identified the conservative momentum of technology, but they use different names for it. Mumford referred to the oppressive character of megamachines; Ellul wrote of the determinism of technological systems. John Kenneth Galbraith, an economist, has described the mammoth "technostructures" into which modern industrial society is organized.[47] The modern historian William McNeill, in *The Pursuit of Power* (1982), described the way in which military-industrial systems interweave politicians, newspaper editors, and a host of other persons and interests. Walter A. McDougall, in . . . *The Heavens and the Earth: A Political History of the Space Age* (1985),[48] related how the resources and authority of the state and the physical and organizational power of technology have been bound, with momentous social consequences, into a technocracy. McDougall believes that the technological enthusiasts and power-hungry among the politicians fuel the technocratic drive. We can see this in the race for supremacy in space.

CONTINGENCIES, CATASTROPHES, AND CONVERSION

In the face of this conservatism and momentum, what might bring the displacement of large-scale, centralized, hierarchically controlled production systems? What forces might counter the tendency of large technological systems to determine social change, even history? Will Toffler's electronic cottages become workplaces? Is promotion of soft technology rational in the face of overwhelmingly large hard-energy systems? What might bring the demise of the nuclear industry, or the breakup of the nuclear-military-industrial complex? In order to bring about a substantial change in the motion and direction of massive systems of production, such as electric light-and-power systems, a counterforce of comparable magnitude becomes imperative. Changes in circumstances comparable to those that cause the demise of organisms well adapted to, even shaping, their environment need to occur. To counter large technological systems, forces analogous to those that killed off the dinosaurs are needed. Like the dinosaurs, some technological systems have embedded in them characteristics that were taken on in times past, characteristics suited for past environments but not for the present. Because these characteristics are often embedded in the hardware of a technological system, they are especially long-lived. These anachronistic characteristics persist despite incremental changes in the environment that favor different characteristics. Only an overpowering change in environmental circumstances can kill off the new dinosaurs.

There are precedents for contingent environmental changes that can alter the course of the evolution of a large technological system. The oil embargo of 1973 and the subsequent rise in gasoline prices ultimately compelled U.S. automobile manufacturers to change substantially an automobile design that had been singularly appropriate to a low-cost-energy environment. Earlier, there had been model changes and alterations in the manufacturing processes, but nothing comparable to the change in product and the characteristics of the system for producing it that the sharp rise in gasoline prices precipitated. Detroit responded to the new market environment when Americans began to buy Japanese and German automobiles designed appropriately for the relatively high gasoline prices in those countries. Because the market is a major environment for systems, changes in the market might bring changes in the character of systems of production. As we have observed, systems of mass production in the United States, both military and nonmilitary, expanded rapidly during a century when there was a growing market for consumer goods and, because of war and cold war, an expanding market for military goods. If these markets should contract, then the circumstances that promoted and sustained mass production would become less influential and from such changed market circumstances systems of production less hierarchical and less large might arise.

In recent decades, well-publicized technological catastrophes have caused public reactions, precipitating changes in attitudes and goals that might also counter the momentum of large technological systems. The large coastal oil spills and the urban smog alerts of the 1960s stimulated a rising tide of public sentiment that brought passage of environmental laws modifying industrial practices. Environmental protection agencies, private and public, have gathered momentum as ecological accidents occur. The nuclear-reactor catastrophe at Three-Mile Island in 1979 heightened public concern and stimulated regulations and controls that have dampened the spread of nuclear power. The *Challenger* space shuttle tragedy of 1986 temporarily sidetracked, if not derailed, NASA, a huge government-industrial system infiltrated by the military.

Fortunately, the remedial actions taken, along with a chance convergence of circumstances at the nuclear power plant at Three-Mile Island, avoided the ultimate disaster, but the nation and the world were introduced as never before to the complexity and accident-prone character of large and complex technological systems. The Three-Mile Island disaster has been labeled a "normal accident." If normal accidents continue, they may mobilize public opinion against large, centralized, hierarchically controlled systems. Sociologist Charles Perrow, who served as a consultant to the U.S. presidential commission that investigated the Three-Mile Island episode, wrote:

> As our technology expands, as our wars multiply, and as we invade more and more of nature, we create systems—organizations, and the organization of organizations—that increase the risks for the operators, passengers, innocent bystanders, and for future generations. . . . Most of these risky enterprises have catastrophic potential. . . . Every year there are more such systems.[49]

In the interaction of multiple failures within a matter of seconds at Three-Mile Island, and in the incomprehensibility of their interaction, we can interpret the episode as representative of past and future failures among large, tightly knit technological systems. Such multiple interactive failures are most likely to occur among tightly coupled systems with a high level of interconnectedness. These include nuclear plants; high-voltage electric-transmission systems, or grids; nuclear weapons systems;

EXPLODING BALL OF GAS FROM SPACE SHUTTLE *CHALLENGER* DISASTER, 28 JANUARY 1986

***CHALLENGER* FRAGMENT**

space missions; aircraft-control systems; and chemical plants.[50] The managers of these often acknowledge that they cannot comprehend the systems in all of their complexity, a complexity that tends to increase incrementally.[51]

Three-Mile Island was presented to the public and is remembered as a nuclear-reactor accident, but in fact it was an accident in an electric-power system. Nuclear reactors, after Shippingport, became the steam-generating boilers in electric-power systems, replacing coal- and oil-fired boilers with uranium-fired ones. A private electrical utility, Metropolitan Edison, operated the power station, and a manufacturer of electrical-power-station equipment, Babcock and Wilcox, designed and manufactured the steam generator. Designed to function in an electric-power system, the reactor had technical characteristics that harmonized with the characteristics of other components in the system. A century earlier, Thomas Edison had written that each component in an electrical system must be designed with all the others in mind. The interaction of multiple failures at Three-Mile Island was not simply the malfunctioning of a nuclear reactor. The disaster occurred within a system that was one of the major technological achievements of the twentieth century, an electric light-and-power system. The Three-Mile Island failure and catastrophe followed by only a few years a previous massive power-system failure, the New York blackout of 1965. The shuttle disaster, the Three-Mile Island catastrophe, and the blackout were all, in Perrow's terminology, "normal accidents" in large technological systems.

The space-shuttle tragedy revealed, in the network of responsibility and action, the complex array of astronauts, O rings, engineers, managers, government departments, and industrial corporations involved in the NASA system. Similarly, the Chernobyl disaster of 1986 showed that the Soviet nuclear-power plants and electrical-supply network are parts of large and complex technological systems that include political and economic institutions organized to supply power for industrial and other purposes. Societal values calling for frequent, economic, and newsworthy public-support-garnering launching, in the case of NASA, and, in the case of the Soviet energy system, values calling for a rapid increase of low-cost energy in the name of national achievement, are integral parts

THREE-MILE ISLAND NUCLEAR POWER PLANT DURING THE ACCIDENT ON THE NIGHT OF 28 MARCH 1979

of these technological systems. Earlier we encountered the misuse and overuse of tractors and other equipment in the Soviet Union so that production goals could be reached. The failure of the Soviets to place containment vessels over reactors at Chernobyl provides another instance of an effort to obtain the largest possible output from a system with as low an input of resources as possible. The Chernobyl and *Challenger* disasters reflect national values, not simply technical inadequacies.

What might break the momentum besides contingencies and catastrophes, or "normal accidents"? Mumford in *Pentagon of Power* and Roszak in *Where the Wasteland Ends* called for a change in belief, attitudes, and intentions comparable to a religious upsurge, or a religious conversion. Mumford, we should recall, argued that values shape technology. Writing early in the 1970s, Mumford and Roszak believed that society, especially young people, might transcend materialism and reject power

UNIT 4 OF THE CHERNOBYL ATOMIC POWER STATION AFTER THE DISASTER ON 26 APRIL 1986

relations of megamachines and technological systems. We have already suggested that counterculture books and articles, like Mumford's, sowed the seeds of change. Attitudes that might generate a technology dedicated to objectives other than economic growth and national prestige may spread, especially if the mass market for consumer goods and services should become saturated. Values may change because activists-become-enthusiasts-for-soft-and-alternate-technology, we we have noted, have provided models for a new technology. There is evidence of institutionalization of attitudes and objectives other than those associated with mass production and mass destruction by means of large technological systems.

Influential changes in attitudes are discernible among managers and system builders, the present-day counterparts of Frederick W. Taylor, Henry Ford, and Insull. During the past decade or so a number of books by young American professors of management, professional writers and journalists interested in management, and young entrepreneurs have called for an end to the massive, centrally controlled (hierarchical), rigid systems of management and the production systems over which they preside. Instead, the new wave of managers advocates small-scale, decentralized (autonomous), flexible management and production complexes. They argue that the steady and predictable market growth that characterized the era of technological enthusiasm is passing. If they are American managers, they hold up more flexible Japanese, Swedish, and Italian models of decentralized modes of production that permit supple changes in response to market conditions. In addition, they maintain that, since the 1960s and the emergence of the counterculture, both white-collar and blue-collar workers, in order to fulfill their potential, seek—and need—an environment free of the Taylorite attitude that "the system must be first."[52]

Because of widespread anxiety about weapons of mass destruction, a change in attitudes and values that would counter the momentum of large military systems of production seems more likely, however, than one that would challenge systems for the mass production of consumer goods and services. The oppressive, even authoritarian character of massive systems of nonmilitary production that Mumford, Ellul, and others believed existed is less obvious to the general public than the dangers arising from the military-industrial system. The final congressional testimony of Admiral Rickover before his retirement in January 1982 suggests the kind of conversion envisaged by Roszak and Mumford. The father of the nuclear navy and of the first commercial nuclear reactor told congressmen that the country was not taking into account the potential danger of the release of radiation from nuclear-power plants. Then he added in his informal testimony:

> I do not believe that nuclear power is worth it if it creates radiation. Then you might ask me why do I have nuclear powered ships. That's a necessary evil. I would sink them all. Have I given you an answer to your question?

To which Senator William Proxmire responded:

> You've certainly given me a surprising answer. I didn't expect it and it's very logical.

When the senator added that it surprised him because he did not expect somebody so close to and so expert in nuclear power to point out its destructiveness, the admiral responded simply, "I'm not proud. . . ."[53]

The idiosyncratic reaction of the admiral should be set in the context of a large-scale public reaction to the destructiveness of modern military technology. Ineradicable memories of the horror of Hiroshima and Nagasaki continue to move people to protest. The use of high technology by the U.S. military during the Vietnam War revolted a large segment of the population, especially the young, and precipitated a sustained effort to end the war. Paul Goodman, the social critic who called himself a neolithic conservative,[54] made one of the most articulate and passionate attacks on the role of military-industrial systems in Vietnam. He spoke before the National Security Industrial Association—an unexpected invitation to address a most unlikely audience, one composed of representatives of the giant U.S. corporations.[55] The horrendous destructiveness of nuclear energy and the arguments of Goodman and others have affected attitudes; the ultimate effect of this change on the development of nuclear energy is difficult to see.

The history we have considered offers some support for Mumford, Roszak, and others who argue that changes in the intentions and goals of individuals and society can alter the course of technological development or can counter technological momentum. They are pointing out that society can shape, or socially construct, technology, and not only be shaped, or determined, by technology. This history has offered numerous examples of the construction of technology by social forces—economic, political, and other. When we considered the invention of technological systems, we observed how the widespread commitment to mass production of consumer goods—materialistic goals—shaped their design. The drive for national prestige and power embodied in the pre–World War I armaments race deeply influenced the inventive activity and the inventions of Elmer Sperry, the Wright brothers, and other independent inventors. An influential group in society can use technology to promote its interests.[56] In the case of Taylorism, or scientific management, managers often sponsored the invention and development of technology that increased their control of the workplace. There are more recent examples.[57] Workmen have struggled to prolong the use of machinery that embodies a rhythm and style of work that they find appropriate.[58] Historians and sociologists referring to these influences of society on technology now speak of "the social construction of technological systems."[59] History reveals an endless interaction of the forces, including attitudes and objectives, which socially construct technology with the forces of technological momentum which shape society. The question we are asking about the possibilities of change can be seen as a conflict between technological momentum and the social construction of technology. We should also note that technological momentum incorporates attitudes and objectives that shaped the technology in times past.

CONFLUENCE AND REVOLUTION

Despite the indications of change today, we have witnessed during the century of technological enthusiasm a steady increase in the momentum of large systems of production. The systems mature, grow large and rigid, then resist further social construction. These systems of production, military and nonmilitary, have, we have seen, become characteristic of modern society and have shaped a modern culture. There is a widespread unexamined assumption, especially in the United States, that the modern will project endlessly into the future despite contingency, catastrophe, and attitudinal change. Americans have coupled technological creativity with economic growth and mass production and assume that this relationship will endure. New systems emerge from centers of creativity linking universities and industry, such as Route 128 outside Boston and Silicon Valley in California, but the new systems, be they electronic and military or computer and industrial, tend to display the same patterns of growth and momentum. In the modern era, old systems do sometimes fade away, but even larger and more complex ones often displace them.

The most likely cause of a displacement of large, centrally controlled systems would seem to be a confluence of contingency, catastrophe, and conversion that would break the technological momentum and socially construct a new style of technology that would not be coupled with the mass production of private consumer and of military goods. Technology linked to mass consumption is a modern American hallmark, but not necessarily one that will survive the century of technological enthusiasm. In the United States of expanding markets and population and of democratic values, technological change and economic growth were tightly intertwined. Economic growth was a "policy-determining ideology."[60] Yet technology can be created and can be deployed that would heighten the quality of life, not simply the quantity of goods. Technology might be deployed more often to provide public services and common goods on a smaller scale and be especially designed for local circumstances and preferences. If this should happen, then massively large systems of production might not be appropriate. With the demise of the large systems so characteristic of modern America, a postmodern era would arrive.

A confluence of circumstances sufficient to break the momentum and change the characteristics of modern technology and modern culture needs to be comparable to the confluence that brought the first industrial revolution in Britain and the second that we have considered as it occurred in the United States. It is important to note that these momentous changes took place in different countries. Britain was unable to break the momentum of the first industrial revolution and become the seat of the second. The momentum of the modern may be so great in the United States that the next great technological and cultural change may occur among other peoples in another nation.

On the other hand, the next sea change might transcend nations and at the same time eclipse the forces tending to diminish the size and influence of large technological systems. In recent years we have seen the continued growth of multinational corporations and consortia. For example, Swedish firms such as ASEA, the electrical manufacturer, have merged with or bought out industrial giants in other countries and formed multinational systems of management, production, and marketing. Japanese government agencies, manufacturing enterprises, and financial institutions coordinate and spread systematically throughout the world.[61] Economic principles like economy of scale, economies of scope, and diversity of load that Insull, Ford, and other system builders applied primarily in the region and nation are being implemented by their system-building successors in an international arena. It will be fascinating to see if supranational systems will embody the controls, tight coupling, and hierarchical characteristics of modern systems or whether they will incorporate in ingenious and subtle ways values of the counter-, or post-modern, culture of recent decades.

NOTES

INTRODUCTION ■ THE TECHNOLOGICAL TORRENT

1. Perry Miller, "The Responsibility of Mind in a Civilization of Machines," *The American Scholar*, XXXI (Winter 1961–1962), 51–69.
2. I am indebted to Elaine Scarry of the University of Pennsylvania for the notion of the invention of the nation's "material constitution" and to Jaroslav Pelikan of Yale University for reminding me of the Genesis story as "Urmythus," of God as the first to practice technology (Genesis III: 21), and of Cain bearing the curse of work and creation (Genesis III: 23).
3. Thomas P. Hughes, "The Order of the Technological World," *History of Technology*, V (1980), 1–16.
4. Casey Blake, "Lewis Mumford: Values over Technique," *Democracy* (Spring 1983), pp. 125–37.
5. Martin Heidegger, *The Question Concerning Technology and Other Essays*, trans. W. Lovitt (New York: Harper & Row, 1977), p. 19. In *American Genesis* I have concentrated on the means of production, especially the mechanical and electrical. Also of surpassing importance during the age of technological enthusiasm were the great works of civil engineering. See, for instance, David McCullough, *The Great Bridge* (New York: Simon & Schuster, 1972), and McCullough, *The Path Between the Seas: The Creation of the Panama Canal, 1870–1914* (New York: Simon & Schuster, 1977).
6. Sidney W. Mintz, "Culture: An Anthropological View," *The Yale Review*, 71 (1982), 499–512.

7. Alfred North Whitehead, *Science and the Modern World* (London: Free Association Books, 1985; first ed. 1926), p. 120.
8. Charles A. Beard and Mary R. Beard, *The Rise of American Civilization* (New York: Macmillan, 1930), pp. 52–121.
9. Lewis Mumford, *Technics and Civilization* (New York: Harcourt, Brace, 1934), pp. 215–21. More recently Louis Galambos has described "The Emerging Organizational Synthesis in Modern American History" (*Business History Review*, XLIV [1970], 279–90), and discerned three major characteristics of modern America as "Technology, Political Economy, and Professionalization: Central Themes of the Organizational Synthesis" (*Business History Review*, LVII [1983], 471–93).
10. "Soviet power + Prussian railroad administration + American technology and monopolistic industrial organization . . . = Socialism," *Leninskij Sbornik*, XXXVI: 37, cited in Eckhart Gillen, "Die Sachlichkeit der Revolutionäre," in *Wem gehört die Welt: Kunst und Gesellschaft in der Weimarer Republik* (Berlin: Neue Gesellschaft für Bildende Kunst, 1977), p. 214.
11. Marcel Duchamp quoted in Stanislaus von Moos, "Die Zweite Entdeckung Amerikas," afterword to Sigfried Giedion, *Die Herrschaft der Mechanisierung* (Frankfurt am Main: Europäische Verlagsanstalt, 1982), p. 807.
12. F. T. Marinetti, "The Founding and Manifesto of Futurism 1909," in *Futurist Manifestos*, ed. and intro. by Umbro Apollonio (London: Thames and Hudson, 1973), p. 22.
13. C. Vann Woodward, *The Burden of Southern History* (Baton Rouge: Louisiana State University Press, 1968), pp. 187–211.

1 ■ A GIGANTIC TIDAL WAVE OF HUMAN INGENUITY

1. Edward W. Byrn, "The Progress of Invention During the Past Fifty Years," *Scientific American*, 75 (25 July 1896), 82–83. Earlier I have written on invention and independent inventors in "The Era of Independent Inventors," *Science in Reflection: The Israel Colloquium: Studies in History, Philosophy, and Sociology of Science* III, ed. Edna Ullmann-Margalit (Dordrecht: Kluwer Academic Publishers, 1988), 151–68.
2. Daniel J. Boorstin, *The Democratic Experience* (New York: Vintage Books, 1974), p. 525.
3. W. S. Woytinsky and E. S. Woytinsky, *World Population and Production: Trends and Outlook* (New York: Twentieth-Century Fund, 1953), pp. 868–

69, 1067–68, 1117–19, 1188; Thomas P. Hughes, *Networks of Power: Electrification in Western Society, 1880–1930* (Baltimore: Johns Hopkins University Press, 1983), pp. 227–61. At the turn of the century, the notable exception to the trend was the British success in holding the lead in the production of textiles.

4. Thomas Parke Hughes, *Elmer Sperry: Inventor and Engineer* (Baltimore: Johns Hopkins University Press, 1971), pp. 293–94.
5. Robert V. Bruce, *Bell: Alexander Graham Bell and the Conquest of Solitude* (Boston: Little, Brown, 1973), pp. 120–50; David A. Hounshell, "Bell and Gray: Contrasts in Style, Politics, and Etiquette," *Proceedings of the IEEE*, 64 (September 1976), 1305–14.
6. Fessenden to Hay Walker, Jr., 21 January 1907; Hay Walker, Jr., to Fessenden, 2 January 1907; Fessenden to Hay Walker, Jr., 17 August 1906; Fessenden to Hay Walker, Jr., 8 May and 19 May 1906; and Fessenden to Hay Walker, Jr., 12 January 1906. Box P.C. 1140.6, Reginald Fessenden Papers, North Carolina State Archives, Raleigh, N.C.
7. Margaret Cheney, *Tesla: Man Out of Time* (Englewood Cliffs, N.J.: Prentice-Hall, 1981), p. 51.
8. Tesla to a *New York Times* reporter, quoted in 1942 in Gordon D. Friedlander, "Tesla: Eccentric Genius," *IEEE Spectrum*, IX (June 1972), 29.
9. Frank Lewis Dyer and Thomas Commerford Martin, *Edison: His Life and Inventions* (New York: Harper & Brothers, 1910), I: 102.
10. Matthew Josephson, *Edison* (New York: McGraw-Hill, 1959), p. 64.
11. Thomas A. Edison Papers, (Frederick, Md.: University Publications of America Microfilm), Part I, reel 12, item 30. Hereafter cited as Edison Microfilm.
12. Robert Conot, *A Streak of Luck* (New York: Seaview Books, 1979), p. 41.
13. Josephson, *Edison*, p. 90.
14. Ibid., p. 87.
15. Ibid., p. 132.
16. Ibid., p. 134.
17. Ibid., pp. 133–34.
18. W. Bernard Carlson, "Thomas Edison's Laboratory at West Orange, New Jersey: A Case Study in Using Craft Knowledge for Technological Innovation, 1886–1888," revision of paper presented at the Edison National Historic Site, West Orange, N.J., 25 April 1987, p. 7.
19. Ibid., p. 13.
20. Ibid., pp. 10–11.
21. Ibid., p. 9.

22. Ibid., p. 16.
23. Lee de Forest, *Father of Radio: The Autobiography of Lee de Forest* (Chicago: Wilcox & Follett, 1950), p. 299.
24. "The Research Laboratory of Mr. Edward Weston," *Scientific American*, LVII (5 November 1887), 287; David O. Woodbury, A *Measure for Greatness: A Short Biography of Edward Weston* (New York: McGraw-Hill, 1949), pp. 150–53.
25. Cheney, *Tesla*, p. 96.
26. Ibid., pp. 133–51.
27. Ibid., p. 77.
28. Hughes, *Sperry*, p. 37.
29. Sperry to Helen Willett, 11 February 1919. Elmer A. Sperry Papers, Hagley Museum and Library, Wilmington, Del. Hereafter cited as Sperry Papers.
30. Hughes, *Sperry*, p. 41.
31. Thomas C. Martin, "William Stanley," chap. VII, p. 6, unpublished book manuscript in Stanley Library of the General Electric Company, Pittsfield, Mass. I am indebted to Samuel Sass, GE librarian, for calling my attention to this source.
32. Ibid., chap. VII, p. 8.
33. W. Bernard Carlson, "Invention, Science, and Business: The Professional Career of Elihu Thomson, 1870–1900," doctoral dissertation, University of Pennsylvania, 1984 (Ann Arbor, Mich.: University Microfilms, 1984), p. 410.
34. Ibid., pp. 369–70.
35. Ibid., pp. 370, 407–10, 425–28, 459.
36. David O. Woodbury, *Elihu Thomson: Beloved Scientist* (Boston: Museum of Science, 1960), p. 205.
37. Thorstein Veblen, *The Instinct of Workmanship and the State of the Industrial Arts* (New York: W. W. Norton, 1964), p. 25. Portions of this section on model builders and the following on experimentation first appeared in Thomas P. Hughes, "Model Builders and Instrument Makers," *Science in Context*, II (1988), 59–75.
38. Dyer and Martin, *Edison*, I: 276.
39. *Harper's Weekly*, XII (4 April 1868), 209–10. See also William and Marlys Ray, *The Art of Invention: Patent Models and Their Makers* (Princeton: Pyne Press, 1974); Eugene Ferguson and Christopher Baer, *Little Machines: Patent Models in the Nineteenth Century* (Greenville, Del.: The Hagley Museum, 1979).
40. Robert V. Bruce, author of *Bell: Alexander Graham Bell and the Conquest of*

Solitude (Boston: Little, Brown, 1973), is an exception in that he duly notes the role of Charles Williams.

41. Edison Microfilm, Part I, reel 12, items 34 and 225.
42. Thomas Watson, Alexander Bell's technical assistant, quoted in Lillian Hoddeson, "The Emergence of Basic Research in the Bell Telephone System, 1875–1915," *Technology and Culture*, 22 (July 1981), 517.
43. Bruce, *Bell*, p. 134.
44. Ibid., p. 356.
45. "Gyroscope Experiments of Elmer Sperry," two-page typewritten report of experiments of 18 November 1908 signed by EAS and witnessed. Sperry Papers.
46. Chas. E. Dressler & Bro. brochure. Sperry Papers.
47. Memorandum, Francis Jehl to Francis Upton, 22 April 1913, p. 4. Hammer Collection, Smithsonian Institution.
48. Folder labeled "Biographical—Upton, Francis," item E-6285-11, Edison Archives, Edison National Historic Site, West Orange, N.J.
49. Lawrence Lessing, *Man of High Fidelity: Edwin Howard Armstrong* (New York: Bantam Books, 1969), p. 162.
50. Ibid., p. 163.
51. Edwin Howard Armstrong, "Mathematical Theory vs. Physical Concept," *FM and Television* (August 1944).
52. David P. Billington, "The Rational and the Beautiful: Maillart and the Origins of Reinforced Concrete," paper presented at the New Materials and the Modern World Seminar at the Hagley Museum and Library, Wilmington, Del., 4 March 1988.
53. Edison Microfilm, reel 12, item 120.
54. Francis Jehl, *Menlo Park Reminiscences* (Dearborn, Mich.: Edison Institute, 1937), I: 51.
55. Ibid., I: 115–23.

2 ■ CHOOSING AND SOLVING PROBLEMS

1. John Jewkes, David Sawers, and Richard Stillerman, *The Sources of Invention* (New York: W. W. Norton, 1969), pp. 79–103.
2. Sperry address to Brooklyn YMCA, 1922, quoted in Thomas Parke Hughes, *Elmer Sperry: Inventor and Engineer* (Baltimore: Johns Hopkins University Press, 1971), p. 63.
3. David Hounshell, "Bell and Gray: Contrasts in Style, Politics, and Etiquette," *Proceedings of the IEEE*, 64 (September 1976), 1305–14.

4. Orville Wright, *How We Invented the Airplane*, Fred C. Kelly, ed. (New York: David McKay, 1953), pp. 18–57.
5. Ibid., reprint, p. 5.
6. Fred Howard, *Wilbur and Orville: A Biography of the Wright Brothers* (New York: Alfred A. Knopf, 1987), pp. 337–38.
7. *The Papers of Wilbur and Orville Wright*, ed. Marvin W. McFarland (New York: Arno Press, 1972), I: 142.
8. Ibid., pp. 420–21, 431.
9. Ibid., p. 235. The Wrights' basic patent was Orville and Wilbur Wright, "Flying Machine," application, 23 March 1903; patent no. 821,393, 22 May 1906.
10. Lee de Forest, *Father of Radio: The Autobiography of Lee de Forest* (Chicago: Wilcox & Follett, 1950), p. 105.
11. Article count on early wireless detectors (1900–1909) done by Tony Mount of Southern Methodist University, 1969, unpublished.
12. Hugh G. J. Aitken, *The Continuous Wave: Technology and American Radio, 1900–1932* (Princeton: Princeton University Press, 1985), pp. 40, 50, 52.
13. Ibid., p. 180.
14. Frdk. K. Vreeland to T. H. Given and Hay Walker, 4 February 1904. Box P.C. 1140.5, Reginald A. Fessenden Papers, North Carolina State Archives, Raleigh, N.C.
15. Hay Walker, Jr., to R. A. Fessenden, 8 February 1904. Box P.C. 1140.5, Reginald A. Fessenden Papers, North Carolina State Archives, Raleigh, N.C.
16. De Forest, *Father of Radio*, p. 161.
17. Aitken, *Continuous Wave*, pp. 187–91.
18. Junius Edwards, *The Immortal Woodshed: The Story of the Inventor Who Brought Aluminum to America* (New York: Dodd, Mead, 1955), p. 38.
19. Thomas P. Hughes, *Networks of Power: Electrification in Western Society, 1880–1930* (Baltimore: Johns Hopkins University Press, 1983), pp. 109–17.
20. Ibid., pp. 117–18.
21. Harold C. Passer, *The Electrical Manufacturers, 1875–1900* (Cambridge, Mass.: Harvard University Press, 1953), pp. 277–82.
22. Margaret Cheney, *Tesla: Man Out of Time* (Englewood Cliffs, N.J.: Prentice-Hall, 1981), pp. 55, 95.
23. Joseph W. Slade, "The Man Behind the Killing Machine," *American Heritage of Invention and Technology*, II (Fall 1986), 18–25.
24. Hughes, *Sperry*, pp. 64–70.
25. Sperry testifying, 1886, in patent interference 10,426, *Van Depoele* v. *Henry* v. *Sperry*, "Electric Railway," U.S. National Archives, Box 1262.
26. Hughes, *Sperry*, p. 67 n. 7.

27. Edison public testimony in folder, "Electric Light Histories Written by Thomas A. Edison for Henry Ford, 1926," pp. 3128–34. Edison Papers, Edison National Laboratory, U.S. National Park Service, West Orange, N.J. Hereafter cited as Edison Papers.
28. Arthur A. Bright, Jr., *The Electric-Lamp Industry: Technological Change and Economic Development from 1800 to 1947* (New York: Macmillan, 1949), pp. 39–40.
29. Hughes, *Networks of Power*, pp. 18–46.
30. Telegram from Edison to Puskas, 22 September 1878. Edison Papers. See also Frank Lewis Dyer and Thomas Commerford Martin, *Edison: His Life and Inventions* (New York: Harper & Brothers, 1910), I: 247–49.
31. Edison to Puskas, 13 November 1878. Edison Papers.
32. D. O. Edge, "Technological Metaphor," in *Meaning and Control*, ed. Edge and Wolfe (London: Tavistock, 1973), p. 31. Mary Hesse, *Models and Analogies in Science* (Notre Dame, Ind.: University of Notre Dame Press, 1966).
33. Max Black, *Models and Metaphors: Studies in Language and Philosophy* (Ithaca, N.Y.: Cornell University Press, 1962), p. 33.
34. Silvano Arieti, *Creativity: The Magic Synthesis* (New York: Basic Books, 1976), pp. 136–37.
35. Ibid., pp. 69–71.
36. Earl R. MacCormac, "Men and Machines: The Computational Metaphor," *Technology in Society*, VI (1984), 209–10.
37. Theodore M. Edison, "Diversity Unlimited: The Creative Work of Thomas A. Edison," condensation of paper given before the MIT Club of Northern New Jersey, 24 January 1969, p. 2.
38. Robert Friedel and Paul Israel with Bernard S. Finn, *Edison's Electric Light: Biography of an Invention* (New Brunswick, N.J.: Rutgers University Press, 1986), pp. 63–64.
39. Hughes, *Sperry*, p. 173.
40. Ibid., p. 112.
41. Ibid., p. 291.
42. Ibid., p. 64.
43. Ibid., p. 116.
44. De Forest, *Father of Radio*, p. 114.
45. Ibid., p. 116.
46. Ibid., p. 116.
47. Ibid., p. 116.
48. Aitken, *Continuous Wave*, p. 199.
49. De Forest, *Father of Radio*, pp. 469–76.

50. Ibid., p. 119.
51. Aitken, *Continuous Wave*, pp. 194–205.
52. Robert A. Chipman, "De Forest and the Triode Detector," *Scientific American*, 212 (March 1965), 92–100.
53. Goethe, *Faust*, pt. 1, in *Goethes sämtliche Werke*, 36 vols., intro. by Karl Goedeke (Stuttgart: Gotta, 1867–82), 10: 44–45 (lines 737–38). Author's translation.
54. Nikola Tesla, "Some Personal Recollections," *Scientific American* (June 1915), reprinted in *Nikola Tesla: Lectures, Patents, and Articles*, comp. V. Popović, R. Horvat, and N. Nikolić (Belgrade: Nikola Tesla Museum, 1956), p. A 198.
55. Aitken, *Continuous Wave*, pp. 33–34, 58.
56. Brooke Hindle, *Emulation and Invention* (New York: New York University Press, 1982), pp. 133–38; Arthur I. Miller, *Imagery in Scientific Thought: Creating 20th-Century Physics* (Boston: Birkhäuser, 1984); Thomas J. Misa, "Visualizing Invention and Development: Henry Bessemer and Alexander Holley," paper presented to the Society for the History of Technology, 24 October 1986; Betty Edwards, *Drawing on the Right Side of the Brain* (Los Angeles: Tarcher, 1979).
57. The Polhem "alphabet" is displayed at the Technology Museum in Stockholm, Sweden.
58. Quoted in Hindle, *Emulation and Invention*, p. 135.
59. Eugene Ferguson quoted in ibid., p. 133.
60. Edwards, *Right Side of the Brain*.
61. Hughes, *Sperry*, pp. 52 (quote), 291.
62. Tesla to John Pierpont Morgan, 1 March 1901, and Charles Steele to Tesla, 4 March 1901. Tesla Collection, Library of Congress, Washington, D.C. (microfilm from Tesla Museum in Belgrade, Yugoslavia), title 7227, reel 3.
63. Gordon D. Friedlander, "Tesla: Eccentric Genius," *IEEE Spectrum*, IX (June 1972), 29.
64. Cheney, *Tesla*, pp. 98–100, 133, 139, 157–58, 164–67.
65. Tesla to Andrew W. Robertson, 22 May 1941. Tesla Collection, Library of Congress, Washington, D.C. (microfilm from Tesla Museum in Belgrade, Yugoslavia), title 7229, reel 6.
66. Lowrey to Edison, 10 October 1878. Edison Papers.
67. Hughes, *Networks of Power*, p. 43.
68. De Forest, *Father of Radio*, pp. 126–28.
69. Ibid., p. 130.
70. Ibid., p. 131.
71. Ibid., pp. 184–85.

72. Ibid., pp. 217–18.
73. M. A. Rosanoff, "Edison in His Laboratory," *Harper's Magazine*, 988 (September 1932), 409.
74. Memorandum, Francis Jehl to Francis Upton, 22 April 1913, pp. 10–11. Hammer Collection, Smithsonian Institution.
75. Susan J. Douglas, *Inventing American Broadcasting, 1899–1922* (Baltimore: Johns Hopkins University Press, 1987), pp. 165–66.
76. Slade, "Man Behind the Killing Machine," p. 22.
77. Frank L. Dyer and T. C. Martin, *Edison: His Life and Inventions* (New York: Harper & Brothers, 1910), II: 499.
78. Bernard Carlson, "Edison in the Mountains: The Magnetic Ore Separation Venture, 1879–1900," *History of Technology*, VIII (1983), 37–59.
79. Matthew Josephson, *Edison* (New York: McGraw-Hill, 1959), p. 309.
80. Ibid., p. 87.
81. Memorandum on the "Experimental Laboratory," prepared by Naval Consulting Board Secretary Thomas Robins. Sperry Papers.

3 ■ BRAIN MILL FOR THE MILITARY

1. William H. McNeill, *The Pursuit of Power: Technology, Armed Force, and Society since A.D. 1000* (Chicago: University of Chicago Press, 1982), pp. 89–94, 278–79.
2. Ibid., pp. 262–304.
3. Richard Hough, *A History of the Modern Battleship: Dreadnought* (London: Michael Joseph, 1964), p. 15.
4. Admiral Sir R. A. Bacon quoted in ibid., p. 21.
5. Admiral Sir R. A. Bacon quoted in Arthur J. Marder, *From the Dreadnought to Scapa Flow: The Royal Navy in the Fisher Era, 1904–19*, I: *The Road to War, 1904–1914* (London: Oxford University Press, 1961), 43.
6. Elting Morison, *Admiral Sims and the Modern American Navy* (Boston: Houghton Mifflin, 1942), chaps. 8, 9.
7. Bradley A. Fiske, "Naval Power," *United States Naval Institute Proceedings*, XXXVII (1911), 683ff; William Atherton Du Puy, "Inventors and the Army and Navy," *Scientific American*, CVII (14 September 1912), 227.
8. Elting E. Morison, *Men, Machines, and Modern Times* (Cambridge, Mass.: MIT Press, 1966), pp. 27–28.
9. Wilbur Wright to Octave Chanute, 1 June 1905, in *The Papers of Wilbur and Orville Wright*, ed. Marvin W. McFarland (New York: Arno Press, 1972), I: 495.

10. Wilbur Wright to Octave Chanute, 28 May 1905, in ibid., I: 493.
11. Octave Chanute to Wilbur Wright, 6 June 1905, in ibid., I: 496–97.
12. Fred C. Kelly, *The Wright Brothers* (New York: Ballantine Books, 1975), p. 95.
13. Ibid., pp. 95–97.
14. Ibid., p. 139.
15. Joseph W. Slade, "The Man Behind the Killing Machine," *American Heritage of Invention & Technology*, II (Fall 1986), 22.
16. John Ellis, *The Social History of the Machine Gun* (New York: Pantheon Books, 1975), p. 18.
17. Slade, "Man Behind Killing Machine," p. 21.
18. Frederic Manning, *The Life of Sir William White* (New York: Dutton, 1923).
19. Thomas Parke Hughes, *Elmer Sperry: Inventor and Engineer* (Baltimore: Johns Hopkins University Press, 1971), pp. 123–24.
20. Ibid., p. 128.
21. Ibid., pp. 190–91.
22. *New York Times*, 17 June 1930.
23. Susan J. Douglas, *Inventing American Broadcasting, 1899–1922* (Baltimore: Johns Hopkins University Press, 1987), pp. 110–12.
24. *Electrical World and Engineer*, 36 (1900), 157, quoted in ibid., p. 119.
25. Lee de Forest, *Father of Radio: The Autobiography of Lee de Forest* (Chicago: Wilcox & Follett, 1950), p. 191.
26. Douglas, *Inventing American Broadcasting*, pp. 127–31.
27. Fessenden to Cleland Davis, 30 November 1906, quoted in ibid., p. 131.
28. De Forest to Francis X. Butler, 14 October 1905, quoted in ibid., p. 133.
29. Raymond Aron, *The Century of Total War* (Garden City, N.Y.: Doubleday, 1954).
30. Thomas Parke Hughes, "Technological Momentum in History: Hydrogenation in Germany 1898–1933," *Past & Present*, 44 (August 1969), 106–32.
31. Ulrich Trumpener, "The Road to Ypres: The Beginnings of Gas Warfare in World War I," *Journal of Modern History*, 47 (September 1975), 467.
32. Fritz Haber, letter in *Nature*, 109 (1922), 40. See also L. F. Haber, *The Poisonous Cloud: Chemical Warfare in the First World War* (Oxford: Clarendon Press, 1986), pp. 22–40.
33. Winston Churchill, *The World Crisis, 1915* (London: T. Butterworth, 1923), p. 22.
34. Alex Roland, *Model Research: The National Advisory Committee for Aeronautics 1915–1958* (Washington, D.C.: NASA, 1985), I: 51.

35. *The Impact of Air Power*, ed. and with intro. by Eugene M. Emme (Princeton: Van Nostrand, 1959), p. 39.
36. Waldemar Kaempffert, "The Inventors' Board and the Navy," *American Review of Reviews*, LII (1915), 298.
37. Hughes, *Sperry*, pp. 243–46.
38. *New York World*, 30 May 1915.
39. *New York Times*, 16 October 1915, p. 4.
40. Ibid.
41. Edward Marshall, "Edison's Plan for Preparedness," *New York Times*, 30 May 1915, sect. V, pp. 6–7.
42. Daniels to Edison, 31 May 1915. Josephus Daniels Papers, Library of Congress, Washington, D.C.
43. Edward Marshall, interview with Secretary Daniels, *New York Times*, 8 August 1915, magazine section.
44. Hughes, *Sperry*, pp. 248–50.
45. *New York Times*, 14 July 1915, p. 1.
46. Editorial, *New York Times*, 13 September 1915, p. 8.
47. Daniel J. Kevles, *The Physicists: The History of a Scientific Community in Modern America* (New York: Vintage Books, 1979), p. 109.
48. Ibid.
49. Millikan quoted in A. Hunter Dupree, *Science in the Federal Government* (Cambridge, Mass: Belknap Press, 1957), p. 308.
50. Kevles, *Physicists*, p. 111.
51. Ibid., p. 113.
52. Millikan quoted in Dupree, *Science in the Federal Government*, p. 312.
53. Robert H. Kargon, *The Rise of Robert Millikan: Portrait of a Life in American Science* (Ithaca, N.Y.: Cornell University Press, 1982), pp. 86–87.
54. Hughes, *Sperry*, p. 258.
55. Kevles, *Physicists*, p. 120.
56. Dupree, *Science in the Federal Government*, p. 318.
57. Kevles, *Physicists*, p. 121.
58. Lloyd N. Scott, *Naval Consulting Board of the United States* (Washington, D.C.: U.S. Government Printing Office, 1920), p. 125.
59. Hughes, *Sperry*, p. 255.
60. Matthew Josephson, *Edison* (New York: McGraw-Hill, 1959), p. 454.
61. Memorandum on the "Experimental laboratory," prepared by the Board secretary, Thomas Robins. Sperry Papers.
62. Quotes from Hughes, *Sperry*, pp. 252–53.

63. Kevles, *Physicists*, p. 138; Dupree, *Science and the Federal Government*, pp. 306–8.
64. Hughes, *Sperry*, pp. 250, 256.
65. Ibid., p. 258.
66. Josephus Daniels, *The Cabinet Diaries* (Lincoln: University of Nebraska Press, 1963), p. 149.
67. EAS to Admiral Earle, 19 December 1918. Sperry Papers.
68. Hughes, *Sperry*, p. 261.
69. Grover Cleveland Loening, *Our Wings Grew Faster* (Garden City, N.Y.: Doubleday, Doran, 1935), p. 93.
70. Rear Admiral Delmar S. Fahrney and Robert Strobell, "America's First Pilotless Aircraft," *Aero Digest*, LXIX (1954), 28ff.
71. Hughes, *Sperry*, p. 267.
72. Stuart W. Leslie, *Boss Kettering* (New York: Columbia University Press, 1983), p. 81.
73. Elmer A. Sperry, recollection to Commander John H. Towers, 1 October 1918. Sperry Papers.
74. Leslie, *Boss Kettering*, p. 83.
75. Colonel William Mitchell, "Lawrence Sperry and the Aerial Torpedo," *U.S. Air Service* (January 1926), p. 16.
76. Hughes, *Sperry*, pp. 179–81, 272–73, 322–24.
77. Ibid., p. 231.
78. Ibid., pp. 232–33.
79. Ibid., p. 233.

4 ■ NO PHILANTHROPIC ASYLUM FOR INDIGENT SCIENTISTS

1. Lawrence Lessing, *Man of High Fidelity: Edwin Howard Armstrong* (New York: Bantam Books, 1969), pp. 45–47. Thomas S. W. Lewis, "Radio Revolutionary," *American Heritage of Invention and Technology*, I (Fall 1985), 36.
2. Ibid., p. 127.
3. Herbert H. Thompson (Elmer Sperry's patent lawyer), "A Patent's Value," *Sperryscope*, I (November-December 1919), 12–14.
4. Lessing, *Man of High Fidelity*, p. 135.
5. Lee de Forest, *Father of Radio: The Autobiography of Lee de Forest* (Chicago: Wilcox & Follett, 1950), p. 352.
6. Ibid., p. 319.

7. Ibid., p. 377.
8. Ibid., p. 379.
9. *Radio Corporation of America, American Telephone & Telegraph Company, and De Forest Radio Company* vs. *Radio Engineering Laboratories, Inc., Respondent* (293 U.S. 1-14, 79 L. Ed. 164), quoted in De Forest, *Father of Radio*, p. 380.
10. Lessing, *Man of High Fidelity*, p. 159.
11. Edwin Armstrong, "Mathematical Theory vs. Physical Concept," *FM and Television*, IV (August 1944), 11–13, 36.
12. Lessing, *Man of High Fidelity*, pp. 163–64.
13. Ibid., p. 213; Lewis, "Radio Revolutionary," p. 10.
14. Ibid., p. 236.
15. Ibid., p. 248.
16. Ibid., p. 268.
17. Neil H. Wasserman, *From Invention to Innovation: Long-Distance Telephone Transmission at the Turn of the Century* (Baltimore: Johns Hopkins University Press, 1985), p. 91.
18. Lillian Hoddeson, "The Emergence of Basic Research in the Bell Telephone System, 1875–1915." *Technology and Culture*, 22 (July 1981), 521–22.
19. "Engineering Department Annual Report for 1906," pp. 4–5, quoted in Leonard S. Reich, *The Making of American Industrial Research: Science and Business at GE and Bell, 1876–1926* (Cambridge: Cambridge University Press, 1985), p. 149.
20. A. Michal McMahon, *The Making of a Profession: A Century of Electrical Engineering in America* (New York: IEEE Press, 1984), pp. 53–54; Reich, *American Industrial Research*, p. 149.
21. Reich, *American Industrial Research*, p. 177.
22. Wasserman, *From Invention to Innovation*, p. 98.
23. Ibid., p. 97.
24. Ibid., pp. 79–88.
25. On Morgan, see Vincent P. Carosso, *The Morgans, Private International Bankers, 1854–1913* (Cambridge, Mass.: Harvard University Press, 1987).
26. Reich, *American Industrial Research*, p. 140.
27. Ibid., p. 158.
28. *The Autobiography of Robert A. Millikan* (London: Macdonald, 1951), p. 134.
29. Reich, *American Industrial Research*, p. 162.
30. *Autobiography of Robert Millikan*, p. 136.
31. Maurice Holland, author of *Industrial Explorers* (New York: Harper & Brothers, 1928), quoted in George Wise, *Willis R. Whitney: General Electric and the*

Origins of U.S. Industrial Research* (New York: Columbia University Press, 1985), p. 209.

32. Wise, *Willis R. Whitney*, pp. 34–35.
33. Georg Meyer-Thurow, "The Industrialization of Invention: A Case Study from the German Chemical Industry," *ISIS*, 73 (1982), 363–81.
34. Wise, *Willis R. Whitney*, p. 58.
35. Ibid., p. 76.
36. Ibid., p. 66.
37. Ibid., p. 77.
38. Ibid., p. 88.
39. Arthur A. Bright, Jr., *The Electric-Lamp Industry: Technological Change and Economic Development from 1800 to 1947* (New York: Macmillan, 1949), p. 168.
40. Ibid., pp. 172–73; Wise, *Willis R. Whitney*, p. 116.
41. Wise, *Willis R. Whitney*, p. 123.
42. William Coolidge quoted in ibid., p. 123.
43. Quote from ibid., p. 150. See also Michael Aaron Dennis, "Accounting for Research: New Histories of Corporate Laboratories and the Social History of American Science," *Social Studies of Science*, 17 (1987), 490.
44. Wise, *Willis R. Whitney*, p. 135.
45. Ibid., p. 157.
46. Reich, *American Industrial Research*, p. 86.
47. Wise, *Willis R. Whitney*, pp. 147–48.
48. Ibid., p. 3.
49. Ibid., p. 141.
50. Spencer R. Weart, "The Rise of 'Prostituted' Physics," *Nature*, 262 (1 July 1976), 13–17. Quotes from pp. 13, 14.
51. Wise, *Willis R. Whitney*, p. 91.
52. George Wise, "A New Role for Professional Scientists in Industry: Industrial Research at General Electric, 1900–1916," *Technology and Culture*, 21 (July 1980), 413.
53. Wise, *Willis R. Whitney*, p. 276.
54. Ibid., pp. 172–77.
55. Thomas P. Hughes, *Networks of Power: Electrification in Western Society, 1880–1930* (Baltimore: Johns Hopkins University Press, 1983), pp. 379–85.
56. David A. Hounshell and John K. Smith, *Science and Corporate Strategy: Du Pont R&D, 1902–1980* (New York: Cambridge University Press), p. 76.
57. Ibid., p. 89.
58. David A. Hounshell, "Continuity and Change in the Management of Industrial

Research: The Du Pont Company, 1902–1980," paper presented at the Second International Conference on the History of Enterprise, Terni, Italy, 2 October 1987, p. 11.

59. Hounshell and Smith, *Science and Corporate Strategy*, pp. 77, 88–89.
60. Ibid., pp. 92–95.
61. Ibid., p. 90.
62. Hounshell, "Continuity and Change," p. 14.
63. Stine quoted in John K. Smith and David A. Hounshell, "Wallace H. Carothers and Fundamental Research at Du Pont," *Science*, 229 (2 August 1985), 437.
64. Carothers quoted in ibid.
65. Carothers quoted in ibid., p. 438.
66. Ibid., pp. 436, 439.
67. John K. Smith, "The Ten-Year Invention: Neoprene and Du Pont Research, 1930–1939," *Technology and Culture*, 26 (January 1985), 34.
68. Carothers quoted in Smith and Hounshell, "Wallace H. Carothers," p. 440.
69. Hounshell, "Continuity and Change," p. 20.
70. Wise, *Willis R. Whitney*, p. 215.
71. C. L. Edgar, "An Appreciation of Mr. Edison Based on Personal Acquaintance"; F. B. Jewett, "Edison's Contributions to Science and Industry"; R. A. Millikan, "Edison as a Scientist," *Science*, 75 (1932), 59–71.
72. Thomas P. Hughes, "Edison's Method," in *Technology at the Turning Point*, ed. William B. Pickett (San Francisco: San Francisco Press, 1977), p. 5.
73. Temporary National Economic Committee, *Concentration of Economic Power Hearings, 1938–39*. (Washington, D. C.: U.S. Government Printing Office, 1939), pts. I–IV, pp. 871–72.
74. Ibid., pts. I–III, pp. 971–76.
75. Henk van den Belt and Arie Rip, "The Nelson-Winter-Dosi Model and Synthetic Chemistry," in *The Social Construction of Technological Systems*, eds. W. Bijker, T. Hughes, and T. Pinch (Cambridge, Mass.: MIT Press, 1987), p. 155.
76. John Jewkes, David Sawers, and Richard Stillerman, *The Sources of Invention* (New York: W. W. Norton, 1969), p. 73.

5 ■ THE SYSTEM MUST BE FIRST

1. Alan Trachtenberg, *The Incorporation of America: Culture and Society in the Gilded Age* (New York: Hill and Wang, 1982), p. 38. Cecelia Tichi prefers "gears and girders" to "machine" as a symbol of early-twentieth-century tech-

nology (*Shifting Gears: Technology, Literature, Culture in Modernist America* [Chapel Hill: University of North Carolina Press, 1987], p. xii).

2. *Yankee Enterprise: The Rise of the American System of Manufactures*, eds. Otto Mayr and Robert C. Post (Washington, D.C.: Smithsonian Institution Press, 1982).
3. David A. Hounshell, *From the American System to Mass Production: The Development of Manufacturing Technology in the United States* (Baltimore: Johns Hopkins University Press, 1984), pp. 331–36.
4. The concept of technological system used in this essay is less elegant but more useful to the historian who copes with messy complexity than the system concepts used by engineers and many social scientists. Several works on systems, as defined by engineers, scientists, and social scientists, are: Ludwig von Bertalanffy, *General System Theory* (New York: George Braziller, 1968); Günter Ropohl, *Eine Systemtheorie der Technik* (Munich, Carl Hanser, 1979); *The Social Theories of Talcott Parsons: A Critical Examination*, ed. Max Black (Carbondale: Southern Illinois University Press, 1961); C. West Churchman, *The Systems Approach* (New York: Dell, 1968); Herbert Simon, "The Architecture of Complexity," in *General Systems Yearbook*, X (1965), 63–76. For further references to the extensive literature on systems, the reader should refer to the Ropohl and Bertalanffy bibliographies. Among historians, Bertrand Gille has used the systems approach explicitly and applied it to the history of technology. See, for instance, *The History of Techniques*, ed. B. Gille (New York: Gordon & Breach, 1986, 2 vols.)
5. John Dos Passos, *U.S.A.* (New York: Viking Penguin, 1986), p. 746.
6. Frederick W. Taylor, *The Principles of Scientific Management* (New York: Harper & Brothers, 1911), p. 7. This publication has been combined with Taylor's *Shop Management*, a long paper presented in 1903, and with his testimony in *Hearings Before Special Committee of the House of Representatives to Investigate the Taylor and Other Systems of Shop Management Under Authority of House Resolution 90* (1912) into a single volume, Frederick Winslow Taylor, *Scientific Management* (New York: Harper & Brothers, 1947). Taylor, *Scientific Management* of 1911, hereafter cited as Taylor, *Scientific Management*. The testimony of 1912 will be cited as Taylor, *Testimony*.
7. Frank Barkley Copley, *Frederick W. Taylor: Father of Scientific Management* (New York: Harper & Brothers, 1923), II: 372. (Reprint: New York: Augustus M. Kelley, 1969). Hereafter cited as Copley, *Taylor*. For a recent appraisal of Taylor's achievements and a criticism of Copley's biography, see Daniel Nelson, *Frederick W. Taylor and the Rise of Scientific Management* (Madison: University of Wisconsin Press, 1980), esp. pp. 193–97.

8. Copley, *Taylor*, I: 108, 110.
9. Sudhir Kakar, *Frederick Taylor: A Study in Personality and Innovation* (Cambidge, Mass.: MIT Press, 1970), p. 95.
10. Taylor, *Scientific Management*, p. 79.
11. Taylor, *Testimony*, pp. 8, 115.
12. Kakar, *Frederick Taylor*, pp. 70–71.
13. Nelson, *Taylor*, p. 202.
14. Taylor, *Scientific Management*, p. 44.
15. Ibid., pp. 44–46ff. Charles Wrege and Amedeo Perroni raise serious doubts about several details in the Schmidt story and about Taylor's version of the pig-iron experiments. See their "Taylor's Pig-Tale: A Historical Analysis of Frederick W. Taylor's Pig-Iron Experiments," *Academy of Management, Journal*, XVII (1974), 6–27.
16. Taylor, *Scientific Management*, p. 46.
17. Copley, *Taylor*, II: 150.
18. Kakar, *Frederick Taylor*, p. 183.
19. Hugh G. J. Aitken, *Scientific Management in Action: Taylorism at Watertown Arsenal, 1908–15* (Princeton: Princeton University Press, 1985), p. 150.
20. Ibid., pp. 229–35.
21. Copley, *Taylor*, II: 407.
22. Ibid., II: 418.
23. Ibid., II: 370.
24. Kenneth Trombley, *The Life and Times of a Happy Liberal* (New York: Harper & Brothers, 1954), p. 9.
25. Martha M. Trescott in "Women Engineers in History: Profiles in Holism and Persistence," in *Women in Scientific and Engineering Professions*, eds. V. Haas and C. Perrucci (Ann Arbor: University of Michigan Press, 1984), pp. 192–204, has alerted us to Lillian Gilbreth's contributions.
26. Raymond C. Miller, *Kilowatts at Work: A History of the Detroit Edison Company* (Detroit: Wayne State University Press, 1957), pp. 63–65, 161–62.
27. Henry Ford, "Mass Production," in *Encyclopaedia Britannica*, 13th ed., suppl. vol. 2 (1926), 821–23.
28. Hounshell, *From the American System*, p. 229.
29. Ibid., p. 222.
30. Allan Nevins and Frank Ernest Hill, *Ford: Expansion and Challenge 1915–33*. (New York: Charles Scribner's Sons, 1957), pp. 206–7.
31. Grant Hildebrand, *Designing for Industry: The Architecture of Albert Kahn* (Cambridge, Mass.: MIT Press, 1974), p. 121.

32. Anne Jardim, *The First Henry Ford: A Study in Personality and Business Leadership* (Cambridge, Mass.: MIT Press, 1970), p. 62.
33. Nevins and Hill, *Ford: Expansion and Challenge*, p. 257.
34. David Halberstam, *The Reckoning* (New York: William Morrow, 1986), pp. 90ff.
35. Jardim, *First Henry Ford*, p. 234.
36. Henry Ford with Samuel Crowther, *My Life and Work* (Garden City, N.Y.: Doubleday, Page, 1922), p. 86.
37. Jardim, *First Henry Ford*, p. 227.
38. Ibid., pp. 227–28.
39. Halberstam, *Reckoning*, pp. 95–96, 101.
40. Jardim, *First Henry Ford*, pp. 217–19. See also Robert Lacey, *Ford: The Men and the Machine* (New York: Ballantine Books, 1986), pp. 309–11.
41. Jardim, *First Henry Ford*, p. 72.
42. Ibid., p. 35.
43. Halberstam, *Reckoning*; Jardim, *First Henry Ford*, passim.
44. Jardim, *First Henry Ford*, pp. 65–68.
45. Ford with Crowther, *My Life and Work*, p. 43.
46. Ibid., p. 17.
47. Ibid., p. 22.
48. Hounshell, *From the American System*, p. 259.
49. Ibid., p. 296.
50. Jardim, *First Henry Ford*, p. 73.
51. Ford with Crowther, *My Life and Work*, p. 56.
52. James J. Flink, *America Adopts the Automobile, 1895–1910* (Cambridge, Mass.: MIT Press, 1970), esp. chaps. 6, 7.
53. John Lawrence Enos, *Petroleum Progress and Profits: A History of Process Innovation* (Cambridge, Mass.: MIT Press, 1962), p. 2.
54. Ibid., p. 61.
55. Ibid., pp. 61–130.
56. Lynwood Bryant, "The Problem of Knock in Gasoline Engines," paper presented at the annual meeting of the American Society of Mechanical Engineers, 19 November 1974, p. 1.
57. "How We Found Ethyl Gas," *Motor* (January 1925), p. 93.
58. T. A. Boyd, *Professional Amateur: The Biography of Charles Franklin Kettering* (New York: Dutton, 1957), pp. 184–85.
59. T. A. Boyd, "Pathfinding in Fuels and Engines: Horning Memorial Lecture," paper presented at the annual meeting of the Society of Automotive Engineers, 11 January 1950.

60. Harold F. Williamson, Ralph L. Andreano, Arnold Daum, and Gilbert C. Klose, *The American Petroleum Indusry: The Age of Energy, 1899–1959* (Westport, Conn.: Greenwood Press, 1981), pp. 409–15.
61. R. R. Sayers, A. C. Fieldner, W. P. Yant, B. G. H. Thomas, and W. J. McConnell, "Exhaust Gases from Engines Using Ethyl Gasoline," *Bureau of Mines Reports of Investigations* (I 28.23:2661), December 1924.
62. "The Use of Tetraethyl Lead Gasoline in Its Relation to Public Health," U.S. Public Health Bulletin no. 163, comp. office of the Surgeon General of the United States (Washington, D.C.: U.S. Government Printing Office, 1926), pp. 117–23.
63. Enos, *Petroleum Progress and Profits*, p. 133.
64. Ibid., p. 196.
65. Thomas K. McCraw, *TVA and the Power Fight, 1933–1939* (Philadelphia: J. B. Lippincott, 1971), p. 12.
66. Forrest McDonald, *Insull* (Chicago: University of Chicago Press, 1962), p. 3.
67. Bernard Weisberger, "The Forgotten Four: Chicago's First Millionaires," *American Heritage*, 38 (November 1987), 43.
68. Thomas P. Hughes, *Networks of Power: Electrification in Western Society, 1880–1930* (Baltimore: Johns Hopkins University Press, 1983), pp. 203–4.
69. Samuel Insull, "Memoirs of Samuel Insull," typescript written in 1934–35, p. 7. Samuel Insull Papers, Loyola University, Chicago, Illinois.
70. McDonald, *Insull*, pp. 53–54.
71. Hughes, *Networks of Power*, p. 250.
72. McDonald, *Insull*, p. 274.
73. Harold L. Arnold, "Ford's Methods and the Ford Shops," *Engineering Magazine* 47 (1914).
74. Ford, "Mass Production," pp. 821–23.
75. Samuel Insull, *Central-Station Electric Service: Its Commercial Development and Economic Significance as Set Forth in the Public Addresses (1897–1914) of Samuel Insull*, ed. William E. Keily (Chicago: privately printed, 1915).
76. Hughes, *Networks of Power*, pp. 175–200, 227–61.
77. Ibid.
78. Stephen Kern, *The Culture of Time and Space, 1880–1918* (Cambridge, Mass.: Harvard University Press, 1983).
79. McDonald, *Insull*, p. 279.
80. Ibid., p. 281.
81. Ibid., p. 291.
82. Ibid., p. 296.

83. Ibid., p. 309.
84. Hughes, *Networks of Power*, pp. 175–200, 227–61.
85. McDonald, *Insull*, p. 314.
86. Ibid., p. 331.
87. Ibid., p. 337.
88. David F. Noble, *America by Design: Science, Technology, and the Rise of Corporate Capitalism* (New York: Alfred A. Knopf, 1977), pp. 24, 39.
89. Ibid., p. 44.
90. Edwin Layton, *The Revolt of the Engineers* (Cleveland: Case Western Reserve University Press, 1971), pp. 154, 156, 170, 185.
91. Noble, *America by Design*, p. 44.
92. Ibid., p. 46.
93. Ibid., pp. 34–35.
94. Ibid., p. 41.
95. Ibid., p. 51.
96. Samuel Haber, *Efficiency and Uplift* (Chicago: University of Chicago Press, 1964), pp. 110–16.
97. Samuel P. Hays, *Conservation and the Gospel of Efficiency: The Progressive Movement, 1890–1920* (Cambridge, Mass.: Harvard University Press, 1959), pp. 3–4.
98. Taylor, *Scientific Management*, p. 8.
99. Daniel Bell's introduction to Thorstein Veblen, *The Engineers and the Price System* (New York: Harcourt, Brace & World, 1963), pp. 2–35.
100. Ibid., p. 13.
101. Joseph Dorfman, *Thorstein Veblen and His America* (New York: Viking Press, 1934), p. 423.
102. Veblen, *Engineers and Price System*, quotes from pp. 72, 74.
103. Ibid., pp. 77, 88.

6 ■ TAYLORISMUS + FORDISMUS = AMERIKANISMUS

1. Edward Hallett Carr and R. W. Davies, *A History of Soviet Russia: Foundations of a Planned Economy, 1926–1929* (New York: Macmillan, 1969), I: 433.
2. Harold Dorn, "Hugh Lincoln Cooper and the First Détente," *Technology and Culture*, XX (1979), 336.
3. Antony C. Sutton, *Western Technology and Soviet Economic Development, 1917–1930* (Stanford: Stanford University Press, 1968), pp. 345–47.

4. A. B. Dibner, "Russia as an Electrical Market," *Electrical World*, 95 (1930), 485.
5. John P. McKay, *Pioneers for Profit: Foreign Entrepreneurship and Russian Industrialization, 1885–1913* (Chicago: University of Chicago Press, 1970), pp. 3–5.
6. Judith A. Merkle, *Management and Ideology: The Legacy of the International Scientific Management Movement* (Berkeley: University of California Press, 1980), pp. 106–7.
7. V. I. Lenin, *Selected Works* (London: Lawrence & Wishard, 1937), VII: 332, cited in Merkle, *Management and Ideology*, p. 113.
8. Merkle, *Management and Ideology*, pp. 115–20.
9. Ibid., p. 122.
10. Ibid., pp. 122–23.
11. Kendall E. Bailes, "Alexei Gastev and the Soviet Controversy over Taylorism," *Soviet Studies*, XXIX (July 1977), 374.
12. Rainer Traub, "Lenin and Taylor: The Fate of 'Scientific Management' in the (Early) Soviet Union," *Telos*, XXXVII (Fall 1978), 87.
13. Bailes, "Alexei Gastev," p. 377.
14. Traub, "Lenin and Taylor," p. 88.
15. Bailes, "Alexei Gastev," p. 384.
16. Ibid., p. 385.
17. Traub, "Lenin and Taylor," p. 86.
18. Gleb M. Krzhizhanovsky, *The Basis of the Technological Economic Plan of Reconstruction of the U.S.S.R.* (Moscow: Cooperative Publishing Society of Foreign Workers in the USSR, 1931), pp. 16, 32 (quote). On Krzhizhanovsky see Raissa L. Berg, *Acquired Traits: Memoirs of a Geneticist from the Soviet Union*, trans. D. Lowe (New York: Viking, 1988), pp. 92–109; Alexander Vucinich, *Empire of Knowledge: The Academy of Sciences of the USSR (1917–1970)* (Berkeley: University of California Press, 1984), pp. 130, 180.
19. Jonathan Charles Coopersmith, "The Electrification of Russia, 1880 to 1925," doctoral dissertation, Oxford University, 1985, p. 143.
20. *The Letters of Lenin*, trans. and ed. Elizabeth Hill and Doris Mudie (New York: Harcourt, Brace, 1937), p. 470.
21. Coopersmith, "Electrification of Russia," pp. 144, 145.
22. Anne Dickason Rassweiler, "Dnieprostroy, 1927–1932: A Model of Soviet Socialist Planning and Construction, doctoral dissertation, Princeton University, 1980, pp. 59, 61, 67.
23. Coopersmith, "Electrification of Russia," pp. 169–70.

24. "An American Engineering Firm in the USSR: Extract from an Address by Albert Kahn, President of Albert Kahn, Inc., of Detroit, before the Cleveland Engineering Society, December 15, 1930," *Economic Review of the Soviet Union*, VI (15 January 1931), 41.
25. Carr and Davies, *History of Soviet Russia*, II: 898–99.
26. Ibid., II: 908.
27. H. R. Knickebocker, *The Soviet Five-Year Plan and Its Effect on World Trade* (London: John Lane The Bodley Head, 1931), p. 161.
28. Margaret Bourke-White, *Eyes on Russia* (New York: Simon & Schuster, 1931), p. 80.
29. Carr and Davies, *History of Soviet Russia*, II: 908.
30. Ibid., II: 911.
31. K. A. Pohl, "Das neue Wasserkraftwerk am Dnjepr," *Elektrotechnische Zeitschrift*, XXXI (4 August 1932), 746.
32. Dorn, "Hugh Lincoln Cooper," p. 337.
33. Carr and Davies, *History of Soviet Russia*, II: 900.
34. A. Bonwetsch, "Die Gross-Wasserkraftanlage Dnjeprostroj," *Zeitschift des Vereines Deutscher Ingenieure*, 76 (20 August 1932), 814–15.
35. Dorn, "Hugh Lincoln Cooper," p. 335.
36. Hugh Lincoln Cooper, "Address Prepared for the 6th General Congress of the International Chamber of Commerce, Washington, D.C., 5 May 1931," in *Trade with Russia* (New York: American-Russian Chamber of Commerce, 1931); "Address Delivered before the Institute of Politics, Williamstown, Mass., 1 August 1930," in *Soviet Russia* (New York: [place?, 1930?]).
37. Dorn, "Hugh Lincoln Cooper," pp. 341–47.
38. Maurice Hindus quoted in Mira Wilkins and Frank Ernest Hill, *American Business Abroad: Ford on Six Continents* (Detroit: Wayne State University Press, 1964), p. 216.
39. Walter Duranty, "Talk of Ford Favors Thrills Moscow," *New York Times*, 17 February 1928, p. 7.
40. Wilkins and Hill, *American Business Abroad*, p. 209.
41. Henry Ford with Samuel Crowther, *My Life and Work* (Garden City, N.Y.: Doubleday, Page, 1922), pp. 4–5.
42. Ibid., p. 5.
43. Allan Nevins and Frank Ernest Hill, *Ford: Expansion and Challenge, 1915–1933* (New York: Charles Scribner's Sons, 1957), p. 673.
44. Dana G. Dalrymple, "The American Tractor Comes to Soviet Agriculture: The Transfer of a Technology," *Technology and Culture*, 5 (1964), 191.

45. Ibid., p. 194.
46. Christine White, "Ford in Russia: In Pursuit of the Chimeral Market," *Business History*, XXVII (October 1986), 92–93.
47. Ibid., p. 93.
48. Grant Hildebrand, *Designing for Industry: The Architecture of Albert Kahn* (Cambridge, Mass.: MIT Press, 1974), pp. 43–54, 92–100.
49. Dalrymple, "American Tractor," p. 199.
50. Ibid., p. 201.
51. Ibid., p. 202.
52. Ibid., pp. 206–7.
53. Wilkins and Hill, *American Business Abroad*, p. 209.
54. Dalrymple, "American Tractor," pp. 212–13.
55. White, "Ford in Russia," p. 91.
56. Nevins and Hill, *Ford: Expansion and Challenge*, p. 677.
57. Ibid., p. 679.
58. Wilkins and Hill, *American Business Abroad*, pp. 222–24.
59. N. Ossinsky, "Zwei Giganten der Sowjetindustrie," *Moskauer Rundschau*, 34 (19 July 1931), 131.
60. "Magnitostroi," *Sowjet-Russland von Heute*, XI (1932), 8–9.
61. "Magnitostroj und Kusnetzkstroj," *Moskauer Rundschau*, 24 (15 June 1930).
62. John Scott, *Behind the Urals: An American Worker in Russia's City of Steel* (Bloomington: Indiana University Press, 1973), p. 6.
63. Ibid., p. 3.
64. Ibid., p. 65.
65. Ibid., pp. 69–70.
66. *Frankfurter Zeitung*, 30 November 1930, trans. by and included in El Lissitzky, *Russia: An Architecture for World Revolution*, trans. Eric Dluhosch (Cambridge, Mass.: MIT Press, 1984), pp. 175–79.
67. Scott, *Behind the Urals*, p. 209.
68. Ibid., pp. 187, 193. See also Antony C. Sutton, *Western Technology and Soviet Economic Development, 1930 to 1945*, (Stanford: Hoover Institution Press, 1971), pp. 74–77.
69. Scott, *Behind the Urals*, pp. 69–70.
70. Frank Trommler, "The Rise and Fall of Americanism in Germany," in *America and the Germans: The Relationship in the Twentieth Century*, ed. F. Trommler and J. McVeigh (Philadelphia: University of Pennsylvania Press, 1985), p. 335.
71. Winfried Nerdinger, *Walter Gropius* (Berlin: Gebr. Mann Verlag, 1985), p. 11.

72. Gustav Winter, *Der Taylorismus: Handbuch der wissenschaftlichen Betriebs- und Arbeitsweise für die Arbeitenden aller Klassen: Stände und Berufe* (Leipzig: S. Hirzel, 1920), p. vii.
73. Charles S. Maier, "Between Taylorism and Technocracy: European Ideologies and the Vision of Industrial Productivity in the 1920s," *The Journal of Contemporary History*, 5 (1970), 27–61.
74. Merkle, *Management and Ideology*, p. 172.
75. Henry Ford, *Mein Leben und Werk*, trans. Curt and Margerite Thesing (Leipzig: Paul List Verlag, 1923).
76. Peter Berg, *Deutschland und Amerika, 1918–1929: Über das deutsche Amerikabild der zwanziger Jahre* (Lübeck/Hamburg: Matthiesen Verlag, 1963), p. 101.
77. Author's translation of Paul Rieppel, *Ford-Betriebe und Ford-Methoden* (Munich/Berlin: Verlag R. Oldenbourg, 1926), p. 29.
78. Ibid., p. 29. Oswald Spengler, "Preussentum und Sozialismus," in *Politische Schriften* (Munich/Berlin: C. H. Beck'schen Verlag, 1934), pp. 1–105.
79. Friedrich von Gottl-Ottlilienfeld, *Fordismus: Über Industrie und Technische Vernunft* (Jena: Gustav Fisher, 1926).
80. Ibid., pp. 18–20, 23, 45, 55, 62–63.
81. Jost Hermand/Frank Trommler, *Die Kultur der Weimarer Republik* (Munich: Nymphenburger Verlagshandlung, 1978), pp. 54–55.
82. J. Walcher, *Ford Oder Marx: Die praktische Lösung der sozialen Frage* (Berlin: Neuer Deutscher Verlag, 1925).
83. Ibid., pp. 43, 67, 84, 92, 107, 111.
84. Gustav Winter, *Der falsche Messias Henry Ford: Ein Alarmsignal für das gesamte deutsche Volk* (Leipzig: Verlag "Freie Meinung," 1924).
85. Helmut Lethen, *Neue Sachlichkeit, 1924–1932: Studien zur Literatur des "Weissen Sozialismus"* (Stuttgart: J. B. Metzlersche, 1970), p. 20.
86. Ibid., pp. 21–24.
87. Ibid., p. 22.
88. Rieppel, *Ford-Betriebe und Ford-Methoden*, pp. 41–42.
89. Nevins and Hill, *Ford: Expansion and Challenge*, pp. 311–23.
90. Berg, *Deutschland und Amerika*, p. 98.
91. Otto Moog, *Drüben steht Amerika: Gedanken nach einer Ingenieurreise durch die Vereinigten Staaten* (Braunschweig: G. Westermann, 1927).
92. Author's translation of ibid., p. 72.
93. Ibid., p. 76.
94. Franz Westermann, *Amerika wie ich es sah: Reiseskizzen eines Ingenieurs* (Halberstadt: H. Meyer's Buchdruckerei, 1926), p. 5.

95. Author's translation of ibid., p. 99.

96. Jeffrey Herf, *Reactionary Modernism: Technology, Culture, and Politics in Weimar and the Third Reich* (Cambridge: Cambridge University Press, 1984), pp. 12, 31.

97. Oswald Spengler, *Decline of the West*, trans. Charles Francis Atkinson (New York: Modern Library, 1965), pp. 24–27.

98. Trommler, "Americanism in Germany," p. 337.

99. Thomas P. Hughes, "Technology," in *The Holocaust: Ideology, Bureaucracy, and Genocide*, ed. Henry Friedlander and Sybil Milton (Millwood, N.Y.: Kraus International, 1980), pp. 165–81.

7 ■ THE SECOND DISCOVERY OF AMERICA

1. Lewis Mumford, "Machinery and the Modern Style," *The New Republic*, XXVII (3 August 1921), 264–65. Portions of the section on the second industrial revolution in this chapter first appeared in Thomas P. Hughes, "Visions of Electrification and Social Change," in *Histoire de l'électricité: 1880–1980, un siècle d'électricité dans le monde*, ed. Fabienne Cardot (Paris: Presses Universitaires de France, 1987), pp. 327–40.

2. Lewis Mumford, "The City," in *Civilization in the United States: An Inquiry by 30 Americans*, ed. Harold E. Stearns (New York: Harcourt, Brace, 1922), p. 12. Quoted in Richard G. Wilson, Dianne H. Pilgrim, and Dickran Tashjian, *The Machine Age in America, 1918–1941* (New York: The Brooklyn Museum, 1986), p. 30.

3. Robert Hughes, *The Shock of the New* (New York: Alfred A. Knopf, 1981), p. 9.

4. Alfred D. Chandler, Jr., *The Visible Hand: The Managerial Revolution in American Business* (Cambridge, Mass.: Belknap Press, 1977).

5. James Beniger, *The Control Revolution: Technological and Economic Origins of the Information Society* (Cambridge, Mass.: Harvard University Press, 1986).

6. Leslie Hannah, *Electricity Before Nationalisation* (Baltimore: Johns Hopkins University Press, 1979), pp. 65–66, 82, 85; R. Blanchard, *Les Forces hydroélectriques pendant la guerre* (Paris: Presses Universitaires de France, 1924); Thomas P. Hughes, "Technology as a Force for Change in History: The Effort to Form a Unified Electric Power System in Weimar Germany," in *Industrielles System und politische Entwicklung in der Weimarer Republik*, eds. H. Mommsen, D. Petzina, and B. Weisbrod (Düsseldorf: Droste, 1974), pp. 153–66; Thomas P. Hughes, *Networks of Power: Electrification in Western Society, 1880–1930* (Baltimore: Johns Hopkins University Press, 1983), pp. 285–323.

7. Vladimir Lenin, *Collected Works*, XXV: 490–91 (in Russian), trans. and quoted in U.S.S.R. Committee for International Scientific and Technical Conferences, *Electric Power Development in the U.S.S.R.* (Moscow: INRA Publishing Society, 1936), p. 11.
8. Great Britain, Ministry of Reconstruction, Reconstruction Committee, Coal Conservation Sub-Committee, *Interim Report on Electric Power Supply in Great Britain* (London: HMSO, 1918), pp. 4, 7, 8.
9. For various post–World War I plans for regional and all-German grids, see Georg Boll, *Entstehung und Entwicklung des Verbundbetriebs in der deutschen Elektrizitätswirtschaft bis zum europäischen Verbund* (Frankfurt am Main: VWEW, 1969), pp. 56–67.
10. For more on these postwar schemes, see Hughes, *Networks of Power*, pp. 285–323.
11. Lewis Mumford, *Technics and Civilization* (New York: Harcourt, Brace, 1934).
12. Lewis Mumford, "Regionalism and Irregionalism," *The Sociology Review*, XIX (1927), 288.
13. Mumford, *Technics*, p. 215.
14. Mumford, "Regionalism and Irregionalism," p. 286.
15. Mumford, *Technics*, pp. 215–21.
16. Ibid., p. 223.
17. Lewis Mumford, "The Theory and Practice of Regionalism," *The Sociology Review*, XX (1928), 23–24.
18. Gifford Pinchot to Morris L. Cook, 15 January 1927. Morris L. Cooke Papers, Franklin D. Roosevelt Library, Hyde Park, N.Y.
19. For more about Giant Power, see Hughes, *Networks of Power*, pp. 297–313.
20. Governor Gifford Pinchot's message to the General Assembly of the Commonwealth of Pennsylvania on 17 February 1925, published as Pinchot, "Introduction" to "Giant Power: Large Scale Electrical Development as a Social Factor," *The Annals*, CXVIII (March 1925), viii.
21. Pinchot, "Introduction," pp. xi–xii.
22. Ibid., p. xi.
23. Ibid.
24. Ibid.
25. Joseph K. Hart, "Power and Culture," *The Survey: Graphic Number*, LI (1 March 1924), 625–28, reprinted in Thomas Parke Hughes, ed., *Changing Attitudes Toward American Technology* (New York: Harper & Row, 1975), pp. 241–52. Pages cited in subsequent references are in Hughes.
26. Hart, "Power," pp. 243–45.
27. Ibid., p. 246.

28. Ibid., p. 250.
29. Ibid., p. 251.
30. Ibid.
31. Paul Kellogg, "The Play of a Big Man with a Little River," *The Survey: Graphic Number*, LI (1 March 1924), 637ff.
32. Henry Ford with Samuel Crowther, *My Life and Work* (Garden City, N.Y.: Doubleday, Page, 1922), quoted (without page number) in Kellogg, "Play of a Big Man," p. 640.
33. Kellogg, "Play of a Big Man," p. 641.
34. Ford with Crowther, *My Life and Work*, quoted in Kellogg, "Play of a Big Man," p. 641.
35. Allan Nevins and Frank Ernest Hill, *Ford: Expansion and Challenge, 1915–1933* (New York: Charles Scribner's Sons, 1957), p. 228.
36. Ibid., p. 226.
37. Kellogg, "Play of a Big Man," p. 641; Nevins and Hill, *Ford: Expansion and Challenge*, pp. 229–30.
38. Le Corbusier, "Architecture, the Expression of the Materials and Methods of Our Times," *The Architectural Record*, LXVI (August 1928), 123, 128. Le Corbusier's theory of the useful object, however, implied a kind of Darwinian evolution—even among American designers; Reyner Banham, *Theory and Design in the First Machine Age* (Cambridge, Mass.: MIT Press, 1980), pp. 211–12. See also Robert Twombly, *Louis Sullivan: His Life and Work* (New York: Viking, 1986), pp. 130–32, 217–18; Louis H. Sullivan, *The Autobiography of an Idea* (New York: Dover, 1956), pp. 247–50; Carl W. Condit, "Sullivan's Skyscrapers as the Expression of Nineteenth Century Technology," *Technology and Culture*, I (1959), 78–93; Colin Rowe, "Chicago Frame: Chicago's Place in the Modern Movement," *The Architectural Review*, 128 (November 1956), 285–89; Frank Lloyd Wright, "In the Cause of Architecture," *The Architectural Record*, 61 (May and June 1927), 394–96, 478–80; 62 (July and October 1927), 163–66, 318–21.
39. Peter Behrens, "Art and Technology," lecture delivered at the 18. Jahresversammlung des Verbandes Deutscher Electrotechniker, Braunschweig, 26 May 1910, reprinted in *Industriekultur: Peter Behrens and the* AEG, *1907–14*, ed. Tilmann Buddensieg, trans. Iain Boyd Whyte (Cambridge, Mass.: MIT Press, 1984), pp. 212–19.
40. Behrens quoted by Buddensieg, "Industriekultur," in *Industriekultur*, p. 8.
41. Karin Wilhelm, "Fabrikenkunst: The Turbine Hall and What Came of It," in *Industriekultur*, p. 142.
42. Behrens quoted by Buddensieg, "Industriekultur," in *Industriekultur*, p. 17.

43. Joan Campbell, *The German Werkbund* (Princeton: Princeton University Press, 1978), pp. 11–17.
44. Hermann Muthesius, "Handarbeit und Massenerzeugnis," *Technischer Abend im Zentralinstitut für Erziehung und Unterricht*, IV (1917).
45. On Gropius, see Reginald R. Isaacs, *Walter Gropius: Der Mensch und sein Werk* (Berlin: Gebr. Mann Verlag, 1983).
46. Walter Gropius, "Monumentale Kunst und Industriebau," typescript (37 pages) of address given in the Folkwang Museum, Hagen, Westfalen, 29 January 1911, p. 12. Bauhaus Archiv, West Berlin.
47. Ibid., p. 9. Gropius was not taking into account that a bridge builder like Robert Maillart was coming to terms with the new aesthetic as well as with the new technology. I am indebted to David Billington of Princeton University for this observation.
48. Ibid., p. 13.
49. Ibid.
50. "From Americanism to the New World," in *Walter Gropius*, ed. Winfried Nerdinger (Berlin: Gebr. Mann Verlag, 1985), p. 9.
51. Thomas P. Hughes, "Gropius, Machine Design, and Mass Production," *Wissenschaftskolleg Jahrbuch 1983–84* (Berlin: Siedler Verlag, 1985), p. 172.
52. The Gropius papers are at the Busch-Reisinger Museum, Harvard University, Cambridge, Mass. Copies with additional Gropius materials are at the Bauhaus Archiv, West Berlin, Federal Republic of Germany.
53. Walter Gropius, "Wohnhaus-Industrie," *Berlin Tageblatt*, 24 September 1924; Gropius, *Scope of Total Architecture* (New York: Harper & Row, 1955), pp. 128–35; Gropius to Schumann Haus- und Küchengeräte, Berlin, 11 July 1925, Korrespondenz W. Gropius über Rationalisierung im Bauwesen, Bauhaus Archiv, Berlin; "Arbeitsprogrammskizze für einige Vesuchswohnhäuser," six-page typescript dated 29 December 1924, Korrespondenz 1924–26, Dewog, Bauhaus Archiv, West Berlin.
54. The expression came from the architectural historian Sigfried Giedion (Gilbert Herbert, *The Dream of the Factory-Made House: Walter Gropius and Konrad Wachsmann* [Cambridge, Mass.: MIT Press, 1984], p. 4).
55. Gropius, "Wohnhaus-Industrie."
56. "Arbeitsprogrammskizze für einige Versuchswohnhäuser."
57. Herbert, *Factory-Made House*, p. 60. See also Hughes, "Gropius, Machine Design, and Mass Production," passim.
58. "Weissenhof, 1927–87," *Info Bau* 83, X (1983), 15–35.

59. Henry-Russell Hitchcock and Philip Johnson, *The International Style* (New York: W. W. Norton, 1932).
60. Le Corbusier-Saugnier, "Trois Rappels à MM. Les Architectes," *L'Esprit nouveau*, I: 2 (November 1920), 196.
61. Mary McLeod, "Architecture or Revolution: Taylorism, Technocracy, and Social Change," *Art Journal*, 43 (Summer 1983), 133.
62. See Le Corbusier-Saugnier, "Les Maisons 'Voisin,' " *L'Esprit nouveau*, I: 2 (November 1920), 214.
63. McLeod, "Architecture," pp. 143–45.
64. *L'Esprit nouveau*, I: 1 (October 1920), title page.
65. Thomas P. Hughes, "Appel aux Industriels," in *L'Esprit nouveau: Le Corbusier und die Industrie 1920–25*, ed. Stanislaus von Moos (Berlin: Wilhelm Ernst & Sohn Verlag, 1987), pp. 26–31.
66. Le Corbusier-Saugnier, "Trois Rappels," p. 199.
67. Stanislaus von Moos, "Le Corbusier und Gabriel Voison," in *Avant Garde und Industrie*, ed. S. von Moos and C. Smeenk (Delft: Delft University Press, 1983), pp. 80–82.
68. Stanislaus von Moos, "Standard und Elite: Le Corbusier, die Industrie und der Esprit nouveau," in *Die nützlichen Künste*, ed. T. Buddensieg and Henning Rogge (Berlin: Quadriga, 1981), pp. 311–13. Von Moos, "Im Vorzimmer des 'Machine Age,' " in *L'Esprit mouveau: Le Corbusier und die Industrie 1920–25*, pp. 20–22.
69. Le Corbusier-Saugnier, "Esthétique de l'ingénieur: Maison en série," *L'Esprit nouveau*, III: 13 (December 1921), 1525–42.
70. Brian Brace Taylor, *Le Corbusier et Pessac 1914–1928* (Paris: Fondation Le Corbusier and Harvard University, 1972).
71. William J. R. Curtis, *Modern Architecture Since 1900* (Oxford: Phaidon, 1982), pp. 156–57.
72. Barbara Miller Lane, *Architecture and Politics in Germany, 1918–1945* (Cambridge, Mass.: Harvard University Press, 1985), pp. 125–40.
73. *Futurist Manifestos*, ed. Umbro Apollonio (London: Thames and Hudson, 1973), pp. 19–20.
74. Gerald Silk, "Automobile," in *Futurismo & Futurismi*, catalogue of an exhibition organized by Pontus Hulten (Milan: Bompiani, 1986), p. 421.
75. Ester Coen, "City," in *Futurismo & Futurismi*, p. 452.
76. Enrico Crispolti, "Sant'Elia," in *Futurismo & Futurismi*, pp. 563–64.
77. Coen, "City," in *Futurismo & Futurismi*, p. 452.
78. Gail Levin, "Joseph Stella," in *Futurismo & Futurismi*, p. 579.

79. Dorothy Norman, *Alfred Stieglitz: An American Seer* (New York, Random House, 1973), p. 45.
80. Marius De Zayas quoted in Dickran Tashjian, *Skyscraper Primitives: Dada and the American Avant-Garde, 1910–25* (Middletown, Conn.: Wesleyan University Press, 1975), p. 205.
81. William Innes Homer, *Alfred Stieglitz and the American Avant-Garde* (Boston, Mass.: New York Graphic Society, 1977), pp. 4, 37–38, 192–94.
82. Association of American Painters and Sculptors, *Catalogue of International Exhibition of Modern Art* (New York: Vreeland Press, 1913), pp. 29, 36.
83. *New York Tribune*, 24 October 1915, sect. IV, p. 2. I am indebted to Agatha H. Hughes for calling my attention to this source.
84. Ibid.
85. Ibid.
86. Tashjian, *Skyscraper Primitives*, pp. 42–43.
87. Calvin Tomkins, *The World of Marcel Duchamp*, (New York: Time-Life Books, 1966), p. 37.
88. Katia Samaltanos, *Apollinaire: Catalyst for Primitivism, Picabia and Duchamp* (Ann Arbor: UMI Research Press, 1984), p. 70.
89. K. G. Pontus Hultén, *The Machine as Seen at the End of the Mechanical Age* (New York: Museum of Modern Art, 1968), p. 84.
90. Francis Naumann, "Walter Conrad Arensberg: Poet, Patron, and Participant in the New York Avant-Garde, 1915–20," *PMA Bulletin*, 76 (Spring 1980), 4–11.
91. Martin L. Friedman, *The Precisionist View in American Art* (Minneapolis: Walker Art Center, 1960), pp. 24, 25 n. 16.
92. Wilson, Pilgrim, and Tashjian, *Machine Age in America*, pp. 213–14. Many Duchamp scholars today doubt that his interest in readymades was aesthetic. I am indebted to Stanislaus von Moos, University of Zurich, for this observation.
93. Pontus Hultén, *Machine*, pp. 80–83.
94. Ileana B. Leavens, *From "291" to Zurich: The Birth of Dada* (Ann Arbor: UMI Research Press, 1983), p. 137. Werner Spies, "Die Klischeedrucke—Resultat der Verarbeitung typographischen Materials," in *Max Ernst in Köln: Die rheinische Kunstszene bis 1922*, ed. Wulf Herzogenrath (Cologne: Rheinland-Verlag, 1980), pp. 197–205.
95. Robert Alden Sanborn, "A Champion in the Wilderness," *Broom*, III (October 1922), 175.
96. Coady quoted in ibid.
97. Ibid.

98. Coady quoted in ibid., p. 176.
99. Ibid., p. 177.
100. Susan Fillin Yeh, *The Precisionist Painters 1916–1949: Interpretations of a Mechanical Age* (Huntington, N.Y.: Heckscher Museum, 1978), p. 12.
101. Wilson, Pilgrim, and Tashjian, *Machine Age in America*, p. 225.
102. *Broom*, V (August 1923).
103. Matthew Josephson, "Made in America," *Broom*, II: 3 (June 1922), 269–70.
104. Yeh, *Precisionist Painters*, p. 9.
105. Wilson, Pilgrim, and Tashjian, *Machine Age in America*, p. 213.
106. Tashjian, *Skyscraper Primitives*, p. 204.
107. Ibid., p. 212.
108. Ibid., p. 209.
109. Sheeler quoted in ibid., p. 221.
110. "Comments," *The Little Review*, IX: 2 (Winter 1922), 22, as cited in Wilson, Pilgrim, and Tashjian, *Machine Age in America*, p. 231.
111. Yeh, *Precisionist Painters*, p. 12.
112. Jean Heap, "Machine-Age Exposition," in *Machine-Age Exposition Catalogue: The Little Review*, XII (1926–29), 36.
113. Wilson, Pilgrim, and Tashjian, *Machine Age in America*, p. 234.
114. Louis Lozowick, "The Americanization of Art," in *Machine-Age Exposition Catalogue: The Little Review*, XII (1926–29), 18.
115. Ingeborg Güssow, "Malerei der Neuen Sachlichkeit," in *Kunst und Technik in den 20er Jahren: Neue Sachlichkeit und gegenständlicher Konstruktivismus* (Munich: Städtische Gallerie im Lenbachhaus, München, 1980), p. 49.
116. "700 Intellektuellen beten einen Öltank an" is from Bertolt Brecht, *Gesammelte Werke* (Frankfurt am Main: Suhrkamp Verlag, 1967), VIII: 316–17. On Grossberg see Hessisches Landesmuseum, *Carl Grossberg, Gemälde, Aquarelle, Zeichnungen und Druckgrafik (1914–1940)* (Darmstadt: Druckerei Anthes, 1976), p. 19.
117. Waldemar Augustiny, *Franz Radziwill* (Göttingen: Musterschmidt, 1964).

8 ■ TENNESSEE VALLEY AND MANHATTAN ENGINEER DISTRICT

1. Lewis Mumford, *The Culture of Cities* (New York: Harcourt Brace Jovanovich, 1970), p. 378.
2. The regional plan number of *The Survey: Graphic Number*, LIV (1 May 1925).
3. I am indebted to Professors Howard Segal of the University of Maine, Allen Tullos of Emory University, and John Thomas of Brown University for calling

my attention to these regional developments. Segal, "Mumford's Alternatives to the Megamachine"; Tullos, "The Politics of Regional Development, Lewis Mumford, and Howard W. Odum"; and Thomas, "Lewis Mumford, Benton MacKaye, and the Regional Vision." Papers presented at the International Symposium on Lewis Mumford, University of Pennsylvania, Philadelphia, 5–7 November 1987.

4. Robert W. Bruère, "Pandora's Box," *The Survey: Graphic Number*, LI (1 March 1924), 557. Portions of this section on regional planners first appeared in Thomas P. Hughes, "Visions of Electrification and Social Change," in *Histoire de l'électricité: 1880–1980, un siècle d'électricité dans le monde*, ed. Fabienne Cardot (Paris: Presses Universitaires de France, 1987), pp. 327–40.
5. *The Survey: Graphic Number*, LI (1 March 1924) and LIV (1 May 1925).
6. "Giant Power," *The Annals*, CXVIII (March 1925). Editor in charge of this volume was Morris Llewellyn Cooke.
7. "The Fourth Migration," *The Survey: Graphic Number*, LIV (1 May 1925), 129–33.
8. Mumford to Patrick Geddes, 4 December 1924. Lewis Mumford Papers, University of Pennsylvania.
9. "The Regional Community," *The Survey: Graphic Number*, LIV (1 May 1925), 129. This preface to the issue is unsigned, but a Mumford letter to Patrick Geddes indicated Mumford as the author (Mumford to Geddes, 4 December 1924. Lewis Mumford Papers, University of Pennsylvania).
10. On 12 July 1893, the American historian Frederick Jackson Turner read his now classic essay, "The Significance of the Frontier in American History," at the meeting of the American Historical Association in Chicago. It was first published in the *Proceedings of the State Historical Society of Wisconsin* (14 December 1893).
11. Lewis Mumford, *The Golden Day: A Study of American Literature and Culture* (Boston: Beacon Hill, 1957; first published in 1926).
12. Mumford, *The Culture of Cities*, pp. 307–8.
13. Ibid., p. 344.
14. Ibid., p. 301.
15. Ibid., p. 463.
16. Ibid., p. 381.
17. Allan Nevins and Frank Ernest Hill, *Ford: Expansion and Challenge, 1915–1933* (New York: Charles Scribner's Sons, 1957), p. 310. See also Preston J. Hubbard, *Origins of the TVA: The Muscle Shoals Controversy, 1920–1932* (New York: W. W. Norton, 1968), pp. 62–71.

18. Hubbard, *Origins of the* TVA, pp. 28–47, 37 (quotes). On Ford's anti-Semitism see also Robert Lacey, *Ford: The Men and the Machine* (New York: Ballantine Books, 1986), pp. 215–31.
19. Richard Lowitt, *George W. Norris: The Persistence of a Progressive, 1913–33* (Urbana: University of Illinois Press, 1971), p. 338.
20. George W. Norris, *Fighting Liberal: The Autobiography of George W. Norris* (New York: Macmillan, 1945), p. 248.
21. Thomas K. McCraw, TVA *and the Power Fight, 1933–1939* (Philadelphia: J. B. Lippincott, 1971), p. 5.
22. Ibid., p. 12.
23. Ibid., p. 35.
24. Nathan Reingold, "Vannevar Bush's New Deal for Research: Or the Triumph of the Old Order," *Historical Studies in the Physical Sciences*, 17 (1987), 323.
25. Roy Talbert, Jr., *FDR's Utopian: Arthur Morgan of the* TVA (Jackson: University Press of Mississippi, 1987), p. 52.
26. Arthur E. Morgan, "Sociology in the TVA," *American Sociological Review*, II (April 1937), 159.
27. Ibid., p. 158.
28. Talbert, *FDR's Utopian*, p. 24.
29. Thomas K. McCraw, *Morgan vs. Lilienthal: The Feud Within the* TVA (Chicago: Loyola University Press, 1970), p. 33.
30. Talbert, *FDR's Utopian*, p. 129.
31. McCraw, *Morgan vs. Lilienthal*, p. 34.
32. Talbert, *FDR's Utopian*, p. 128.
33. McCraw, TVA *and the Power Fight*, p. 55.
34. Franklin D. Roosevelt, *The Public Papers and Addresses of Franklin D. Roosevelt*, II: *The Year of Crisis* (New York: Random House, 1938), 122–23.
35. C. Herman Pritchett, *The Tennessee Valley Authority: A Study in Public Administration* (New York: Russell & Russell, 1971), pp. 116–17.
36. Franklin D. Roosevelt, *On Our Way* (New York: John Day, 1934), pp. 53–56, and quoted in Talbert, *FDR's Utopian*, p. 128.
37. Talbert, *FDR's Utopian*, pp. 128–29.
38. Alfred Lief, *Democracy's Norris: The Biography of a Lonely Crusade* (New York: Stackpole Sons, 1939), p. 415; Norman Zucker, *George W. Norris: Gentle Knight of American Democracy* (Urbana: University of Illinois Press, 1966), p. 124.
39. Morgan, "Sociology in the TVA," p. 160.
40. Pritchett, *Tennessee Valley Authority*, p. 122.

41. Ibid., pp. 121–31; Morgan, "Sociology in the TVA," pp. 157–65.
42. Otis L. Graham, Jr., *Toward a Planned Society: From Roosevelt to Nixon* (New York: Oxford University Press, 1976), pp. 1–68.
43. Pritchett, *Tennessee Valley Authority*, pp. 131–40.
44. McCraw, *TVA and the Power Fight*, p. 143.
45. McCraw, *Morgan vs. Lilienthal*, p. 18; McCraw, *TVA and the Power Fight*, p. 44.
46. McCraw, *TVA and the Power Fight*, pp. 57–63.
47. Thomas P. Hughes, *Networks of Power: Electrification in Western Society, 1880–1930* (Baltimore: Johns Hopkins University Press, 1983), pp. 224–35.
48. McCraw, *Morgan vs. Lilienthal*, p. 29.
49. David E. Lilienthal, *The Journals of David E. Lilienthal: The TVA Years, 1939–1945* (New York: Harper & Row, 1964), I:79–80.
50. McCraw, *TVA and the Power Fight*, p. 142.
51. McCraw, *Morgan vs. Lilienthal*, p. 32.
52. Philip Selznick, *TVA and the Grass Roots: A Study in the Sociology of Formal Organization* (New York: Harper & Row, 1966), p. 92.
53. Pritchett, *Tennessee Valley Authority*, pp. 193, 202.
54. Lilienthal, *Journals*, I:66.
55. Pritchett, *Tennessee Valley Authority*, pp. 205–6.
56. Richard Lowitt, *George W. Norris: The Triumph of a Progressive, 1933–44* (Urbana: University of Illinois Press, 1978), p. 210.
57. David E. Lilienthal, *TVA: Democracy on the March* (Chicago: Quadrangle Books, 1953 ed.), p. 53.
58. Selznick, *TVA and the Grass Roots*, p. 93.
59. McCraw, *Morgan vs. Lilienthal*, p. x.
60. Lilienthal, *TVA: Democracy on the March*, pp. 2–5.
61. Thomas K. McCraw, "Triumph and Irony—The TVA," *Proceedings of the IEEE*, 64 (September 1976), 1375.
62. Ibid., p. 1376.
63. McCraw, *TVA and the Power Fight*, p. 159.
64. John Emerich Edward Dalberg, Lord Acton: Letter to Bishop Mandell Creighton, 24 April 1887. John Bartlett, *Familiar Quotations*, ed. E. M. Beck (Boston: Little, Brown, 1980), p. 615.
65. J. E. Pearce, "The Creeping Conservatism of the TVA," *The Reporter*, 26:4 (January 1962), 34, quoted in McCraw, "Triumph and Irony," p. 1378.
66. Richard G. Hewlett and Oscar E. Anderson, Jr., *The New World, 1939–1946* (University Park: Pennsylvania State University Press, 1962), p. 81 (vol. 1 in *A History of the United States Atomic Energy Commission*); Richard Hewlett,

"Beginnings of Development in Nuclear Technology," *Technology and Culture,* 17 (1976), 470.

67. Stéphane Groueff, *Manhattan Project: The Untold Story of the Making of the Atomic Bomb* (Boston: Little, Brown, 1967), p. 31.
68. Hewlett and Anderson, *New World,* p. 82.
69. Groueff, *Manhattan Project,* p. 34.
70. Richard Rhodes, *The Making of the Atomic Bomb* (New York: Simon & Schuster, 1986), p. 313.
71. Ibid., pp. 413, 423.
72. Leslie R. Groves, *Now It Can Be Told: The Story of the Manhattan Project* (New York: Harper & Row, 1962), p. 39.
73. Ibid., p. 40.
74. Hewlett and Anderson, *New World,* p. 179.
75. Ibid., p. 181.
76. Groves, *Now It Can Be Told,* pp. 46–51.
77. Crawford Greenewalt diary quoted in David A. Hounshell and John K. Smith, "Science and Corporate Strategy: Du Pont R&D, 1902–1980," uncut manuscript, Hagley Museum and Library, Wilmington, Del., chap. 11, p. 30.
78. Groves, *Now It Can Be Told,* p. 51.
79. Hounshell and Smith, *Science and Corporate Strategy,* p. 339.
80. Hewlett and Anderson, *New World,* pp. 193–98.
81. Groves, *Now It Can Be Told,* p. 44.
82. Groueff, *Manhattan Project,* pp. 28–29.
83. Ibid., pp. 29–30.
84. Arthur Holly Compton, *Atomic Quest: A Personal Narrative* (New York: Oxford University Press, 1956), p. 169.
85. Ibid., p. 109.
86. Ibid., pp. 113–14.
87. Hewlett and Anderson, *New World,* p. 199.
88. Ibid., p. 201.
89. Rhodes, *Making of the Atomic Bomb,* p. 502.
90. Ibid., p. 503.
91. Ibid., pp. 20–21.
92. Ibid., pp. 504–8.
93. Greenewalt quoted in Hounshell and Smith, *Science and Corporate Strategy,* p. 340.
94. Greenewalt quoted in Hounshell and Smith, "Science and Corporate Strategy," uncut manuscript, chap. 11, p. 34.
95. Greenewalt quoted in ibid., p. 340.

96. Groves, *Now It Can Be Told*, p. 101.
97. Groueff, *Manhattan Project*, pp. 152–53.
98. Ibid., pp. 306–7.
99. Ibid., pp. 302–9; Hewlett and Anderson, *New World*, pp. 304–8.
100. Hewlett and Anderson, *New World*, p. 148.
101. Herbert Childs, *An American Genius: Ernest Orlando Lawrence* (New York: E. P. Dutton, 1968), p. 138.
102. Compton, *Atomic Quest*, p. 5.
103. Childs, *An American Genius*, pp. 139–40.
104. Ibid., p. 168; Daniel J. Kevles, *The Physicists: The History of a Scientific Community in Modern America* (New York: Vintage Books, 1979), p. 229.
105. Nuel Pharr Davis, *Lawrence and Oppenheimer* (New York: Simon & Schuster, 1968), pp. 42–44.
106. Luis W. Alvarez, "Berkeley: A Lab Like No Other," *Bulletin of the Atomic Scientists*, 30 (April 1974), 18–21, quoted in Richard Hewlett, "Nuclear Physics in the United States During World War II," unpublished manuscript, p. 4.
107. Compton, *Atomic Quest*, pp. 6–7.
108. Ibid., p. 8.
109. Groueff, *Manhattan Project*, pp. 35–39.
110. Childs, *An American Genius*, p. 335.
111. Ibid., p. 341.
112. Ibid., p. 345.
113. Rhodes, *Making of the Atomic Bomb*, p. 601.
114. Childs, *An American Genius*, pp. 348–49.
115. Richard G. Hewlett, "Beginnings of Development in Nuclear Technology," *Technology and Culture*, 17 (1976), 473.
116. Ibid., p. 474. I am indebted to Richard Hewlett for the suggestion that the electromagnetic process contributed only marginally to the production of U-235 of sufficient quantity for the bomb.
117. Hewlett and Anderson, *New World*, p. 122.
118. Ibid., p. 127.
119. Ibid., p. 134.
120. Groueff, *Manhattan Project*, p. 267.
121. Ibid., p. 182.
122. Hewlett and Anderson, *New World*, pp. 168–73; Groueff, *Manhattan Project*, pp. 313–19.
123. Groueff, *Manhattan Project*, p. 324.

124. Ibid., pp. 320–22.
125. Rhodes, *Making of the Atomic Bomb*, pp. 728, 733–34, 741–42.
126. Alice Kimball Smith, *A Peril and A Hope: The Scientists' Movement in America: 1945–47* (Chicago: University of Chicago Press, 1965), pp. 128–31, 294–97, 57–63.
127. Ibid., pp. 271–75; Hewlett and Anderson, *New World*, pp.7–8.
128. Hewlett and Anderson, *New World*, pp. 4–5.
129. Hounshell and Smith, *Science and Corporate Strategy*, pp. 341–45.
130. Mark Hertsgaard, *Nuclear Inc.: The Men and Money Behind Nuclear Energy* (New York: Pantheon Books, 1983), p. 22.
131. Hewlett and Anderson, *New World*, pp. 624–38.
132. Hertsgaard, *Nuclear Inc.*, p. 20.
133. Richard G. Hewlett and Francis Duncan, *Atomic Shield: A History of the United States Atomic Energy Commission, 1947–1952* (Washington, D.C.: U.S. Atomic Energy Commission, 1972), pp. 62–63, 76, 85–86, 120, 142–44, 192.
134. Ibid., p. 197.
135. Ibid., p. 191.
136. Richard Hewlett and Francis Duncan, *Nuclear Navy, 1946–1962* (Chicago: University of Chicago Press, 1974), pp. 34–35.
137. Hewlett and Duncan, *Atomic Shield*, p. 75.
138. Hewlett and Duncan, *Nuclear Navy*, pp. 48–50.
139. Ibid., pp. 57–78.
140. Richard G. Hewlett, "Beginnings of Development in Nuclear Technology," *Technology and Culture*, 17 (1976), 476–77.
141. Hertsgaard, *Nuclear Inc.*, p. 21.
142. Hewlett and Duncan, *Nuclear Navy*, p. 120.
143. Richard G. Hewlett and Jack M. Holl, "Atoms for Peace and War: Eisenhower and the Atomic Energy Commission, 1953–1961," manuscript (to be published by University of California Press, Berkeley, in 1989), VII, 6. Subsequent references are also to the manuscript chapters and pages.
144. Rickover in private conversation with David Lilienthal, July 1954, according to *Venturesome Years, 1950–1955: The Journals of David Lilienthal* (New York: Harper & Row, 1966), p. 532.
145. Hewlett and Holl, "Atoms for Peace and War," VII, 8.
146. Hewlett and Duncan, *Nuclear Navy*, p. 220.
147. Ibid., pp. 186–93; Norman Polmar and Thomas B. Allen, *Rickover: Controversy and Genius: A Biography* (New York: Touchstone, 1984), pp. 183–205.

148. Hewlett and Duncan, *Nuclear Navy*, pp. 307–10.
149. Harvey M. Sapolsky, *The Polaris System Development* (Cambridge, Mass: Harvard University Press, 1972), p. 11.
150. Hertsgaard, *Nuclear Inc.*, pp. 25–27.
151. Ibid., pp. 34–35.
152. Hewlett and Duncan, *Nuclear Navy*, pp. 225–30.
153. Ibid., p. 250.
154. Hertsgaard, *Nuclear Inc.*, pp. 34–36.
155. Hewlett and Holl, "Atoms for Peace and War," XVIII, 41.
156. Hertsgaard, *Nuclear Inc.*, pp. 44–45.
157. Hewlett and Holl, "Atoms for Peace and War," XVIII, 11.
158. Hertsgaard, *Nuclear Inc.*, p. 9.
159. Hewlett and Holl, "Atoms for Peace and War," VII, 20.
160. *The Military Industrial Complex*, ed. Carroll W. Pursell, Jr. (New York: Harper & Row, 1972), pp. 2306–7.
161. Mary Kaldor, *The Baroque Arsenal* (New York: Hill and Wang, 1981).

9 ■ COUNTERCULTURE AND MOMENTUM

1. Richard Striner, "The Machine as Symbol: 1920–1939," doctoral dissertation, University of Maryland, 1982.
2. David Dickson, *Alternative Technology and the Politics of Technical Change* (Glasgow: Fontana, 1974).
3. Jonathan Schell, *The Fate of the Earth* (New York: Avon, 1982).
4. John McDermott, "Technology: The Opiate of the Intellectuals," *The New York Review of Books*, XIII (31 July 1969), 25–35.
5. Theodore Roszak, *The Making of a Counter Culture: Reflections on the Technocratic Society and Its Youthful Opposition* (Garden City, N.Y.: Doubleday, 1969), pp. 4–6.
6. Charles A. Reich, *The Greening of America* (New York: Bantam Books, 1971; first published 1970), pp. 92–93.
7. Ibid., p. 281.
8. Roszak, *Making of a Counter Culture*, p. 110.
9. Russell Jacoby, *The last Intellectuals: American Culture in the Age of Academe* (New York: Basic Books, 1987), p. 5.
10. Leo Marx, "Lewis Mumford: Prophet of Organicism," paper presented at the International Symposium on Lewis Mumford, University of Pennsylvania, Philadelphia, 5–7 November 1987, pp. 6–7. Hereafter papers presented at this symposium are cited as "Mumford symposium." The papers will be edited by

Agatha C. and Thomas P. Hughes and published by Oxford University Press.

11. Ibid., p. 9.
12. Andreas Schüler, "Fortschrittsglaube und Kulturpessimismus," *Zeitschrift für Politik*, 33 (1986), 148–63.
13. Lewis Mumford, "If I Were Dictator," *The Nation*, 133 (9 December 1931), 631.
14. Lewis Mumford, *The Pentagon of Power*, vol. 2 of *The Myth of the Machine* (New York: Harcourt Brace Jovanovich, 1970), p. 1.
15. Author's conversation with Lewis and Sophia Mumford in Amenia, New York, 28 December 1985. See also Everett Mendelsohn, "Prophet of Our Discontent: Lewis Mumford Confronts the Bomb," Mumford symposium.
16. Mumford, *Pentagon of Power*, p. 78.
17. Ibid., p. 74.
18. Ibid., p. 37.
19. Donald L. Miller, "The Making of *The Myth of the Machine*," Mumford symposium, pp. 5–9.
20. Ibid., p. 9.
21. Rosalind Williams, "Lewis Mumford as a Historian of Technology in *Technics and Civilization*," Mumford symposium, pp. 12–16.
22. Howard P. Segal, "Mumford's Alternatives to the Megamachine," Mumford symposium, p. 3.
23. Jacques Ellul, *The Technological System*, trans. Joachim Neugroschel (New York: Continuum, 1980); Jacques Ellul, *The Technological Society*, trans. J. Wilkinson (New York: Alfred A. Knopf, 1964); Jacques Ellul, "The Technological Order," in *The Technological Order*, ed. Carl F. Stover (Detroit: Wayne State University Press, 1963), reprinted in *Philosophy and Technology: Readings in the Philosophical Problems of Technology*, eds. Carl Mitcham and Robert Mackey (New York: Free Press, 1983), pp. 86–105. Portions of the sections on Mumford and Ellul first appeared in Thomas P. Hughes, "Machines, Megamachines, and Systems," in *In Context: History and the History of Technology—Essays in Honor of Melvin Kranzberg*, eds. Stephen Cutcliffe and Robert Post (Bethlehem, Pa.: Lehigh University Press, 1988), pp. 106–19.
24. Ellul, *Technological System*, p. 16.
25. Ibid., p. 7.
26. Ibid.
27. Ibid., pp. 56–57, 311.
28. Ibid., pp. 45–46, 48, 312–13.
29. Ibid., p. 318.

30. Ellul, "Technological Order," p. 83.
31. Ibid., pp. 88–92.
32. Ibid., p.96.
33. Langdon Winner, *The Whale and the Reactor: A Search for Limits in an Age of High Technology* (Chicago: University of Chicago Press, 1986), p. 64.
34. Ibid., p. 65.
35. *The Whole Earth Catalog*, ed. Stewart Brand et al. (Menlo Park, Calif.: Nowels, 1968).
36. Winner, *Whale and Reactor*, p. 65.
37. John Hostetler, *Amish Society* (Baltimore: Johns Hopkins University Press, 1980), pp. 369–71.
38. Amory B. Lovins, "Energy Strategy: The Road Not Taken," *Foreign Affairs*, 55 (October 1976), 77–78.
39. Ibid., p. 74.
40. Ibid., p. 93.
41. Barry Commoner, *The Closing Circle: Nature, Man, and Technology* (New York: Alfred A. Knopf, 1971).
42. Winner, *Whale and Reactor*, p. 75.
43. Alvin Toffler, *The Third Wave* (New York: Bantam Books, 1982; first published 1980).
44. Ibid., p. 14.
45. Ibid., pp. 46–60.
46. Thomas P. Hughes, "A Technological Frontier: The Railway," in *The Railroad and the Space Program: An Exploration in Historical Analogy*, ed. Bruce Mazlish (Cambridge, Mass.: MIT Press, 1965), pp. 53–73; Thomas P. Hughes, "Technological Momentum in History: Hydrogenation in Germany, 1898–1933," *Past and Present*, 44 (August 1969), 106–32. On momentum, see also John Staudenmaier, S.J., *Technology's Storytellers* (Cambridge, Mass.: MIT Press, 1985), pp. 148–61.
47. John Kenneth Galbraith, *The New Industrial State* (Boston: Houghton Mifflin, 1971).
48. William H. McNeill, *The Pursuit of Power: Technology, Armed Force, and Society since A.D. 1000* (Chicago: University of Chicago Press, 1982); Walter A. McDougall, . . . *The Heavens and the Earth: A Political History of the Space Age* (New York: Basic Books, 1985).
49. Charles Perrow, *Normal Accidents: Living with High-Risk Technologies* (New York: Basic Books, 1984), p. 3.
50. Ibid., p. 97.
51. I am indebted to Todd R. La Porte of the University of California (Berkeley)

for this observation. On large systems see his "The United States Air Traffic System: Increasing Reliability in the Midst of Rapid Growth," in *The Development of Large Technical Systems*, eds. R. Mayntz and T. P. Hughes (Boulder, Colo.: Westview Press, 1988), pp. 215–44.

52. See, for instance, Shoshana Zuboff, *In the Age of the Smart Machine: The Future of Work and Power* (New York: Basic Books, 1984), pp. 229–35, 392–414.
53. Center for the Study of Responsive Law, *No Holds Barred: The Final Congressional Testimony of Admiral Hyman Rickover* (Washington, D.C.: CSRL, 1982), pp. 70–71.
54. Paul Goodman, *New Reformation: Notes of a Neolithic Conservative* (New York: Random House, 1970).
55. Paul Goodman, "A Causerie at the Military-Industrial," *The New York Review of Books*, IX (23 November 1967), 14–19.
56. Dickson, *Alternative Technology*, pp. 9–10, 183.
57. David F. Noble, *Forces of Production: A Social History of Industrial Automation* (New York: Alfred A. Knopf, 1984); Zuboff, *Age of the Smart Machine*.
58. Merritt Roe Smith, *Harpers Ferry Armory and the New Technology: The Challenge of Change* (Ithaca, N.Y.: Cornell University Press, 1977).
59. *The Social Construction of Technological Systems: New Directions in the Sociology and History of Technology*, ed. W. Bijker, T. Hughes, and T. Pinch (Cambridge, Mass.: MIT Press, 1987).
60. Organization for Economic Cooperation and Development, Ad Hoc Committee on New Concepts of Science Policy, *Science, Growth and Society: A New Perspective* (Paris: OECD, 1971), p. 22. Committee headed by Harvey Brooks.
61. On the general subject of the spread of international enterprises during the recent century, see Alfred D. Chandler, "Scale and Scope: International Comparison of the Dynamics of Managerial Capitalism," forthcoming from Harvard University Press, 1989.

INDEX

INDEX

ILLUSTRATION CREDITS: p. 14—From *Scientific American*, 75 (25 July 1896): 23, Courtesy of National Museum of American History, Smithsonian Institution; p. 26—Courtesy of U.S. Department of the Interior, National Park Service, Edison National Historic Site; pp. 28, 29—Courtesy of National Museum of American History, Smithsonian Institution; p. 31—From *Leslie's Weekly* (1880); p. 32—Courtesy of U.S. Department of the Interior, National Park Service, Edison National Historic Site; p. 33—Courtesy of National Museum of American History, Smithsonian Institution; p. 37—bottom: Courtesy of Burndy Library, Norwalk, Connecticut; p. 40—Courtesy of American Philosophical Society; p. 42—Courtesy of U.S. Department of the Interior, National Park Service, Edison National Historic Site; p. 46—top: Courtesy of Hagley Museum and Library, middle: Author's collection, bottom: Courtesy of Hagley Museum and Library; p. 50—Courtesy of National Museum of American History, Smithsonian Institution; pp. 58, 60—From the collections of the Henry Ford Museum and Greenfield Village, Neg. Nos. p. 188.16081 and p. 188.22433; p. 61—Courtesy of Franklin Institute, Philadelphia; p. 62—Courtesy of AT&T Archives; p. 63—Courtesy of National Museum of American History, Smithsonian Institution; p. 79—Courtesy of Hagley Museum and Library; pp. 87, 90—Courtesy of National Museum of American History, Smithsonian Institution; p. 93—From A. Hickenlooper, *Edison's Incandescent Electric Lights for Street Illumination* (1886); p. 94—Courtesy of U.S. Department of the Interior, National Park Service, Edison National Historic Site; pp. 103, 104—Courtesy of National Air and Space Museum, Smithsonian Institution; p. 105—Courtesy of National Museum of American History, Smithsonian Institution; p. 108—Courtesy of Hagley Museum and Library; p. 111—Author's collection; p. 120—Courtesy of National Museum of American History, Smithsonian Institution; pp. 128, 129, 131—Courtesy of Hagley Museum and Library; p. 133—Courtesy of National Air and Space Museum, Smithsonian Institution; p. 136—Courtesy of Hagley Museum and Library; p. 142—Courtesy of U.S. Army Communications Electronics Museum; pp. 144, 148, 153—Courtesy of National Museum of American History, Smithsonian Institution; pp. 157, 158—Courtesy of AT&T Archives; p. 161—Courtesy of General Electric Company; p. 164—Courtesy of Loyola University of Chicago Archives, Samuel Insull Collection; pp. 165, 167, 168, 172, 173—Courtesy of General Electric Company; p. 179—Courtesy of Hagley Museum and Library; pp. 192, 196, 198, 201, 202—Courtesy of Taylor Collection, Stevens Institute of Technology, Hoboken, N.J.; pp. 206, 207, 209, 211, 213, 215, 218—From the collections of the Henry Ford Museum and Greenfield Village, Neg. Nos. P.833.697, P.O.4475, P.833.65387, P.833.68057.105, P.O.331, P.34.177.21, P.O.833.979; p. 228—Courtesy of Commonwealth Edison Company, Chicago; pp. 233, 234—From Insull, *Central-Station Electric Service* (Chicago: Privately printed, 1915), pp. 461, 56; p. 236—From *Cassiers Magazine*, VIII (1895), 358; p. 237—Courtesy Consolidated Edison Company of New York; p. 238—Courtesy of Loyola University of Chicago Archives, Samuel Insull Collection; p. 242—Copyrighted, Chicago Tribune Company, all rights reserved, used with permission; pp. 252, 253—Lenin Library, Moscow; p. 262—Courtesy of the Burndy Library, Norwalk, Conn.; p. 263—Courtesy of the National Museum of American Art, Smithsonian Institution, Gift of Adele Lozowick; p. 267—Tretjakov Gallery, Moscow; p. 268—Courtesy of the Burndy Library, Norwalk, Conn.; p. 269—Courtesy of Harold Dorn and the late Elizabeth Hardin; p. 270—Courtesy of Harold Dorn; pp. 272, 273—From the collections of the Henry Ford Museum and Greenfield Village, Neg. Nos. P.189.1934, P.189.1933; p. 274—Tretjakov Gallery, Moscow; p. 277—From the collections of the Henry Ford Museum and Greenfield Village, Neg. No. P.P.5827; p. 283—Courtesy of Margaret Bourke-White Papers, George Arents Research Library for Special Collections, Syracuse University; p. 287—Author's collection; p. 311—Courtesy of Akademie der Künste, Berlin; pp. 313, 316, 318, 320—Courtesy of Bauhaus-Archiv, Berlin; p. 328—The Alfred Stieglitz Collection, the National Gallery of Art, Washington, D.C.; pp. 332, 333—From *291*, Nos. 5–6 (July–August 1915); p. 340—top: Courtesy of Columbus Museum of Art, Columbus, Ohio, Gift of Ferdinand Howard, bottom: Courtesy of Regis Collection, Minneapolis, Minn.; p. 341—Courtesy of Metropolitan Museum of Art, The Alfred Stieglitz Collection, 1949; p. 342—top: Courtesy of the Lane Collection, bottom: Courtesy of the Dallas Museum of Art, Gift of Edmund J. Kahn; p. 344—top: Courtesy of The Museum of Modern Art, Gift of Abby Aldrich Rockefeller, bottom: From the Collection of Whitney Museum of American Art, New York; p. 345—Courtesy of Worcester Art Museum, Worcester, Mass.; pp. 347, 348—Courtesy of National Museum of American Art, Smithsonian Institution, Gift of Adele Lozowick; p. 350—top: private collection, bottom: Courtesy of Tilde Grossberg, private collection; p. 351—Courtesy of Radziwill Foundation and Verwertungsgesellschaft Bild-Kunst, Federal Republic of Germany; pp. 363, 365, 370, 371, 380, 382—Courtesy of the Tennessee Valley Authority, Knoxville, Tenn.; p. 384—Courtesy of Niels Bohr Library, American Institute of Physics; p. 386—Courtesy of Hagley Museum and Library; p. 388—Courtesy of Lawrence Radiation Laboratory, Berkeley, Calif.; p. 389—Courtesy of the American Institute of Physics and W. F. Meggers; p. 393—Courtesy of the Hagley Museum and Library; p. 396—Courtesy of American Institute of Physics and G. W. Szilard; p. 400—Courtesy of Hagley Museum and Library; p. 406—Courtesy of Science Service and American Institute of Physics; p. 407—Courtesy of American Institute of Physics and Lawrence Radiation Laboratory; p. 415—Courtesy of History Office, U.S. Department of Energy; p. 418—Courtesy of Niels Bohr Library, American Institute of Physics; p. 420—Courtesy of Los Alamos Scientific Laboratory and American Institute of Physics; p. 427—Courtesy of Argonne National Laboratory; p. 429—From Hewlett and Duncan, *Nuclear Navy*; p. 434—Courtesy of History Office, U.S. Department of Energy and U.S. Navy; p. 464—Courtesy of History Office, National Aeronautics and Space Administration; p. 466—From Allied Pix Service, Inc.; p. 467—From Novosti/Gamma-Liaison.